Advanced Electromagnetic Computation

Second Edition

T0179084

Advanced Electromagnetic Computation

Second Edition

Dikshitulu K. Kalluri

CRC Press
Taylor & Francis Group
Boca Raton London New York

CRC Press is an imprint of the
Taylor & Francis Group, an **informa** business

CRC Press
Taylor & Francis Group
6000 Broken Sound Parkway NW, Suite 300
Boca Raton, FL 33487-2742

First issued in paperback 2019

© 2018 by Taylor & Francis Group, LLC
CRC Press is an imprint of Taylor & Francis Group, an Informa business

No claim to original U.S. Government works

ISBN-13: 978-1-4987-3340-3 (hbk)
ISBN-13: 978-0-367-87386-8 (pbk)

Visit the Taylor & Francis Web site at
http://www.taylorandfrancis.com

and the CRC Press Web site at
http://www.crcpress.com

Contents

Part VI Appendices

Part VII Chapter Problems

Preface

The second edition retains the flow and the flavor of the first edition, which was stated in the preface to the first edition. Item 1 describes this flavor in a few words.

1. The millennium generation of students, immersed in computer culture, responds to teaching techniques that make heavy use of computers. They are very comfortable in writing the code to implement an algorithm. In this connection, I remember a graduate student doing research with me mentioning that his struggle is always in understanding the concept. Thus, my approach in teaching is to definitely include the simplest possible model that brings out the concept to the satisfaction of the student and then show how one can improve on the accuracy of the prediction of the performance of a more realistic model to be solved by using more advanced mathematical and computational techniques.

2. It is important to include material that makes my comprehensive book a resource to be referred to when the student encounters, at some stage, topics that are not covered in one or two courses he/she could take using my book as textbook. Having taken one or two courses, the student should be able to quickly get to the new material in the book as and when needed and proceed in a mode of self-study.

3. In view of the advances in the areas of electromagnetic metamaterials and plasmonics, the discussions of the underlying topics of "electromagnetics and plasmas" are strengthened in the second edition.

4. In addition to correcting the typos and errors in the first edition, and making a few minor changes involving rearrangement of the material in the chapters, the following are the major additions to Volume 2 of the second edition:

   a. Chapter 21 is new and deals with perfectly matched layer and its later variants in Section 21.2. This aspect is essential in terminating the computational domain for exterior problems like radiation and scattering. In Section 21.2, near-field to far-field transformation is discussed to enable the computation of the far fields from the data obtained close to the scatterer. Finally in Section 21.4, the finite element boundary integral method is discussed to take advantage of the merits of the moment method as well as the finite element method.

   b. Chapter 23, "Miscellaneous Topics on Electromagnetic Computation" is new and brings the power of the well-known transform methods (extensively studied in the undergraduate course "Signals and Systems") such as Laplace, Fourier, Z transforms, and other spectral domain methods. Section 23.9 deals with frequency selective surfaces and Section 23.6 deals with rapidly developing computational methods such as WLP-FDTD, which overcomes the CLF limit in the choice of time and space steps. Transverse resonance method is discussed in Section 23.2

   c. Appendix 23A is new and explains the most general three-dimensional FDTD algorithm that can be used to compute when the electromagnetic medium has anisotropic, dispersive, and space- and time-varying parameters. It makes use of the Laplace transform method as well as the location of the current variable at the center of the Yee's cube to implement the constitutive equation of the medium numerically by solving the associated auxiliary vector differential equation. This algorithm and its illustration through the computation of the fields in a time-varying magnetoplasma medium is one of the highlights of Volume 2.

d.   Section 20.5 is new to this edition. A one-dimensional FDTD solution of inter-action of a pulsed wave with a switched plasma slab (given in Section 20.4 in the first edition) is extended to the case of a switched magnetoplasma slab. It is shown that the wiggler magnetic field in Section 20.4 is now converted to a low-frequency whistler wave.

MATLAB® is a registered trademark of The MathWorks, Inc. For product information, please contact:

The MathWorks, Inc.
3 Apple Hill Drive
Natick, MA, 01760-2098 USA
Tel: 508-647-7000
Fax: 508-647-7001
E-mail: info@mathworks.com
Web: www.mathworks.com

Acknowledgment

I thank Dr. Robert Kevin Lade for generating the results in Section 20.5, which are based on his doctoral thesis.

Author

Dikshitulu K. Kalluri, PhD, is professor emeritus of electrical and computer engineering at the University of Massachusetts Lowell. He received his BE in electrical engineering from Andhra University, India; DIISc in high voltage engineering from the Indian Institute of Science, Bangalore; master's degree in electrical engineering from the University of Wisconsin, Madison; and doctorate in electrical engineering from the University of Kansas, Lawrence.

Dr. Kalluri began his career at the Birla Institute of Technology, Ranchi, India, advancing to the rank of professor, heading the Electrical Engineering Department, then serving as (dean) assistant director of the institute.

Since 1984, he had been with the University of Massachusetts Lowell, Lowell, advancing to the rank of full professor in 1987. He coordinated the doctoral program (1986–2010) and codirected/directed the Center for Electromagnetic Materials and Optical Systems (1993–2002/2003–2007). As a part of the center, he established the Electromagnetics and Complex Media Research Laboratory. He had collaborated with research groups at the Lawrence Berkeley Laboratory, the University of California, Los Angeles; the University of Southern California; New York Polytechnic University; and the University of Tennessee and has worked several summers as a faculty research associate at Air Force Laboratories. He retired in May 2010 but continues his association with the university guiding doctoral students. He continues to teach his graduate courses online as professor emeritus. CRC Press published his two recent books: *Electromagnetics of Time Varying Complex Media Second Edition*, in April 2010, and the first edition of this book, in August 2011. He has published extensively on the topic of electromagnetics and plasmas. He supervised doctoral thesis research of 17 students.

Dr. Kalluri is a fellow of the Institute of Electronic and Telecommunication Engineers and a senior member of IEEE.

Part V

Electromagnetic Computation

15

Introduction and One-Dimensional Problems

15.1 Electromagnetic Field Problem: Formulation as Differential and Integral Equations

Maxwell's equations are vector PDEs coupling the electric and magnetic fields (read Chapter 1 for a quick review of Maxwell's equations). The independent variables are the three spatial coordinates and the time variable. For time-harmonic fields where we assume the time variation as harmonic (cosinusoidal), one can use the phasor concepts and replace $\partial/\partial t$ in the time domain with multiplication (by $j\omega$) in the phasor domain. Thus, one can reduce the dimensionality of the problem from four independent variables to three, thus reducing the mathematical complexity of the problem. Further reduction in the dimensionality of the problem is possible for certain applications. Let us review the waveguide problem (see Chapter 3) from this viewpoint. If we consider the TM modes of a rectangular waveguide $\left(\tilde{E}_z \neq 0,\ \tilde{H}_z = 0\right)$, the longitudinal component satisfies the scalar Helmholtz equation

$$\left[\frac{\partial^2}{\partial x^2} + \frac{\partial^2}{\partial y^2} + \frac{\partial^2}{\partial z^2} \right] \tilde{E}_z + k^2 \tilde{E}_z = 0, \tag{15.1}$$

$$k^2 = \omega^2 \mu \varepsilon. \tag{15.2}$$

Since we are interested in studying the propagation or nonpropagation of the wave, we can assume that

$$\tilde{E}_z \left(x, y, z \right) = F\left(x, y \right) e^{-j\beta z} \tag{15.3}$$

and study whether and under what circumstances β is real.

Substituting Equation 15.3 into Equation 15.1, we get a two-dimensional PDE in F:

$$\left[\frac{\partial^2 F}{\partial x^2} + \frac{\partial^2 F}{\partial y^2} \right] + k_c^2 F = 0, \tag{15.4}$$

where

$$k_c^2 = k^2 - \beta^2. \tag{15.5}$$

Equation 15.4 can be written as

$$-\nabla_t^2 F = +k_c^2 F = \lambda F. \tag{15.6}$$

In Equation 15.6, ∇_t^2 is the Laplacian operator in the transverse plane and λ is the eigenvalue. The waveguide problem is the eigenvalue problem involving a differential operator and when a numerical technique is employed, the problem is converted into the familiar form of

$$AX = \lambda X, \tag{15.7}$$

where A is a square matrix and λ's are the eigenvalue and X's are the corresponding eigenvectors. Equation 15.6 is a scalar second-order PDE and we have seen that for a rectangular region $0 < x < a$ and $0 < y < b$. One can solve it analytically by using separation of variable technique and on imposing the PEC boundary conditions (Figure 15.1), one obtains

$$\tilde{E}_z = E_{mn} \sin\frac{m\pi x}{a} \sin\frac{n\pi y}{b} e^{-j\beta z}, \quad m = 1,2,\ldots\infty, \; n = 1,2,\ldots\infty, \tag{15.8}$$

where

$$\lambda_{mn} = \left(k_c\right)_{mn}^2 = k_x^2 + k_y^2 = \left(\frac{m\pi}{a}\right)^2 + \left(\frac{n\pi}{b}\right)^2. \tag{15.9}$$

Recognize in Equations 15.8 and 15.9 that the problem has double infinity number of eigenvalues λ_{mn} and the corresponding eigenvectors are proportional to $\sin(m\pi x/a)$ $\sin(n\pi y/b)$.

Many practical problems do not have such elegant analytical solution. For example, Figure 15.2 shows a ridged waveguide used in practice to increase the bandwidth of single-mode propagation of a rectangular waveguide. Numerical methods (finite difference, finite element and moment, and other methods) or approximate analytical methods (perturbation or variational technique) are used to solve such problems [1–10]. We give next integral formulation of an electromagnetic field problem. Suppose we wish to find the capacitance of a square conducting plate located in free space with reference to sphere at infinity (Figure 15.3).

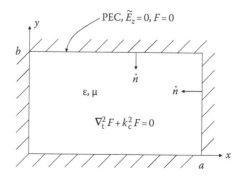

FIGURE 15.1
Rectangular waveguide: TM modes.

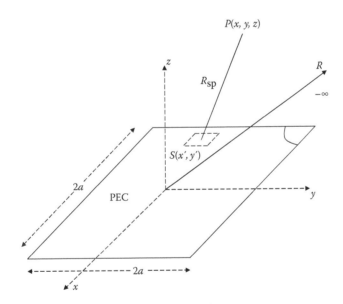

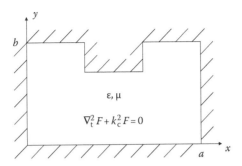

FIGURE 15.2
Ridged waveguide.

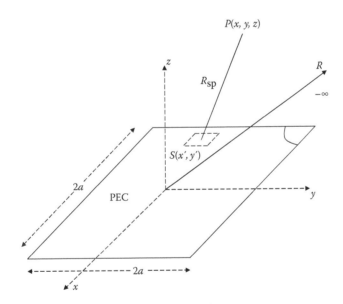

FIGURE 15.3
Conducting plate in free space.

The key step in the solution is determining the surface charge density $\rho_s(x, y)$ on the conducting plate, assuming that the voltage of the plate is V_0 volts. In terms of $\rho_s(x', y')$, the potential at a general point $P(x, y, z)$ is given by

$$\Phi(x,y,z) = \int_{-a}^{a}\int_{-a}^{a} \frac{\rho_s dx' dy'}{4\pi\varepsilon R_{sp}}, \qquad (15.10)$$

where

$$R_{sp} = \sqrt{(x-x')^2 + (y-y')^2 + z^2}. \qquad (15.11)$$

If the point P is on the plate, that is, $z = 0$, $-a < x < a$, and $-a < y < a$, the potential is a constant given by V_0:

$$V_P = V_0 = \int\limits_{-a}^{a}\int\limits_{-a}^{a} \frac{\rho_s \, dx' dy'}{4\pi\varepsilon\sqrt{(x-x')^2+(y-y')^2}} \quad (P \text{ on the plate}). \qquad (15.12)$$

In Equation 15.12, the unknown ρ_s appears under the integral sign and hence Equation 15.12 is called an integral equation.

15.2 Discretization and Algebraic Equations

Electromagnetic field problems are defined on a continuous domain as differential, integral, or integrodifferential equations. A continuous domain has infinite number of points, and an analytical solution, when it can be found, gives the field at every point in the domain as a mathematical expression. The analytical solution often appears as an infinite series and the calculation of fields at a specified point involves summation of an infinite series and requires truncating the series to a specified accuracy. Many numerical methods aim at finding the field at a discrete set of points and obtain the fields at other points by interpolation and extrapolation based on the calculated set of discrete fields. The technique converts the differential, integral, or integrodifferential equations into algebraic equations, the unknowns being the fields at the discrete points, which are finite in number. One can then use algebraic equation solver routines available as mathematical software. Examples are MATLAB®, MATHEMATICA, and MATHCAD. These techniques will be illustrated next using one-dimensional problems.

15.3 One-Dimensional Problems

In the context of this discussion, some examples of one-dimensional problems of electromagnetics, are as follows.

a. Static problems:

$$\frac{d^2V}{dz^2} = -\frac{\rho_v}{\varepsilon} \quad (\text{one-dimensional Poisson's equation}). \qquad (15.13)$$

b. Time-harmonic problem:

$$\frac{d^2V}{dz^2} + k^2V = 0 \quad (\text{transmission line equation}). \qquad (15.14)$$

c. Resonant transmission lines:

$$\frac{d^2V}{dz^2} + k_c^2V = 0. \qquad (15.15)$$

d. Transients on transmission lines

$$\frac{\partial^2 V}{\partial z^2} - \mu\varepsilon \frac{\partial^2 V}{\partial t^2} = 0. \tag{15.16}$$

Problems (a) and (b) are classified as equilibrium problems; the boundary conditions at the endpoints, which bound the one-dimensional domain, are specified. Problem (c) is an eigenvalue problem where the eigenvalue is $\lambda = k_c^2$ of the differential operator with the specified boundary conditions of the potential V on the endpoints of the domain. Problem (d) is called propagation or marching problem. The boundary in the time domain is open and the initial conditions and the boundary conditions at the endpoints are specified.

15.3.1 Finite Differences

Let $y(x)$ be tabulated at equal intervals h and the table of values x_n (independent variable) and y_n (dependent variable) are available, where

$$x_n = x_0 + nh, \tag{15.17}$$

$$y_n = y(x_n), \quad n = 0,1,2,.... \tag{15.18}$$

We can express the first derivative at a tabulated point in terms of the values of y at the neighboring points. For example, if we approximate the tangent to the curve $y(x)$ at P (see Figure 15.4) by the Chord 1:

$$\left.\frac{dy}{dx}\right|_n \approx \frac{y_{n+1} - y_n}{h} + O(h). \tag{15.19}$$

The approximation is called forward-difference approximation and it can be shown that it is approximate to the order of h, denoted by $O(h)$.

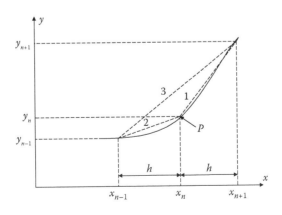

FIGURE 15.4
Approximation of the first derivative.

On the other hand, we can approximate the derivative by the Chord 2 and write

$$\left.\frac{dy}{dx}\right|_n \approx \frac{y_n - y_{n-1}}{h} + O(h).$$

(15.20)

This approximation is called backward-difference approximation of order h. Central-difference approximation is based on approximating the curve by Chord 3:

$$\left.\frac{dy}{dx}\right|_n \approx \frac{y_{n+1} - y_{n-1}}{2h} + O(h^2).$$

(15.21)

Note that the order of approximation in Equation 15.21 is stated to be $O(h^2)$. It would be useful to prove the order of approximation in Equations 15.19 through 15.21 and find a systematic approach to get higher-order approximation formulas; such an approach is available through "finite difference calculus" [1].

We start by defining some operators and find the symbolic relationship between the operators.

Displacement operator E:

$$y_{n+1} = Ey_n,$$
$$y_{n+2} = E(y_{n+1}) = E^2 y_n,$$

(15.22)

$$\therefore y_{n+m} = E^m y_n.$$

(15.23)

Averaging operator μ:

$$\mu y_n = \frac{1}{2}\left[y_{n+1/2} + y_{n-1/2}\right].$$

(15.24)

Forward-difference operator Δ:

$$\Delta y_n = y_{n+1} - y_n,$$
$$\Delta y_n = y_{n+1} - y_n = Ey_n - y_n = (E-1)y_n,$$

(15.25)

$$\therefore \Delta = E - 1.$$

(15.26)

Backward-difference operator ∇:

$$\nabla y_n = y_n - y_{n-1},$$
$$\therefore \nabla = 1 - E^{-1},$$

(15.27)

$$E = (1 - \nabla)^{-1}.$$

(15.28)

Central-difference operator δ:

$$\delta y_n = y_{n+1/2} - y_{n-1/2},$$

(15.29)

$$\delta = E^{1/2} - E^{-1/2}.$$

(15.30)

Note the following:

$$\Delta = \nabla E = \delta E^{1/2}, \tag{15.31}$$

$$\mu^2 = 1 + \frac{1}{4}\delta^2. \tag{15.32}$$

Differential operator D:

$$Dy = \frac{dy}{dx}. \tag{15.33}$$

The relation between D and E can be obtained by considering Taylor series

$$Ey(x) = y(x+h) = y(x) + h\frac{dy}{dx} + \frac{h^2}{2!}\frac{d^2y}{dx^2} + \cdots$$

$$= \left(1 + hD + \frac{h^2D^2}{2!} + \cdots\right)y(x) = e^{hD}y(x), \tag{15.34}$$

$$E = e^{hD},$$

$$hD = \ln E = \ln(1+\Delta), \tag{15.35}$$

$$= \ln(1-\nabla)^{-1} = -\ln(1-\nabla). \tag{15.36}$$

From Equation 15.35,

$$hy'_j = hDy_j = -\left[\ln(1-\nabla)\right]y_j = \left[\nabla + \frac{1}{2}\nabla^2 + \frac{1}{3}\nabla^3 + \cdots\right]y_j,$$

$$\therefore D = \frac{1}{h}\left[\nabla + \frac{1}{2}\nabla^2 + \frac{1}{3}\nabla^3 + \cdots\right]. \tag{15.37}$$

Let us terminate the series in Equation 15.37 with one term only:

$$Dy_j \approx \frac{1}{h}\nabla y_j = \frac{1}{h}\left[y_j - y_{j-1}\right]. \tag{15.38}$$

Equation 15.38 is the same as Equation 15.20. We can improve the accuracy by taking two terms in Equation 15.37:

$$Dy_j \approx \frac{1}{h}\left[\nabla + \frac{1}{2}\nabla^2\right]y_j,$$

$$\nabla y_j = y_j - y_{j-1},$$

$$\nabla^2 y_j = \nabla(\nabla y_j) = \nabla(y_j - y_{j-1}) = (y_j - y_{j-1}) - (y_{j-1} - y_{j-2}),$$

$$\nabla^2 y_j = y_j - 2y_{j-1} + y_{j-2},$$

$$\therefore Dy_j \approx \frac{1}{h}\left[(y_j - y_{j-1}) + \frac{1}{2}(y_j - 2y_{j-1} + y_{j-2})\right]. \tag{15.39}$$

TABLE 15.1

Relations between Symbolic Operators

	E	Δ	δ	∇	hD
E	E	$1 + \Delta$	$1 + \frac{1}{2}\delta^2 + \delta\sqrt{1 + \frac{1}{4}\delta^2}$	$(1-\nabla)^{-1}$	e^{hD}
Δ	$E-1$	Δ	$\frac{1}{2}\delta^2 + \delta\sqrt{1+\frac{1}{4}\delta^2}$	$\nabla(1-\nabla)^{-1}$	$e^{hD}-1$
δ	$E^{1/2} - E^{-1/2}$	$\Delta(1+\Delta)^{-1/2}$	δ	$\nabla(1-\nabla)^{-1/2}$	$2\sinh\frac{1}{2}hD$
∇	$1 - E^{-1}$	$\Delta(1+\Delta)^{-1}$	$-\frac{1}{2}\delta^2 + \delta\sqrt{1+\frac{1}{4}\delta^2}$	∇	$1 - e^{-hD}$
hD	$\log E$	$\log(1 + \Delta)$	$2\sinh^{-1}\frac{1}{2}\delta$	$-\log(1 - \nabla)$	hD
μ	$\frac{1}{2}\left(E^{1/2} + E^{-1/2}\right)$	$\left(1+\frac{1}{2}\Delta\right)(1+\Delta)^{-1/2}$	$\sqrt{1+\frac{1}{4}\delta^2}$	$\left(1-\frac{1}{2}\nabla\right)(1-\nabla)^{-1/2}$	$\cosh\frac{1}{2}hD$

Source: Adapted from Thom, A. and Apelt, C.J., *Field Computations in Engineering and Physics*, Van Nostrand Reinhold, London, U.K., 1961.

Simplifying, we obtain

$$Dy_j \approx \frac{1}{2h}\left[3y_j - 4y_{j-1} + y_{j-2}\right] + O\left(h^2\right).$$ (15.40)

Equation 15.40 is more accurate than Equation 15.38 but note that Equation 15.40 involves more neighboring points than Equation 15.38. It can be shown that, using the forward-difference formula,

$$Dy_j = \frac{1}{2h}\left[-y_{j+2} + 4y_{j+1} - y_j\right] + O\left(h^2\right)$$ (15.41)

and using central-difference formulas

$$\left.\frac{d^2 y}{dx^2}\right|_j = D^2 y_j = \frac{1}{h^2}\left[y_{j+1} - 2y_j + y_{j-1}\right] + O\left(h^2\right).$$ (15.42)

Table 15.1 gives relations between symbolic operators [1].

15.3.2 Method of Weighted Residuals

The underlying concepts of method of weighted residuals can be explained through a simple example [2]. Let us suppose that we wish to obtain the solution of the first-order marching problem:

$$\frac{dx}{dt} + x = 0,$$ (15.43a)

with the initial condition

$$x(0) = 1 \tag{15.43b}$$

in the domain $0 < t < 1$.

The exact solution is

$$x = e^{-t} = 1 - t + \frac{t^2}{2!} - \frac{t^3}{3!} + \cdots. \tag{15.44}$$

The exact solution is an infinite series and one can evaluate the series on a computer to the accuracy desired. Let us suppose that we put the restriction that we cannot use more than three terms and write an approximate solution

$$x_T(t) = 1 - t + \frac{t^2}{2}. \tag{15.45}$$

We will show that we can do better than Equation 15.45 if we write that $x(t)$ is approximately equal to $x_a(t)$:

$$x(t) \approx x_a(t) = 1 + c_1 t + c_2 t^2 \tag{15.46}$$

and adjust c_1 and c_2 so that the residue

$$R(t) = \frac{dx}{dt} + x = 1 + c_1(1 + t) + c_2(2t + t^2) \tag{15.47}$$

is minimized in some sense. Of course, $R(t)$ will be zero for all t in the domain for an exact solution of x. Let us examine several possible ways of defining the minimization of $R(t)$.

15.3.2.1 Collocation (Point Matching)

Since there are two unknowns in Equation 15.46, let us determine them by reducing $R(t)$ to zero at two specific points in the domain, say at $t = 1/3$ and $t = 2/3$:

$$R\left(\frac{1}{3}\right) = 1 + c_1\left(1 + \frac{1}{3}\right) + c_2\left[2 \times \frac{1}{3} + \left(\frac{1}{3}\right)^2\right],$$

giving

$$\frac{4}{3}c_1 + \frac{7}{9}c_2 = -1. \tag{15.48a}$$

Similarly, $R(2/3) = 0$ gives the second equation:

$$\frac{5}{3}c_1 + \frac{16}{9}c_2 = -1. \tag{15.48b}$$

Solving Equation 15.48a and 15.48b, we get

$$c_1 = -0.9310, \quad c_2 = 0.3103 \tag{15.49}$$

and from Equation 15.46, we get

$$x_a(t) = x_c(t) = 1 - 0.9310t + 0.3103t^2. \tag{15.50}$$

15.3.2.2 Subdomain Method

The domain is divided into as many subdomains as there are adjustable constants; in this case two and use the criteria that the average value of the residue over each of the subdomains is zero. Let the subdomains be $0 < t < 1/2$ and $1/2 < t < 1$. The equations for determining c_1 and c_2 are obtained from

$$\int_0^{1/2} R(t)dt = 0, \quad \int_{1/2}^1 R(t)dt = 0. \tag{15.51}$$

The approximate solution $x_a = x_{SD}$ is given by

$$x_{SD} = 1 - 0.9474t + 0.3158t^2. \tag{15.52}$$

15.3.2.3 Galerkin's Method

In Equation 15.46, we have used as our "basis functions" t^0, t^1, t^2 which belong to a series and we let the coefficient of t^0 be 1 to satisfy the initial condition and we let the coefficients c_1 and c_2 adjustable so as to minimize the residue R in some sense. The Russian engineer Galerkin used the basis functions as the weights and defined the criteria as reducing the weighted average of the residue over the entire domain to zero:

$$\int_0^1 tR(t)dt = 0; \quad \int_0^1 t^2R(t)dt = 0. \tag{15.53}$$

From Equation 15.53, it can be shown that $x_a = x_G$ is given by

$$x_G = 1 - 0.9143t + 0.2857t^2. \tag{15.54}$$

15.3.2.4 Method of Least Squares

In this case, the parameters c_1 and c_2 are adjusted in such a way that the integral of the square of the residual is minimized:

$$\frac{\partial}{\partial c_1} \int_0^1 R^2(t)dt = 2 \int_0^1 \frac{\partial R}{\partial c_1} R(t)dt = 0.$$

The criteria may then be written as

$$\int_0^1 \frac{\partial R}{\partial c_1} R dt = \frac{3}{2} + \frac{7}{3}c_1 + \frac{9}{4}c_2 = 0, \tag{15.55a}$$

$$\int_0^1 \frac{\partial R}{\partial c_2} R dt = \frac{4}{3} + \frac{9}{4}c_1 + \frac{38}{15}c_2 = 0. \tag{15.55b}$$

The approximate solution $x_a = x_L$ thus obtained is given by

$$x_L = 1 - 0.9427t + 0.3110t^2. \tag{15.56}$$

Figure 15.5 compares the approximate solutions by calculating the difference between the exact solution $x(t) = e^{-t}$ and the various approximation solutions x_a. The worst approximation is the truncated Talyor-series solution (Equation 15.45) for $t > 0.4$ and the others are with in an error of 0.01 in the entire domain. The computational labor among the methods 1–4 is of increasing order. Collocation is the most direct. The four methods can be described under the category of "method of weighted residuals" since the criterion used to obtain the algebraic equations may be written as

$$\int_0^1 w_j(t)R(t)dt = 0, \quad j = 1,2, \tag{15.57}$$

where w_j are the weight functions. The weight functions are sketched in Figure 15.6. The weight function for the collocation method (also called point-matching technique) is impulse functions. This may be readily seen from

$$\int_0^1 \delta(t - t_j)R(t)dt = R(t_j), \quad j = 1,2. \tag{15.58}$$

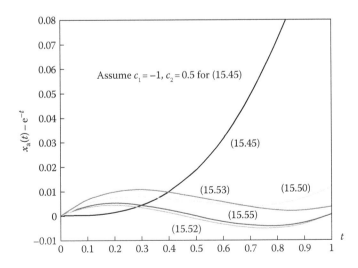

FIGURE 15.5
Comparison of the approximation solution.

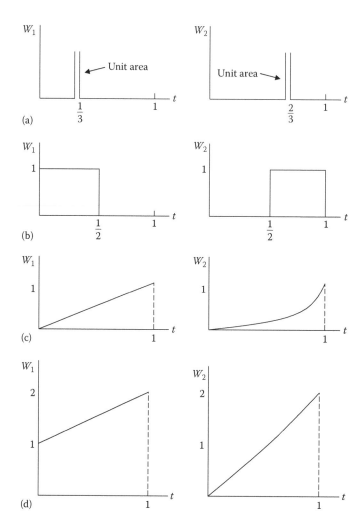

FIGURE 15.6
Weight functions for various approximation techniques: (a) collocation, (b) subdomain method, (c) Galerkin's
method, and (d) least squares.

The weight functions for the subdomain method are pulse functions as shown in
Figure 15.6. The basis functions in this example are called entire domain functions since
they are defined over the entire domain.

One can also use subdomain basis functions that are nonzero over the subdomain and
zero outside the subdomain. More will be said on this topic when we discuss the "moment
method." In the previous section, we discussed finite differences and let us see whether
Equation 15.43 can also be solved using finite differences. Let us develop this algorithm for
the more general case of Equation 15.43 given in Equations 15.59a and 15.59b:

$$\frac{dx}{dt} = f(t,x),\qquad\qquad(15.59a)$$

$$x(0) = x_0.\qquad\qquad(15.59b)$$

Using forward-difference approximation for the LHS of Equation 15.43a at the point j,

$$\frac{dx}{dt}\bigg|_j = \frac{x_{j+1} - x_j}{h} = f(t_j, x_j), \qquad (15.59c)$$

which gives the "Euler algorithm" for differential equation initial value problem:

$$x_{j+1} = x_j + hf(t_j, x_j), \quad j = 0,1,2,\ldots. \qquad (15.60)$$

Note that the above is self-starting in the sense that Equation 15.60 can be used as a recurrence relation starting with the initial value x_0. The solution can be marched in time by a step-by-step process. For the example on hand, $f(t, x) = -x$ and Equation 15.60 becomes

$$\begin{aligned} x_0 &= 1, \\ x_{j+1} &= x_j - hx_j = (1-h)x_j, \quad j = 0,1,2,\ldots, \end{aligned} \qquad (15.61)$$

choosing, $h = 1/3$, $x_1 = 2/3$, $x_2 = (2/3)x_1 = 4/9$, and $x_3 = (2/3)x_2 = 8/27 = x(1)$.
The approximate answer for $x_a = x_{FD}$:

$$x_{FD} = \frac{8}{27} = 0.2963. \qquad (15.62)$$

The results obtained by various approximation techniques are compared in Table 15.2.
From the table, we see that all the methods are better than the two-term approximation based on Talyor expression. The forward-difference approximation based on three steps ($h = 1/3$) is not good but of course the computational time is much less than the methods based on weighted residuals since they involve solution of simultaneous algebraic equations. The FD step-by-step calculation can be made more accurate by taking $h = 0.1$ involving 10-step recursion but one does not have to solve 10 simultaneous-equations. The x_{FD} using $h = 0.1$ will give the approximate result of 0.3487 for the value and 5.2% for the error. One can try to use the central-difference formula for the first derivative (Equation 15.21) since the error is $O(h^2)$ and obtain a higher accuracy for the samenumber of steps.

TABLE 15.2

Approximate Solution of $(dx/dt) + x = 0$, $x(0) = 1$, by Various Methods and Comparison with the Exact Value of $x(t) = e^{-t}$ at $t = 1$ and $x(1) = 0.3679$

Equation #	15.45	15.50	15.52	15.53	15.56	15.60
$x_a(1)$	$x_T(1)$	$x_c(1)$	$x_{SD}(1)$	$x_G(1)$	$x_L(1)$	$x_{FD}(1)$
Value	0.5	0.3793	0.3684	0.3714	0.3683	0.2963
% Error	−35.91	−3.1	−0.14	−0.96	−0.11	19.46
$\frac{x(1)-x_a(1)}{x(1)} \times 100$						

From Equation 15.21,

$$\left.\frac{dx}{dt}\right|_j = \frac{x_{j+1} - x_{j-1}}{2h} = f\left(t_j, x_j\right),$$

$$x_0 = x(0),$$

$$x_{j+1} = x_{j-1} + 2hf\left(t_j, x_j\right), \quad j = 0, 1, 2, \ldots. \tag{15.63}$$

The above does not give a step-by-step procedure since x_{-1} is not known. We can recast Equation 15.63 as $x_0 = x(0)$:

$$x_i = x_{i-2} + 2hf\left(t_{i-1}, x_{i-1}\right), \quad i = 2, 3, \ldots. \tag{15.64}$$

Equation 15.64 requires the same amount of computational work as Equation 15.60 but it is not self-starting. The values of x_1 and $f(t_j, x_j) = f_1$ have to be obtained by some other method to start the recursion going. That other method will also have to be of $O(h^2)$ accuracy to maintain the order of the accuracy of the central-difference methods. Such methods are discussed in books on numerical analysis [3].

15.3.3 Moment Method

Moment method is a version of the method of weighted residuals discussed in the previous section, popularized by Harrington [4] in its application to electromagnetics. It is usually stated in the language of linear spaces [5] and will be explained through an example. Let the problem be stated as

$$Lf = g, \tag{15.65}$$

where L is an operator, f is the unknown function, and g is excitation. The problem is to find f given L and g. The operator L may be a differential, integral, or integrodifferential operator. The domain of the problem and the boundary conditions are a part of the specification of the problem. The formulation requires the following steps. Assume that

$$f \approx \sum_{n=1}^{N} \alpha_n f_n, \tag{15.66}$$

where f_n are basis functions chosen for the problem. They satisfy the boundary conditions. Choose the weight functions w_m for the problem so as to minimize the residual in some sense as explained in the previous section. Choose the inner product for the problem. It is best to explain the choice of the above three through an example [4]. Suppose we wish to solve the following problem:

$$-\frac{d^2 f}{dx^2} = 1 + 4x^2 \tag{15.67a}$$

with the domain given as $0 < x < 1$ and the boundary conditions are given by

$$f(0) = f(1) = 0. \tag{15.67b}$$

This problem is easily solved by integrating Equation 15.67a twice and the constants of integration determined by the boundary conditions. The exact answer to the problem is

$$f(x) = \frac{5x}{6} - \frac{x^2}{2} - \frac{x^4}{3}. \tag{15.67c}$$

Approximate solution based on the "moment method" is formulated as follows. We substitute Equation 15.66 into Equation 15.65, multiply by w_m and find the inner product on both sides thus generalizing the mth algebraic equation for solving α's. A formal way of writing this step is

$$\left\langle w_m, L \sum_{n=1}^{N} \alpha_n f_n \right\rangle = w_m, g, \tag{15.68}$$

where $\langle\ \rangle$ is the symbol for the inner product. A suitable inner product of two functions ψ_m and ψ_n for the problem on hand is

$$\langle \psi_m, \psi_n \rangle = \int_0^1 \psi_m \psi_n \, dx. \tag{15.69}$$

Using Equation 15.69 in Equation 15.68, one obtains

$$\int_0^1 \left[w_m L \sum_{n=1}^{N} \alpha_n f_n \right] dx = \int_0^1 w_m g \, dx. \tag{15.70}$$

Next is to consider the choice of w_m and f_n for the problem. Note that w_m and f_n need not be entire domain functions, and they need not be nonzero over the complete domain. They can be subsectional functions that are nonzero over a section of the domain and zero elsewhere. Many times, use of such functions reduces the work involved in obtaining the algebraic equations. Let us define two such subsectional basis functions.
    Pulse function $P(x)$; Let

$$P(x) = \begin{cases} 1, & |x| < \dfrac{1}{2(N+1)}, \\ 0, & \text{otherwise.} \end{cases} \tag{15.71}$$

This function is sketched in Figure 15.7. The center of the pulse function can be shifted by defining

$$P(x - x_n) = \begin{cases} 1, & |x - x_n| < \dfrac{1}{2(N+1)}. \\ 0, & \text{otherwise.} \end{cases} \tag{15.72}$$

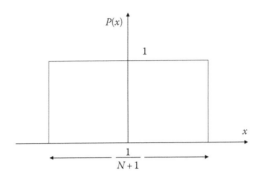

FIGURE 15.7
Pulse function.

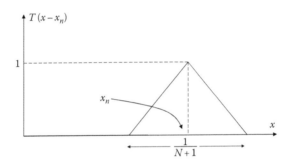

FIGURE 15.8
Triangle function.

Another useful function is a triangle function $T(x - x_n)$ given by

$$T\left(x-x_n\right) = \begin{cases} 1-\left|x-x_n\right|\left(N+1\right), & \left|x-x_n\right| < \dfrac{1}{N+1}, \\ 0, & \left|x-x_n\right| > \dfrac{1}{N+1}. \end{cases} \tag{15.73}$$

This function is sketched in Figure 15.8.

After choosing w_m and f_n in Equation 15.70, one can perform the definite integrals with respect to x and the result may be written as an algebraic equation

$$\sum_{n=1}^{N} \ell_{mn}\alpha_n = g_m, \tag{15.74}$$

where

$$\ell_{mn} = \int_0^1 w_m L f_n \, dx, \tag{15.75}$$

$$g_m = \int_0^1 w_m g \, dx. \tag{15.76}$$

Equation 15.74 yields N equations for the α_n's to be determined by choosing N weight functions w_m, $m = 1, 2, 3, \dots N$. Equations 15.75 and 15.76 give some guidance for choosing w_m and f_n. For hand computation, we would like to have as simple integrals as possible. The integral can be performed only if Lf_n is a well-defined integrable function. We know that integration, when an impulse function is in the integral, is given by

$$\int_a^b \delta(x - x_j) f(x) \, dx = \begin{cases} f(x_j) = f_j, & a < x_j < b, \\ 0, & \text{otherwise.} \end{cases} \tag{15.77}$$

Thus, we can get away by choosing f_n such that Lf_n is a sum of impulses. This will be the result if $f_n = T(x - x_n)$ and L is a second-order derivative. Figure 15.9 shows d^2T/dx^2.

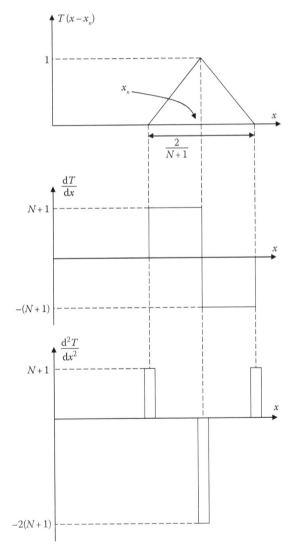

FIGURE 15.9
Sketch of the second derivative of triangular function.

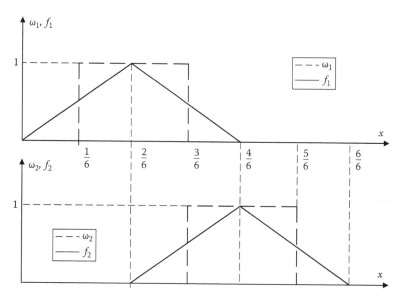

FIGURE 15.10
Weight and basic functions.

Since Lf_n gives impulses, the simplest functions we can choose for w_m are pulse functions. Let us illustrate the calculations by choosing $N = 2$ in Equation 15.66. The weight functions w_1 and w_2 and the basis functions T_1 and T_2 are sketched in Figure 15.10.

Now we can obtain the two algebraic equations for α_1 and α_2. A few details in obtaining the first equation are shown below:

$$Lf_1 = -\frac{d^2 f_1}{dx^2} = -3\left[\delta(x) - 2\delta\left(x - \frac{2}{6}\right) + \delta\left(x - \frac{4}{6}\right)\right],$$

$$\ell_{11} = \int_0^1 w_1 Lf_1 \, dx = \int_{1/6}^{3/6} -3\left[\delta(x) - 2\delta\left(x - \frac{2}{6}\right) + \delta\left(x - \frac{4}{6}\right)\right] dx = 6,$$

$$\ell_{12} = \int_{1/6}^{3/6} -3\left[\delta\left(x - \frac{2}{6}\right) - 2\delta\left(x - \frac{4}{6}\right) + \delta\left(x - \frac{6}{6}\right)\right] dx = -3,$$

$$g_1 = \int_0^1 w_1 g \, dx = \int_{1/6}^{3/6} \left(1 + 4x^2\right) dx = \frac{40}{81}.$$

The first equation is

$$\ell_{11}\alpha_1 + \ell_{12}\alpha_2 = g_1,$$

$$6\alpha_1 - 3\alpha_2 = \frac{40}{81}. \tag{15.78}$$

Using similar technique, one can obtain the second equation:

$$\ell_{21}\alpha_1 + \ell_{22}\alpha_2 = g_2,$$
$$-3\alpha_1 + 6\alpha_2 = \frac{76}{81}. \tag{15.79}$$

Solving for α_1 and α_2, we get

$$\alpha_1 = \frac{52}{243}, \quad \alpha_2 = \frac{64}{243} \tag{15.80}$$

and

$$f \approx \frac{52}{243} T_1\left(x - \frac{2}{6}\right) + \frac{64}{243} T_2\left(x - \frac{4}{6}\right). \tag{15.81}$$

The accuracy can be further improved by choosing a larger value for N but the number of equations will increase.

15.3.4 Finite-Element Method

There are two aspects to the finite-element method:

i. A continuous domain is broken up into a finite number of elements. Figure 15.11 shows a rectangular two-dimensional region broken up into a number of triangular elements (e). The discrete points are the vertices of the triangle. The method aims to find the unknown potentials at the finite number of these discrete points by generating algebraic equations equal in number to the unknown potentials.

ii. The second aspect of the finite-element method is in the technique of generating the algebraic equations.

15.3.5 Variational Principle

A variation principle [5] is used in generating the algebraic equations. Instead of solving the equilibrium Equation 15.65 directly, we try to find the function f that extremizes its "functional" $I(f)$. For a functional, the argument itself is a function. Among all the functions that satisfy the boundary conditions there is one that extremizes the functional and that function is the solution to the equilibrium problem given by Equation 15.65. Before we can use this principle, we have to findout a method to construct a functional for the

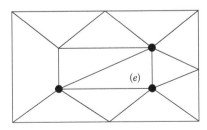

(e)

FIGURE 15.11
Discretization of a continuous domain by finite elements.

equilibrium problem (Equation 15.65). It is known that the functional for Equation 15.65, in the language of linear spaces, is given by

$$I(f) = \langle Lf, f \rangle - \langle 2f, g \rangle \tag{15.82}$$

provided the operator L is positive definite. The operator L is said to be positive definite. If

$$\langle LX_1, X_2 \rangle = \langle X_1, LX_2 \rangle,$$
$$\langle LX, X \rangle \geq 0, \quad \text{for any } X. \tag{15.83}$$

In the above, X, X_1, and X_2 are arbitrary functions that satisfy the same boundary conditions. Table 15.3 lists the common PDEs of electromagnetics and their functionals [10]. The language and the application of Equation 15.82 are best illustrated through a one-dimensional example (Figure 15.12).

TABLE 15.3

Variation Principle Associated with Common PDEs in EM

Name of Equation	PDE	Variational Principle
Inhomogeneous Helmholtz equation	$\nabla^2 \Phi + k^2 \Phi = g$	$I(\Phi) = \dfrac{1}{2} \int_v \left[\left| \nabla \Phi \right|^2 - k^2 \Phi^2 + 2g\Phi \right] dv$
Homogeneous Helmholtz equation	$\nabla^2 \Phi + k^2 \Phi = 0$ or $\nabla^2 \Phi - \dfrac{1}{u^2} \Phi_{tt} = 0$	$I(\Phi) = \dfrac{1}{2} \int_v \left[\left| \nabla \Phi \right|^2 - k^2 \Phi^2 \right] dv$ $I(\Phi) = \dfrac{1}{2} \int\int_v^{t_0} \left[\left| \nabla \Phi \right|^2 - \dfrac{1}{u^2} \Phi_t^2 \right] dv\, dt$
Diffusion equation	$\nabla^2 \Phi - k\Phi_t = 0$	$I(\Phi) = \dfrac{1}{2} \int\int_v^{t_0} \left[\left| \tilde{N}\Phi \right|^2 - k\Phi_t \Phi \right] dv\, dt$
Poisson's equation	$\nabla^2 \Phi = g$	$I(\Phi) = \dfrac{1}{2} \int_v \left[\left| \nabla \Phi \right|^2 + 2g\Phi \right] dv$
Laplace's equation	$\nabla^2 \Phi = 0$	$I(\Phi) = \dfrac{1}{2} \int_v \left[\left| \nabla \Phi \right|^2 \right] dv$

Note: $\Phi_t = \dfrac{\partial \Phi}{\partial t}$; $\Phi_{tt} = \dfrac{\partial^2 \Phi}{\partial t^2}$; $\Phi_x = \dfrac{\partial \Phi}{\partial x}$; $\left| \nabla \Phi \right|^2 = \nabla \Phi \cdot \nabla \Phi = \Phi_x^2 + \Phi_y^2 + \Phi_z^2$.

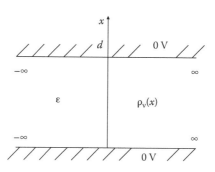

FIGURE 15.12
Parallel plates containing a charge cloud.

Two parallel PEC plates both grounded and separated by a distance $d = 1$ m contain a charge cloud of volume charge density $\rho_v(x)$. The dielectric medium has permittivity ε. The plate dimensions are assumed to be large compared to Poission's equation:

$$-\nabla^2 \Phi = \frac{\rho_v}{\varepsilon} \qquad (15.84)$$

simplifies to the one-dimensional equation

$$-\frac{d^2\Phi}{dx^2} = \frac{\rho_v}{\varepsilon} \qquad (15.85)$$

with the boundary condition:

$$\Phi(0) = \Phi(1) = 0. \qquad (15.86)$$

Defining the inner product for the problem as

$$\langle \psi_1, \psi_2 \rangle = \int_0^1 \psi_1 \psi_2 \, dx. \qquad (15.87)$$

The function for Equation 15.85, from Equation 15.82, can be written as

$$I(\Phi) = \int_0^1 -\frac{d^2\Phi}{dx^2} \Phi \, dx - 2\int_0^1 \Phi(x) \frac{\rho_v(x)}{\varepsilon} \, dx. \qquad (15.88)$$

Integrating the first integral on the RHS of Equation 15.88 and using the boundary conditions (Equation 15.86), we get

$$I(\Phi) = \int_0^1 \left(\frac{d\Phi}{dx} \right)^2 dx - 2\int_0^1 \Phi(x) \frac{\rho_v(x)}{\varepsilon} \, dx. \qquad (15.89)$$

The function $\Phi(x)$ that minimizes (Equation 15.89) is the solution of Equation 15.85 subject to the boundary conditions (Equation 15.86). A physical interpretation of Equation 15.89 may be given if Equation 15.89 is rearranged as

$$I_E(\Phi) = \frac{1}{2}\varepsilon I(\Phi) = \int_0^1 \frac{1}{2}\varepsilon \left| \frac{d\Phi}{dx} \right|^2 dx - \int_0^1 \Phi(x)\rho_v(x) dx. \qquad (15.90)$$

It is easy to see that the functional (Equation 15.90) is the electric potential energy, since $|d\Phi/dx|$ is the magnitude of the electric field $|\bar{E}|$ and $\int_0^1 (1/2)\varepsilon E^2 dx$ is the electric energy stored in the medium. Minimizing the functional I_E minimizes the net potential energy. A similar statement from mechanics is: a gravitational problem can be solved by using Newton's force equation or by minimizing the functional, which is the gravitational potential energy.

FIGURE 15.13
Element (*e*) with endpoints *i* and *j*.

The last step in this explanation is to connect the two aspects and explain the generation of the algebraic equations. The potential function $\Phi(x)$ in each element can be expressed in terms of the potentials of the endpoints of the element (Figure 15.13).

If we use linear interpolation (Figure 15.14),

$$\Phi(x) = \Phi_i + \frac{\Phi_j - \Phi_i}{x_j - x_i}(x - x_i). \tag{15.91}$$

Another way of writing Equation 15.91 is

$$\Phi(x) = N_i(x)\Phi_i + N_j(x)\Phi_j, \tag{15.92}$$

where

$$N_i = \frac{x_j - x}{x_j - x_i}, \tag{15.93a}$$

$$N_j = \frac{x - x_i}{x_j - x_i}. \tag{15.93b}$$

where $N_i(x)$ and $N_j(x)$ are called shape functions. A sketch of these functions is shown in Figure 15.15.

Note that

$$N_i = \begin{cases} 1, & x = x_i, \\ 0, & x = x_j, \end{cases} \quad N_j = \begin{cases} 1, & x = x_j, \\ 0, & x = x_i, \end{cases}$$

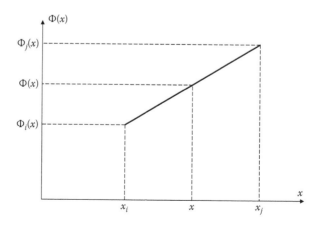

FIGURE 15.14
Linear interpolation.

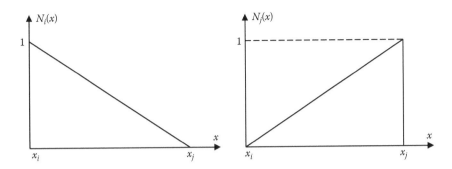

FIGURE 15.15
Sketch of shape functions.

Calling the endpoints as nodes, the shape function N_i is 1 on node i and zero on the other node. In between, it varies linearly. Similar remarks can be made for the shape function N_j, 1 on its own node and zero on the other node. In fact, Equation 15.93a and 15.93b can be derived without resorting to rearranging Equation 15.91, by writing for linear variation

$$N_i(x) = a_i + b_i x \tag{15.94}$$

and since $N_i(x_i) = 1$, $N_i(x_j) = 0$,

$$1 = a_i + b_i x_i, \tag{15.95}$$

$$0 = a_i + b_i x_j, \tag{15.96}$$

$$\begin{bmatrix} 1 & x_i \\ 1 & x_j \end{bmatrix}\begin{bmatrix} a_i \\ b_i \end{bmatrix} = \begin{bmatrix} 1 \\ 0 \end{bmatrix},$$

$$a_i = \frac{\begin{bmatrix} 1 & x_i \\ 0 & x_j \end{bmatrix}}{x_j - x_i} = \frac{x_j}{x_j - x_i},$$

$$b_i = \frac{\begin{bmatrix} 1 & 1 \\ 1 & 0 \end{bmatrix}}{x_j - x_i} = -\frac{1}{x_j - x_i}, \tag{15.97}$$

$$N_i(x) = \frac{x_j - x}{x_j - x_i}.$$

Similar steps will lead to Equation 15.93b. From Equations 15.92 and 15.93,

$$\frac{d\Phi}{dx} = \frac{1}{x_j - x_i}(-1)\Phi_i + \frac{1}{x_j - x_i}(1)\Phi_j = \frac{1}{x_j - x_i}(\Phi_j - \Phi_i). \tag{15.98}$$

The functional $I^{(e)}(\Phi)$ for the element (e) can now be evaluated in terms of Φ_i and Φ_j.

From Equation 15.90,

$$I^{(e)} = \int_{x_i}^{x_j} \frac{1}{2}\varepsilon \frac{1}{(x_j - x_i)^2}(\Phi_j - \Phi_i)^2 \, dx - \int_{x_i}^{x_j}[N_i(x)\Phi_i + N_j(x)\Phi_j]\rho_v(x) dx. \qquad (15.99)$$

Equation 15.99 can further be simplified:

$$I^{(e)} = \frac{1}{2}\varepsilon \frac{(\Phi_j - \Phi_i)^2}{(x_j - x_i)} - \Phi_i \int_{x_i}^{x_j} N_i(x)\rho_v(x) dx - \Phi_j \int_{x_i}^{x_j} N_j(x)\rho_v(x) dx. \qquad (15.100)$$

The total functional can be obtained by summing up the functional for all the elements:

$$I_E(\Phi) = \sum_{\text{elements}} I^{(e)}. \qquad (15.101)$$

The algebraic equations are then obtained by minimizing Equation 15.101 with respect to each of the unknown potentials:

$$\frac{\partial I_E}{\partial \Phi_k} = 0, \quad \Phi_k \ (\text{unknown node potential}). \qquad (15.102)$$

Let us illustrate the computation by taking a simple example. Let $d = 1$, $\rho_v = 1$, and $\varepsilon = 1$ in Figure 15.11 and let the domain be divided into two elements as shown in Figure 15.16. For element 1: $x_i = 0$, $x_j = 1/2$:

$$N_i(x) = \frac{1/2 - x}{1/2 - 0} = 1 - 2x, \quad N_j(x) = \frac{x - 0}{1/2 - 0} = 2x,$$

$$\Phi_i = \Phi_{\boxed{1}} = 0; \quad \Phi_j = \Phi_{\boxed{2}},$$

$$I^{(1)} = \frac{1}{2}\varepsilon \frac{\Phi_{\boxed{2}}^2}{1/2 - 0} - 0\int_0^{1/2}(1 - 2x)dx - \Phi_{\boxed{2}}\int_0^{1/2} 2x \, dx$$

$$= \Phi_{\boxed{2}}^2 - 0 - \Phi_{\boxed{2}} 2 \frac{x^2}{2}\Big|_0^{1/2} = \Phi_{\boxed{2}}^2 - \frac{\Phi_{\boxed{2}}}{4}.$$

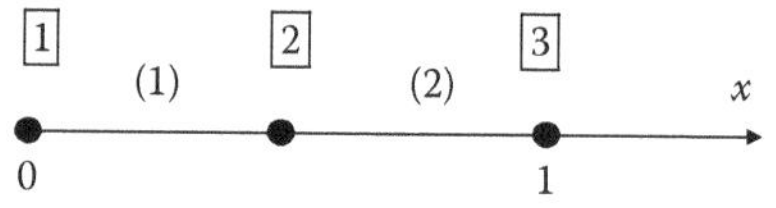

FIGURE 15.16
Domain divided into two elements.

For element 2: $x_i = 1/2$, $x_j = 1$:

$$N_i(x) = \frac{1-x}{1-1/2} = 2(1-x), \quad N_j(x) = \frac{x-1/2}{1-1/2} = 2x-1,$$

$$\Phi_i = \Phi_{\boxed{2}}; \quad \Phi_j = \Phi_{\boxed{3}} = 0,$$

$$I^{(2)} = \frac{1}{2}\varepsilon \frac{(0-\Phi_{\boxed{2}})^2}{1-1/2} - \Phi_{\boxed{2}}\int_{1/2}^{1} 2(1-x)\,dx - 0\int_{1/2}^{1} (2x-1)\,dx$$

$$= \Phi_{\boxed{2}}^2 - 2\Phi_{\boxed{2}}\left(x-\frac{x^2}{2}\right)\Big|_{1/2}^{1} - 0 = \Phi_{\boxed{2}}^2 - \frac{\Phi_{\boxed{2}}}{4},$$

$$I_E = I^{(1)} + I^{(2)} = 2\Phi_{\boxed{2}}^2 - \frac{\Phi_{\boxed{2}}}{2}.$$

For minimum I_E, $\partial I_E/\partial \Phi_{\boxed{2}} = 0$, $4\Phi_{\boxed{2}} - 1/2 = 0$:

$$\Phi_{\boxed{2}} = \frac{1}{8}. \tag{15.103}$$

Let us compare this with the exact answer to this simple problem:

$$-\frac{d^2\Phi}{dx^2} = 1, \quad \begin{array}{l} \Phi(0) = 0, \\ \Phi(1) = 0, \end{array}$$

$$-\frac{d\Phi}{dx} = x + c_1, \tag{15.104}$$

$$-\Phi = \frac{x^2}{2} + c_1 x + c_2.$$

From BC,

$$-0 = 0 + 0 + c_2, \quad c_2 = 0,$$

$$-0 = \frac{1^2}{2} + c_1(1), \quad c_1 = -\frac{1}{2},$$

$$-\Phi = \frac{x^2}{2} - \frac{1}{2}x = \frac{1}{2}x(x-1), \tag{15.105}$$

$$\Phi = \frac{1}{2}x(1-x),$$

$$\Phi\left(\frac{1}{2}\right) = \frac{1}{2}\frac{1}{2}\frac{1}{2} = \frac{1}{8}. \tag{15.106}$$

The exact answer given in Equation 15.105 coincides with the approximate answer by the finite-element method in Equation 15.103; however, it may be noted that this is not true in the entire domain. For example, $\Phi(1/4)$ by exact answer Equation 15.105 is

$$\Phi\left(\frac{1}{4}\right) = \frac{1}{2}\frac{1}{4}\left(1-\frac{1}{4}\right) = \frac{3}{32}. \tag{15.107}$$

FIGURE 15.17
Grid for finite-difference solution.

By the finite-element method, we note that $x = 1/4$ is not an endpoint of an element but it is in the domain of element (1). In this domain,

$$N_i(x) = 1 - 2x, \quad N_j(x) = 2x,$$
$$\Phi(x) = N_i(x)\Phi_i + N_j(x)\Phi_j$$
$$= (1 - 2x)\Phi_{\boxed{1}} + 2x\Phi_{\boxed{2}}, \tag{15.108}$$
$$\Phi\left(\frac{1}{4}\right) = \left(1 - \frac{2}{4}\right)0 + 2\left(\frac{1}{4}\right)\frac{1}{8} = \frac{2}{32}.$$

The source of the error is obvious: the exact solution shows that the potential varies quadratically, whereas the finite-element method we used assumed a linear interpolation. The shape functions are obtained based on Equation 15.94 and are called first-order shape functions. One can define second-order shape functions based on quadratic interpolation. Before we close this topic, let us see what we get if we solve this problem by finite differences. Using central differences (Figure 15.17),

$$-\frac{d^2\Phi}{dx^2}\bigg|_{\boxed{2}} = -\frac{\Phi_{\boxed{1}} + \Phi_{\boxed{3}} - 2\Phi_{\boxed{2}}}{(1/2)^2} = 1.$$

From BC, $\Phi_{\boxed{1}} = \Phi_{\boxed{3}} = 0$, $8\Phi_{\boxed{2}} = 1$, and $\Phi_{\boxed{2}} = 1/8$.

In the one-dimensional case, the first-order finite-element method is equivalent to the finite-difference method.

Before concluding this chapter, let us prove that the operator $L = -d^2/dx^2$ is positive definite [5]. Let Φ_1 and Φ_2 be two functions that satisfy the same boundary conditions as the original problem, namely

$$\Phi_1(0) = 0, \quad \Phi_1(1) = 0, \tag{15.109}$$

$$\Phi_2(0) = 0, \quad \Phi_2(1) = 0, \tag{15.110}$$

$$(LX_1, X_2) = \int_0^1 -\frac{d^2\Phi_1}{dx^2}\Phi_2 dx = -\int_0^1 \Phi_2 d\left(\frac{d\Phi_1}{dx}\right) = -\left[\Phi_2 \frac{d\Phi_1}{dx}\bigg|_0^1 - \int_0^1 \frac{d\Phi_2}{dx}\frac{d\Phi_1}{dx}dx\right],$$

$$(LX_1, X_2) = \int_0^1 \frac{d\Phi_1}{dx}\frac{d\Phi_2}{dx}dx, \tag{15.111}$$

$$(X_1, LX_2) = \int_0^1 \Phi_1\left(-\frac{d^2\Phi_2}{dx^2}\right)dx = -\left[\Phi_1 \frac{d\Phi_2}{dx}\bigg|_0^1 - \int_0^1 \frac{d\Phi_2}{dx}\frac{d\Phi_1}{dx}dx\right] = \int_0^1 \frac{d\Phi_1}{dx}\frac{d\Phi_2}{dx}dx.$$

Thus, we have

$$(LX_1, X_2) = (X_1, LX_2).$$

Also

$$(L\Phi, \Phi) = \int_0^1 -\frac{d^2\Phi}{dx^2}\, \Phi\, dx = -\left[\Phi \frac{d\Phi}{dx}\bigg|_0^1 - \int_0^1 \frac{d\Phi}{dx}\frac{d\Phi}{dx}\, dx \right],$$

$$(L\Phi, \Phi) = \int_0^1 \left| \frac{d\Phi}{dx} \right|^2 dx,$$

which is positive, and we proved $L = -d^2/dx^2$ is positive definite.

References

1. Thom, A. and Apelt, C. J., *Field Computations in Engineering and Physics*, Van Nostr Reinhold, London, U.K., 1961.
2. Crandal, S. H., *Engineering Analysis*, McGraw-Hill, New York, 1956.
3. Conte, S. D., *Elementary Numerical Analysis*, McGraw-Hill, New York, 1965.
4. Harrington, R. F., *Time-Harmonic Electromagnetic Fields*, IEEE Press, New York, 2001.
5. Van Bladel, J., *Electromagnetic Field*, 2nd edn., IEEE Press, New York, 2007.
6. Jin, J. M., *The Finite Element Method in Electromagnetics*, Wiley, New York, 2002.
7. Volakis, J. L., Chatterjee, A., and Kempel, L., *Finite Element Method for Electromagnetics*, Wiley-Interscience, New York, 1998.
8. Pelosi, G., Selleri, S., and Coccioli, R., *Quick Finite Element Method for Electromagnetic Waves*, Artech House, Norwood, MA, 1998.
9. Silvester, P. P. and Ferrari, R. L., *Finite Elements for Electrical Engineers*, 3rd edition, Cambridge University Press, Cambridge, U.K., 1996.
10. Sadiku, N. O. M., *Numerical Techniques in Electromagnetics*, CRC Press, Boca Raton, FL, 1992.

16

Two-Dimensional Problem*

A waveguide problem is a classic two-dimensional problem. Its formulation is discussed in Section 15.1 and we will concentrate on this problem to begin with. For convenience, we will repeat Equation 15.6 and for TM modes:

$$-\nabla_t^2 F = \lambda F, \quad \text{where } F = 0 \text{ on the closed boundary,} \tag{16.1}$$

where the Laplacian operator in the transverse plane is

$$-\nabla_t^2 = -\left[\frac{\partial^2}{\partial x^2} + \frac{\partial^2}{\partial y^2}\right]. \tag{16.2}$$

16.1 Finite-Difference Method

We can approximate the second-order partial derivatives using the following central difference formulas:

$$\left.\frac{\partial^2 F}{\partial x^2}\right|_{j,k} = \frac{F_{j+1,k} + F_{j-1,k} - 2F_{j,k}}{h^2} + O\left(h^2\right),$$

$$\left.\frac{\partial^2 F}{\partial y^2}\right|_{j,k} = \frac{F_{j,k+1} + F_{j,k-1} - 2F_{j,k}}{h^2} + O\left(h^2\right).$$

Equation 16.1 can now be written as

$$\frac{1}{h^2}\left[4F_{j,k} - F_{j+1,k} - F_{j-1,k} - F_{j,k+1} - F_{j,k-1}\right] = \lambda F_{j,k}. \tag{16.3}$$

The application of the "computational molecule" at each of the field points of the rectangular grid will result in an algebraic equation

$$AX = \lambda X. \tag{16.4}$$

By solving Equation 16.4, we obtain the eigenvalues and the corresponding eigenvectors. Let us illustrate this procedure through an example of a rectangular waveguide. We wish to use the FD method to calculate the cutoff wave number and the cutoff frequency.

---

* For chapter appendices, see 16A through 16E in the Appendices section.

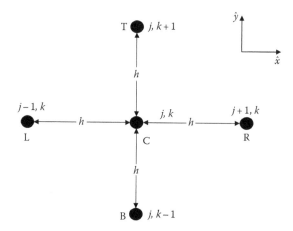

FIGURE 16.1
A central node and the four equidistant surrounding nodes in a Cartesian grid.

Since $F = 0$ on PEC walls, the only unknowns are F_A, F_B, and F_C. We can obtain three equations for the three unknowns by choosing the central node in Figure 16.1 successively at A, B, and C of Figure 16.2.

Since $h = 1/4$,

$$16[4F_A - 0 - 0 - F_B - 0] = \lambda F_A,$$
$$16[4F_B - F_A - 0 - F_C - 0] = \lambda F_B,$$
$$16[4F_C - F_B - 0 - 0 - 0] = \lambda F_C.$$

Arranging them in a matrix form, we obtain

$$AX = \lambda X,$$

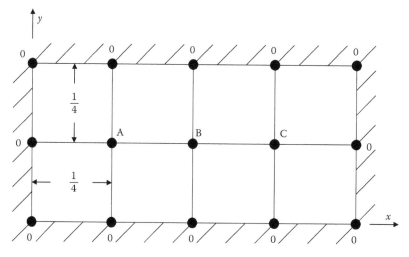

FIGURE 16.2
Rectangular grid for a TM waveguide.

where

$$A = \begin{bmatrix} 64 & -16 & 0 \\ -16 & 64 & -16 \\ 0 & -16 & 64 \end{bmatrix} \qquad (16.5)$$

and

$$X = \begin{bmatrix} F_A \\ F_B \\ F_C \end{bmatrix}. \qquad (16.6)$$

One can now use any standard mathematical software to find λ's and the corresponding eigenvectors. Table 16.1 compares the eigenvalues of Equation 16.5 with the exact answers from the theory of rectangular waveguide (TM modes):

$$\lambda_{mn} = k_c^2 = \left(\frac{m\pi}{1}\right)^2 + \left(\frac{n\pi}{1/2}\right)^2, \quad m = 1,2,\ldots,\infty, \quad n = 1,2,\ldots,\infty. \qquad (16.7)$$

One can obtain better accuracy by reducing h. The size of A will then increase. If we wish to discuss TE modes, the potential is $\tilde{H}_z$ and

$$\tilde{H}_z = G(x,y)e^{-j\beta z}, \qquad (16.8)$$

$$-\nabla_t^2 G = \lambda G. \qquad (16.9)$$

The above equations are similar to Equation 16.1 but the boundary condition is of Neumann type:

$$\frac{\partial G}{\partial n} = 0, \qquad (16.10)$$

on the walls of a PEC and $\hat{n}$ is a unit vector perpendicular to the wall. Referring to Figure 16.3, at a point such as D, $G \neq 0$ but $\partial G/\partial x = 0$, at D. This boundary can be implemented by using the backward-difference formula for the first-order derivative by writing

$$0 = \frac{\partial G}{\partial x}\bigg|_D = \frac{G_D - G_C}{h}, \quad G_D = G_C. \qquad (16.11)$$

However, the backward-difference formula is only accurate to $O(h)$. Even though we used central difference approximation of $O(h^2)$ for the Laplacian operator, we degrade the

TABLE 16.1

Eigenvalues: TM Modes

	Finite Difference	Exact
λ_1	41.3726	$5\pi^2 = 49.3$ (TM$_{11}$)
λ_2	64.00	$8\pi^2 = 78.9$ (TM$_{21}$)
λ_3	86.63	

FIGURE 16.3
Inclusion of an image point E outside the domain.

accuracy to $O(h)$ if we implement Equation 16.11. If we use central difference formula for $\partial G/\partial x$, by considering an image point E outside the domain,

$$\left.\frac{\partial G}{\partial x}\right|_D = \frac{G_E - G_C}{2h} + O\left(h^2\right). \tag{16.12}$$

We can implement the Neumann boundary condition and obtain

$$G_E = G_C. \tag{16.13}$$

The advantage is that the approximations including those for the boundary conditions are of the order of h^2, the price we pay is the increase by one of the unknown potentials, namely G_D is to be calculated. The extra equation needed is obtained by using the computational molecule at D in Figure 16.2:

$$\left.\nabla^2 G\right|_D = \frac{\Phi_C + \Phi_E + \cdots - 4\Phi_D}{h^2} \tag{16.14}$$

$$\left.\nabla^2 G\right|_D = \frac{2\Phi_C + \cdots - 4\Phi_D}{h^2}. \tag{16.15}$$

We eliminated Φ_E by using Equation 16.13. If we decrease h for greater accuracy, the number of equations will increase. However, in any one equation, there are no more than five unknowns (Figure 16.3).

If we arrange the equations as a matrix equation, each row will contain not more than five nonzero elements. Let us say, in Figure 16.2 we make $h = 0.01$ and we are solving the TE problem. The number of unknowns are $100 \times 50 = 5000$ and the matrix size of A is 5000×5000. Any row of A will contain only five nonzero elements. It is extremely inefficient to solve these equations by techniques such as the Gaussian elimination method, since they do not take advantage of the special feature of the matrix, which is "sparse." Numerical analysis books discuss special techniques to invert sparse matrices. A brute force technique such as the "Gaussian elimination method" takes many arithmetic operations and in the process cause "round-off errors," due to rounding off the numbers in floating-point arithmetic. The alternative method is an iterative method described in the following section.

16.2 Iterative Solution

We obtained the algebraic equations by applying the computational molecule for the Laplacian in two dimensions as (see Figure 16.4)

$$\left.\nabla^2 \Phi\right|_C = \frac{\Phi_L + \Phi_B + \Phi_R + \Phi_T + \cdots - 4\Phi_C}{h^2}. \tag{16.16}$$

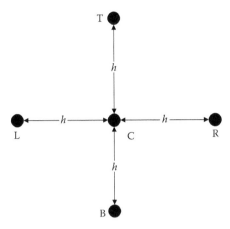

FIGURE 16.4
Computational molecule for the Laplacian operator.

If we are solving the Laplace equation (16.16) becomes

$$\Phi_C = \frac{1}{4}\left(\Phi_L + \Phi_B + \Phi_R + \Phi_T\right). \tag{16.17}$$

Equation 16.17 in words may be stated as: the potential at the center node is the average of the four equidistant surrounding node potentials (left, right, top, and bottom). In the iteration method, we guess a value for the unknowns called zero-iteration values. We calculate new values, first-iteration values using Equation 16.17:

$$\Phi_C^{(1)} = \frac{1}{4}\left(\Phi_L^{(0)} + \Phi_B^{(0)} + \Phi_R^{(0)} + \Phi_T^{(0)}\right). \tag{16.18}$$

After applying Equation 16.18 at a grid point, we move to the next grid point and generate the first-iteration value. It we apply Equation 16.18 at every grid point where the potential is not known, we have a set of first-iteration values. We repeat the process using first-iteration values as guessed values to generate the second set of values. We repeat the process till the set of equations is solved to the desired accuracy and then we terminate the process and accept the values of the last iteration as the approximate solution to the problem:

$$\Phi_C^{(n+1)} = \frac{1}{4}\left(\Phi_L^{(n)} + \Phi_B^{(n)} + \Phi_R^{(n)} + \Phi_T^{(n)}\right). \tag{16.19}$$

Suppose we sweep the grid from left to right and bottom to top, then we notice that we have $(n + 1)$th iteration values as well as nth iteration values for L and B. One could use a new value as soon as it is available rather than waiting till the whole set has new values. Equation 16.19 will be modified as

$$\Phi_C^{(n+1)} = \frac{1}{4}\left(\Phi_L^{(n+1)} + \Phi_B^{(n+1)} + \Phi_R^{(n)} + \Phi_T^{(n)}\right). \tag{16.20}$$

Equation 16.19 describes the "simultaneous displacement method," also called Richardson method [1], whereas Equation 16.20 describes the "successive displace method," also called Liebman method [2]. Intuitively it would appear that the Liebman method should require less number of iterations than the Richardson method. For a linear problem this is true.

For elliptic PDEs such as the Laplace and Helmholtz equations, the solution by FD converges, but the difficulty is its slow convergence. One can speed up the convergence by using additional acceleration techniques. One of them called, extrapolated Liebman method, is easy to code for the computer:

$$\Phi_C^{(n+1)} = \Phi_C^{(n)} + \omega \left[\left\{ \frac{\Phi_L^{(n+1)} + \Phi_B^{(n+1)} + \Phi_R^{(n)} + \Phi_T^{(n)}}{4} \right\} - \Phi_C^{(n)} \right]. \tag{16.21}$$

Note that the value in { } is the value we would have got had we used the Liebman method. The factor ω is called the relaxation factor. Denoting the value in { } as $\Phi_C^{(L)}$, Equation 16.21 may be written as

$$\Phi_C^{(n+1)} = \Phi_C^{(n)} + \omega \left(\Phi_C^{(L)} - \Phi_C^{(n)} \right), \tag{16.22}$$

$$\Phi_C^{(n+1)} = \Phi_C^{(L)} + \alpha \left(\Phi_C^{(L)} - \Phi_C^{(n)} \right), \tag{16.23}$$

where

$$\alpha = \omega - 1. \tag{16.24}$$

The substance of this equation may be stated thus: $\Phi_C^{(n)}$ is the value available from the previous iteration. $\Phi_C^{(L)}$ is the value one would obtain by the Liebman method. $\left(\Phi_C^{(L)} - \Phi_C^{(n)} \right)$ would be the difference between two successive iterations if no relaxation factor were used. By adding a portion of this difference to $\Phi_C^{(L)}$ to obtain the guessed value for the next iteration, we are trying to eliminate at least one iteration in between by predicting the next value from previous experience. Thus, it is only logical that by this relaxation, the convergence rate is improved. The optimum value for ω will give the fastest convergence, while any value $1 < \omega < 2$ would improve the convergence. It is further known that it is preferable to overestimate ω rather than underestimate. This is a useful observation for the reason that the estimation of the optimum relaxation factor, though simple enough in a few cases like a rectangular boundary [1], is in general very difficult. In practice, one could make an experimental study of ω versus the number of iterations for a coarser grid and obtain the optimum ω and use the same optimum ω for a finer grid obtained by reducing the grid spacing h. Kalluri [3], a gist of which is given as Appendix 16E, has more information, examples, and relevant references.

16.3 Finite-Element Method [4–6]

We will illustrate the finite element method (FEM) for a two-dimensional problem by solving the Laplace equation

$$\nabla^2 \Phi = 0, \tag{16.25}$$

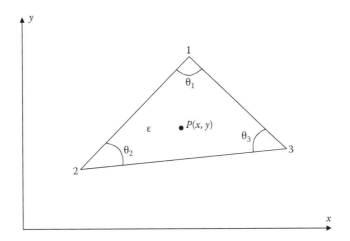

FIGURE 16.5
Triangular finite element.

by minimizing the functional

$$I(\Phi) = \iint \frac{1}{2} \varepsilon \left| \overline{\nabla} \Phi \right|^2 ds. \tag{16.26}$$

We will use a triangle as the finite element as shown in Figure 16.5.

The potential inside the triangle will be expressed in terms of the potentials at vertices 1, 2, 3 of the triangle:

$$\Phi(x,y) = \sum_{k=1}^{3} \zeta_k \Phi_k, \tag{16.27}$$

where ζ_k are the two-dimensional first-order shape functions. These will have the same properties as the one-dimensional first-order shape functions discussed in Section 15.3.5. The properties are ζ_k that can be expressed in terms of x and y as

$$\zeta_k(x,y) = a_k' + b_k' x + c_k' y \tag{16.28}$$

and

$$\zeta_k = \begin{matrix} 1 & \text{on node, } k \\ 0 & \text{on other node.} \end{matrix} \tag{16.29}$$

The ζ's are also called area coordinates, since they have the physical significance of ratios of areas and are useful in integrating an arbitrary function $f(x, y)$ over any area specified by the Cartesian coordinates of its vertices. Figure 16.6 explains the coordinates.

Let

$$\zeta_k = \frac{h_k}{H_k} = \frac{A_k}{A}. \tag{16.30}$$

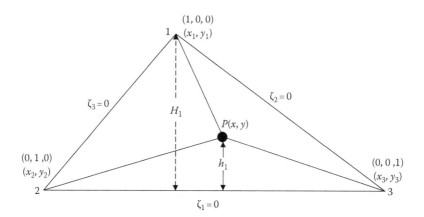

FIGURE 16.6
Area coordinates.

Figure 16.6 shows H_1 which is the height of the node 1 and h_1 is the height of the general point P inside the triangle above the baseline $\overline{23}$. So A_1 is the area of the subtriangle $P23$ and is equal to $(1/2)h_1\left|\overline{23}\right|$. The area of the triangle 123 is $(1/2)H_1\left|\overline{23}\right|$.

Thus, ζ_1 is the ratio of the areas of the subtriangle $P23$ and the triangle 123. It is immediately obvious that

$$\zeta_1 + \zeta_2 + \zeta_3 = 1. \tag{16.31}$$

The node point 1 has the Cartesian coordinates (x_1, y_1) but its ζ coordinates are $(1,0,0)$ since ζ_1 of node 1 is 1 ($h_1 = H_1$ if P is node 1) and $\zeta_2 = \zeta_3 = 0$. In Figure 16.6, the ζ coordinates of each node are written as triplet of numbers $(\zeta_1, \zeta_2, \zeta_3)$. From those numbers, it is obvious that the requirement of shape functions specified in Equation 16.29 are met. The next step in determining the shape function is to solve for a_k' in Equation 16.28 in terms of the node coordinates of the triangle. This process can be initiated by expressing the Cartesian coordinate x in terms of the area coordinates ζ_1, ζ_2, and ζ_3:

$$x = d_1\zeta_1 + d_2\zeta_2 + d_3\zeta_3. \tag{16.32}$$

If the point P is brought to node 1, $x = x_1$ and $\zeta_1 = 1$, $\zeta_2 = 0$, $\zeta_3 = 0$, and thus,

$$x_1 = d_1(1) + d_2(0) + d_3(0),$$

$$x_1 = d_1. \tag{16.33}$$

Similar argument will show $x_2 = d_2$ and $x_3 = d_3$. Equation 16.32 becomes

$$x = x_1\zeta_1 + x_2\zeta_2 + x_3\zeta_3. \tag{16.34}$$

Similarly,

$$y = y_1\zeta_1 + y_2\zeta_2 + y_3\zeta_3. \tag{16.35}$$

Equations 16.31, 16.34, and 16.35 may be written as

$$\begin{bmatrix} 1 & 1 & 1 \\ x_1 & x_2 & x_3 \\ y_1 & y_2 & y_3 \end{bmatrix} \begin{bmatrix} \zeta_1 \\ \zeta_2 \\ \zeta_3 \end{bmatrix} = \begin{bmatrix} 1 \\ x \\ y \end{bmatrix}. \tag{16.36}$$

Solving Equation 16.36 by Kramer's rule, we obtain

$$\zeta_1 = \frac{\begin{vmatrix} 1 & 1 & 1 \\ x & x_2 & x_3 \\ y & y_2 & y_3 \end{vmatrix}}{\begin{vmatrix} 1 & 1 & 1 \\ x_1 & x_2 & x_3 \\ y_1 & y_2 & y_3 \end{vmatrix}}. \tag{16.37}$$

It can be recognized that the denominator determinant is twice the area of the triangle A. Thus, we obtain

$$\zeta_1 = \frac{1}{2A} \left[(x_2 y_3 - x_3 y_2) + (y_2 - y_3) x + (x_3 - x_2) y \right]. \tag{16.38}$$

Comparing with Equation 16.28 and denoting $a_k' = a_k/2A$, $b_k' = b_k/2A$, $c_k' = c_k/2A$, we obtain

$$a_1 = x_2 y_3 - x_3 y_2, \tag{16.39a}$$

$$b_1 = y_2 - y_3, \tag{16.39b}$$

$$c_1 = x_3 - x_2, \tag{16.39c}$$

and

$$A = \frac{1}{2} \begin{bmatrix} 1 & 1 & 1 \\ x_1 & x_2 & x_3 \\ y_1 & y_2 & y_3 \end{bmatrix}. \tag{16.40}$$

Similarly, we get

$$\zeta_2 = \frac{1}{2A} \left[a_2 + b_2 x + c_2 y \right], \tag{16.41}$$

$$a_2 = x_3 y_1 - x_1 y_3, \tag{16.42a}$$

$$b_2 = y_3 - y_1, \tag{16.42b}$$

$$c_2 = x_1 - x_3, \tag{16.42c}$$

$$\zeta_3 = \frac{1}{2A}\left[a_3 + b_3 x + c_3 y\right], \tag{16.43}$$

$$a_3 = x_1 y_2 - x_2 y_1, \tag{16.44a}$$

$$b_3 = y_1 - y_2, \tag{16.44b}$$

$$c_3 = x_2 - x_1. \tag{16.44c}$$

It can be shown that the area A can also be written as (see Appendix 16B)

$$A = \left|\frac{1}{2}\left(b_2 c_3 - b_3 c_2\right)\right|. \tag{16.45}$$

Having found ζ_k in terms of the coordinates of the nodes, we can now express $\bar{\nabla}\Phi$ in terms of the gradient of $\zeta_k(x, y)$. From Equation 16.27,

$$\bar{\nabla}\Phi\left(x,y\right) = \sum_{k=1}^{3} \Phi_k \bar{\nabla}\zeta_k\left(x,y\right) \tag{16.46}$$

$$\left|\bar{\nabla}\Phi\right|^2 = \bar{\nabla}\Phi \cdot \bar{\nabla}\Phi = \sum_{j=1}^{3}\sum_{k=1}^{3} \Phi_j \Phi_k \bar{\nabla}\zeta_j\left(x,y\right) \cdot \bar{\nabla}\zeta_k\left(x,y\right). \tag{16.47}$$

From Equation 16.26, the functional for an element (e) can be written as

$$I^{(e)} = \iint_{(e)} \frac{\varepsilon}{2}\left|\bar{\nabla}\Phi\right|^2 ds = \frac{\varepsilon}{2}\sum_{j=1}^{3}\sum_{k=1}^{3} \Phi_j \Phi_k \iint_{(e)} \bar{\nabla}\zeta_j \cdot \bar{\nabla}\zeta_k \, ds. \tag{16.48}$$

Denoting the last integral by $S_{jk}^{(e)}$, the general element of a square matrix, called stiffness matrix, is given by

$$S_{jk}^{(e)} = \iiint_{(e)} \bar{\nabla}\zeta_j \cdot \bar{\nabla}\zeta_k \, ds. \tag{16.49}$$

Now the functional for the element (e) can be written as

$$I^{(e)} = \frac{\varepsilon}{2}\sum_{j=1}^{3} \Phi_j \sum_{k=1}^{3} \Phi_k S_{jk}^{(e)}. \tag{16.50}$$

Let us evaluate $S_{jk}^{(e)}$ to show how it works:

$$\bar{\nabla}\zeta_j = \hat{x}\frac{\partial\zeta_j}{\partial x} + \hat{y}\frac{\partial\zeta_j}{\partial y} = \frac{1}{2A}\left(\hat{x}b_j + \hat{y}c_j\right),$$

$$\bar{\nabla}\zeta_k = \hat{x}\frac{\partial\zeta_k}{\partial x} + \hat{y}\frac{\partial\zeta_k}{\partial y} = \frac{1}{2A}\left(\hat{x}b_k + \hat{y}c_k\right),$$

$$\bar{\nabla}\zeta_j \cdot \bar{\nabla}\zeta_k = \frac{1}{4A^2}\left(b_j b_k + c_j c_k\right), \tag{16.51}$$

$$\therefore S_{jk}^{(e)} = \frac{1}{4A^2}\left(b_j b_k + c_j c_k\right)\iint\limits_{(e)} ds, \tag{16.52}$$

$$S_{jk}^{(e)} = \frac{1}{4A}\left(b_j b_k + c_j c_k\right).$$

Given the coordinates of the vertices of the triangle b's and c's can be determined from Equations 16.39 through 16.44 and $[S^{(e)}]$ can be determined from Equation 16.52.

An alternative expression for $[S^{(e)}]$ is in terms of vertex angles θ_1, θ_2, and θ_3 of Figure 16.5 (see Appendix 16B):

$$\left[S^{(e)}\right] = \frac{1}{2}\begin{bmatrix} \cot\theta_2 + \cot\theta_3 & -\cot\theta_3 & -\cot\theta_2 \\ -\cot\theta_3 & \cot\theta_1 + \cot\theta_3 & -\cot\theta_1 \\ -\cot\theta_2 & -\cot\theta_1 & \cot\theta_1 + \cot\theta_2 \end{bmatrix}. \tag{16.53}$$

Let us illustrate the technique by a simple example. Given $\Phi_1 = 10$, $\Phi_2 = -5$, determine Φ_3 if the Laplace equation is satisfied inside the isosceles triangle shown in Figure 16.7.

Using Equation 16.53, we get the S matrix:

$$[S] = \frac{1}{2}\begin{bmatrix} 2 & -1 & -1 \\ -1 & 1 & 0 \\ -1 & 0 & 1 \end{bmatrix}. \tag{16.54}$$

The functional I is given by

$$I = \frac{\varepsilon}{2}\sum_{j=1}^{3}\sum_{k=1}^{3}\Phi_j\Phi_k S_{jk}$$

$$= \frac{\varepsilon}{2}\left[S_{11}\Phi_1^2 + S_{12}\Phi_1\Phi_2 + S_{13}\Phi_1\Phi_3 + S_{21}\Phi_2\Phi_1 + S_{22}\Phi_2^2 + S_{23}\Phi_2\Phi_3 + S_{31}\Phi_3\Phi_1 + S_{32}\Phi_3\Phi_2 + S_{33}\Phi_3^2\right]$$

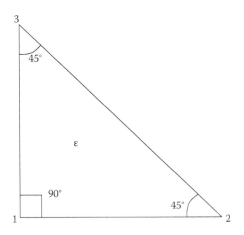

FIGURE 16.7
Isosceles triangle.

The algebraic equations are generated by minimizing the functional with respect to the unknown potential. In this example, Φ_3 is unknown and the equation generated is

$$\frac{\partial I}{\partial \Phi_3} = 0 = \frac{\varepsilon}{2}\left[S_{13}\Phi_1 + S_{23}\Phi_2 + S_{31}\Phi_1 + S_{32}\Phi_2 + 2S_{33}\Phi_3\right].$$ (16.55)

Noting

$$S_{jk} = S_{kj},$$ (16.56)

$$S_{31}\Phi_1 + S_{32}\Phi_2 + S_{33}\Phi_3 = 0.$$ (16.57)

For the example on hand, since $S_{31} = -1/2$, $S_{32} = 0$, $S_{33} = 1/2$, $\Phi_1 = 10$, and $\Phi_2 = -5$,

$$\left(-\frac{1}{2}\right)(10) + (0)(-5) + \left(\frac{1}{2}\right)\Phi_3 = 0,$$

$$\Phi_3 = 10.$$

In this example, nodes 1 and 2 have fixed potential and are called prescribed nodes. Node 3 is called free node. If all the nodes are free, we get the matrix equation

$$\begin{bmatrix} S_{11} & S_{12} & S_{13} \\ S_{21} & S_{22} & S_{23} \\ S_{31} & S_{32} & S_{33} \end{bmatrix}\begin{bmatrix} \Phi_1 \\ \Phi_2 \\ \Phi_3 \end{bmatrix} = 0.$$ (16.58)

One important advantage of FEM based on minimization of the functional is that the Neumann-type boundary conditions are automatically satisfied (see Appendix 16C) and are called natural boundary conditions. So nodes on such boundaries are free nodes. Dirichlet-type (potential specified) boundary conditions are essential boundary conditions and the nodes on such boundaries are prescribed nodes.

16.3.1 Two Elements

To generalize the procedure, let us consider that the problem domain is divided into two elements (Figure 16.8). Let the stiffness matrices computed for each element be denoted by $[S^{(1)}]$ and $[S^{(2)}]$. The functional $I(\Phi)$ for the problem is given by

$$\begin{aligned} I(\Phi) &= I^{(1)} + I^{(2)} \\ &= \frac{\varepsilon_1}{2}\left[S_{11}^{(1)}\Phi_1^2 + S_{12}^{(1)}\Phi_1\Phi_2 + S_{13}^{(1)}\Phi_1\Phi_3 + S_{21}^{(1)}\Phi_2\Phi_1 + S_{22}^{(1)}\Phi_2^2\right. \\ &\quad \left. + S_{23}^{(1)}\Phi_2\Phi_3 + S_{31}^{(1)}\Phi_3\Phi_1 + S_{32}^{(1)}\Phi_3\Phi_2 + S_{33}^{(1)}\Phi_3^2\right] \\ &\quad + \frac{\varepsilon_2}{2}\left[S_{11}^{(2)}\Phi_1^2 + S_{13}^{(2)}\Phi_1\Phi_3 + S_{14}^{(2)}\Phi_1\Phi_4 + S_{31}^{(2)}\Phi_3\Phi_1 + S_{33}^{(2)}\Phi_3^2\right. \\ &\quad \left. + S_{34}^{(2)}\Phi_3\Phi_4 + S_{41}^{(2)}\Phi_4\Phi_1 + S_{43}^{(2)}\Phi_4\Phi_3 + S_{44}^{(2)}\Phi_4^2\right]. \end{aligned}$$ (16.59)

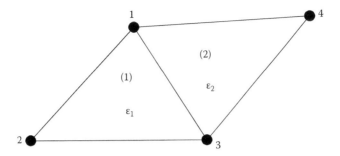

FIGURE 16.8
Two elements.

The algebraic equations are obtained by differentiating $I(\Phi)$ partially with respect to Φ_j and equating to zero. For example,

$$\frac{\partial I}{\partial \Phi_1} = 0 = \frac{\varepsilon_1}{2}\left[2S_{11}^{(1)}\Phi_1 + S_{12}^{(1)}\Phi_2 + S_{13}^{(1)}\Phi_3 + S_{21}^{(1)}\Phi_2 + S_{31}^{(1)}\Phi_3\right]$$
$$+ \frac{\varepsilon_2}{2}\left[2S_{11}^{(2)}\Phi_1 + S_{13}^{(2)}\Phi_3 + S_{14}^{(2)}\Phi_4 + S_{31}^{(2)}\Phi_3 + S_{41}^{(2)}\Phi_4\right]. \tag{16.60}$$

From

$$S_{ij}^{(e)} = S_{ji}^{(e)}, \tag{16.61}$$

the above equation can be written as

$$\left(\varepsilon_1 S_{11}^{(1)} + \varepsilon_2 S_{11}^{(2)}\right)\Phi_1 + \varepsilon_1 S_{12}^{(1)}\Phi_2 + \left(\varepsilon_1 S_{13}^{(1)} + \varepsilon_2 S_{13}^{(2)}\right)\Phi_3 + \varepsilon_2 S_{14}^{(2)}\Phi_4 = 0. \tag{16.62}$$

In matrix form,

$$\left[\varepsilon_1 S_{11}^{(1)} + \varepsilon_2 S_{11}^{(2)} \quad \varepsilon_1 S_{12}^{(1)} \quad \varepsilon_1 S_{13}^{(1)} + \varepsilon_2 S_{13}^{(2)} \quad \varepsilon_2 S_{14}^{(2)}\right]\begin{bmatrix} \Phi_1 \\ \Phi_2 \\ \Phi_3 \\ \Phi_4 \end{bmatrix} = 0. \tag{16.63}$$

When all the equations are included and arranged in matrix form, we get

$$\begin{bmatrix} \varepsilon_1 S_{11}^{(1)} + \varepsilon_2 S_{11}^{(2)} & \varepsilon_1 S_{12}^{(1)} & \varepsilon_1 S_{13}^{(1)} + \varepsilon_2 S_{13}^{(2)} & \varepsilon_2 S_{14}^{(2)} \\ \varepsilon_1 S_{21}^{(1)} & \varepsilon_1 S_{22}^{(1)} & \varepsilon_1 S_{23}^{(1)} & 0 \\ \varepsilon_1 S_{31}^{(1)} + \varepsilon_2 S_{31}^{(2)} & \varepsilon_1 S_{32}^{(1)} & \varepsilon_1 S_{33}^{(1)} + \varepsilon_2 S_{33}^{(2)} & \varepsilon_2 S_{34}^{(2)} \\ \varepsilon_2 S_{41}^{(2)} & 0 & \varepsilon_2 S_{43}^{(2)} & \varepsilon_2 S_{44}^{(2)} \end{bmatrix}\begin{bmatrix} \Phi_1 \\ \Phi_2 \\ \Phi_3 \\ \Phi_4 \end{bmatrix} = 0, \tag{16.64}$$

$$[S][\Phi] = 0. \tag{16.65}$$

Let us see if we can formulate any rule for obtaining the elements of the "global *S*-matrix" from the individual $S^{(e)}$ matrices. Let us look at node 1. This node is common to both the elements. There will be contributions from the stiffness coefficients $S_{11}^{(1)}$ and $S_{11}^{(2)}$. By multiplying by the permittivity of the element and adding them up, we obtain

$$S_{11} = \varepsilon_1 S_{11}^{(1)} + \varepsilon_2 S_{11}^{(2)}. \tag{16.66}$$

Let us look at the global element S_{13}. The line $\overline{13}$ is common to both the elements:

$$S_{13} = \varepsilon_1 S_{13}^{(1)} + \varepsilon_2 S_{13}^{(2)}. \tag{16.67}$$

Following this rule,

$$S_{14} = \varepsilon_2 S_{14}^{(2)}. \tag{16.68}$$

We also note

$$S_{24} = 0, \tag{16.69}$$

since there is no direct connection between nodes 2 and 4.

Example

We wish to calculate the potential at the point $P(1,(1/3))$ if the Laplace equation is satisfied and the boundary conditions are as shown in Figure 16.9. Symmetry allows us to consider half the field region, which has three elements and five nodes. Nodes 2, 3, 4, and 5 are the prescribed nodes.

$$\Phi_2 = \Phi_5 = 0; \quad \Phi_3 = 100. \tag{16.70}$$

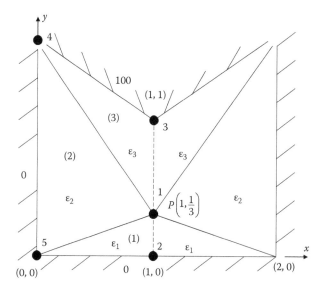

FIGURE 16.9
Example of FEM solution of the Laplace equation for a region with different dielectrics. Node 4 has coordinates (0, 2).

There is a bit of ambiguity about the potential of the node 4, because of the infinitesimal gap. For the purpose of this calculation, let us assume that node 4 is on the 100 V electrode:

$$\Phi_4 = 100. \tag{16.71}$$

Assuming that we calculated the global S-matrix elements from the given coordinates of the nodes, we have the equation

$$S_{11}\Phi_1 + S_{12}\Phi_2 + S_{13}\Phi_3 + S_{14}\Phi_4 + S_{15}\Phi_5 = 0.$$

Substituting the prescribed node potential values,

$$S_{11}\Phi_1 + S_{12}(0) + S_{13}(100) + S_{14}(100) + S_{15}(0) = 0,$$

$$\Phi_1 = \frac{-100S_{13} - 100S_{14}}{S_{11}}. \tag{16.72}$$

Note that the line 213 is the line of symmetry and the boundary condition on this line is

$$\left.\frac{\partial \Phi}{\partial x}\right|_{x=1} = 0. \tag{16.73}$$

The Neumann boundary condition is automatically satisfied and the node 1 is a free node. See Appendix 16C. A general solution of Equation 16.65, where Φ is a column of potentials of free nodes, and prescribed nodes may be obtained by writing the matrix equation in a partitioned form:

$$\begin{bmatrix} S_{ff} & S_{fp} \\ S_{pf} & S_{pp} \end{bmatrix}\begin{bmatrix} \Phi_f \\ \Phi_p \end{bmatrix} = 0, \tag{16.74}$$

$$\therefore \left[S_{ff}\right]\left[\Phi_f\right] + \left[S_{fp}\right]\left[\Phi_p\right] = 0,$$

$$\left[\Phi_f\right] = -\left[S_{ff}\right]^{-1}\left[S_{fp}\right]\left[\Phi_p\right]. \tag{16.75}$$

16.3.2 Global and Local Nodes

We generated the global S-matrix by visually observing the connection of various nodes. In a practical problem, there will be hundreds of elements and it is necessary to generate a scheme (a computer program) to describe the interconnections through inputting the data and obtain as output the assembled global matrix. It can be done in an elegant way by considering local and global node numbering. Let us illustrate this through element (2) of Figure 16.8. See Figure 16.10.

The local node numbers are written inside the triangle and the global node numbers are written outside the triangle enclosed by a box ☐. The local node numbering is arbitrary except that the node sequence indicates one sense of progression, say counterclockwise. Then we create a table describing for each element the correspondence between the local

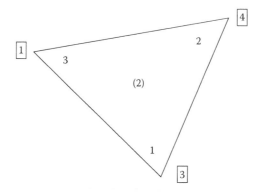

FIGURE 16.10
Local and global node numbering.

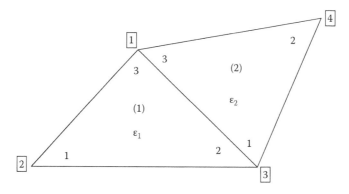

FIGURE 16.11
Local and global node numbering for two elements.

node number and the global node number. Figures 16.10 and 16.11 show the two elements of Figure 16.8 and the local node array table for the figure is given in Table 16.2.

We then write a function program that takes as input the relevant data describing the node connections, and so on, and gives as output the global matrix S. The function statement can be function $[S]$ = GLANT (Nn, Ne, $n1L$, $n2L$, $n3L$, X_n, Y_n). The meanings of the input arguments and their values for Figure 16.11 are

Nn: number of global nodes: 4

Ne: number of elements: 2

n1L: the array of global node numbers corresponding to local node number 1: ②, ③

TABLE 16.2

Local Node Array Table

Element # (e)	ε_e	n1L	n2L	n3L
(1)	ε_1	②	③	①
(2)	ε_2	③	④	①

n2L: the array of global node numbers corresponding to local node number 2: $\boxed{3}$, $\boxed{4}$

n3L: the array of global node numbers corresponding to local node number 3: $\boxed{1}$, $\boxed{1}$

The function program will be so written that the output will be the *S*-matrix.

16.3.2.1 Example of a Main Program

The details of the function program will be explained later but let us first illustrate the data inputting through an example given in Appendix 16A.1, which solves the electrostatic problem (the Laplace equation) shown in Figure 16.12; given $\Phi_{\boxed{2}} = V_2 = 10$, $\Phi_{\boxed{4}} = V_4 = -10$, determine $\Phi_{\boxed{1}}$ and $\Phi_{\boxed{3}}$. Most of the statements are obvious. After obtaining the *S*-matrix, we proceed to describe the free nodes and prescribed nodes:

$$frn = [1,3],$$

$$prn = [2,4],$$

then we give the potentials of the prescribed nodes:

$$Vprn = [10,-10].$$

The other computer input will be the *x*- and *y*-coordinates of the global nodes arranged as in Tables 16.3 through 16.5.

Then we obtain the submatrices $[S_{ff}]$ and $[S_{fp}]$ of $[S]$:

$$Sff = S(frn,frn), Sfp = S(frn,prn),$$

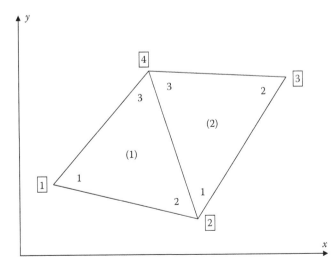

FIGURE 16.12
Two-element example problem.

TABLE 16.3

Table for the Coordinates of the Global Nodes

Global Node	x	y
1		
2		
3		
4		

TABLE 16.4

Node Location for the Example in Figure 6.12

Global Node #	x	y
1	0.8	1.8
2	1.4	1.4
3	2.1	2.1
4	1.2	2.7

TABLE 16.5

Local Node Array Table for the Example in Figure 16.12

Element # e	$n(1, e)$	$n(2, e)$	$n(3, e)$
	$n1L$	$n2L$	$n3L$
(1)	1	2	4
(2)	2	3	4

then we compute the potentials of the free nodes:

$$Vf = -inv(Sff) * S_{fp} * (Vprn)',$$ (16.76)

and get the answers

$$\Phi_{\boxed{1}} = 2.6012,$$
$$\Phi_{\boxed{3}} = -1.0983.$$

Equation 16.75 is implemented by the MATLAB® statement (Equation 16.76). Note that the prime (') in $(Vprn)'$ converts the row matrix into column matrix. It is worth noting the sizes of the matrices in Equation 16.76:

$$S_{ff}: (f_{rn}, f_{rn}) \quad V_f: (f_{rn}, 1),$$
$$S_{fp}: (f_{rn}, p_{rn}) \quad (V_{prn})': (p_{rn}, 1).$$

16.3.2.2 *Example of a Function Program (GLANT) to Assemble the Global S-Matrix*

GLANT is the name used for the function program as an abbreviation of the Global Assembly of Node-based Triangular elements. The output S is the global S-matrix. This program is written assuming $\varepsilon = 1$ for all elements. We need to modify it in a minor way if the elements have different permittivity or we will be using it to solve Poisson's equation. This program is written to obtain as output another global matrix called T-matrix. This matrix will be useful to study the waveguide problem and will be explained later. The key steps are as follows:

1. Initialize to zero all elements of $[S]$.
2. For each element, calculate the local $[S]$ matrix $= [S^{(e)}]$.
3. For each element, locate the global node numbering of the local nodes 1, 2, 3, and Convert $S^{(e)}$ elements to the corresponding S elements with global node subscripts. Add the value to the element in the global S-matrix. To illustrate, suppose for element (2) local node number 1 corresponds to global node number $\boxed{2}$ and the local node number 3 corresponds to the global node number $\boxed{4}$. Then $S_{13}^{(2)}$ will be added to the global element $S_{\boxed{2}\boxed{4}}$.
4. When step 3 is completed for all the elements, the global $[S]$ matrix is assembled.

The table of correspondence of local node numbers of various elements is stored in two-dimensional arrays, $n(1, e)$, $n(2, e)$, and $n(3, e)$. The for loop in MATLAB that achieves this is given by

```
for e = 1: Ne;
      n(1,e)  = n1L(e);
      n(2,e)  = n2L(e);
      n(3,e)  = n3L(e);
end
```

The initialization of the $[S]$ matrix is important. The global element $S_{\boxed{1}\boxed{3}} = 0$ since global nodes $\boxed{1}$ and $\boxed{3}$ are not directly connected (see Figure 16.12). The process of Step 3 will never get a value for $S_{\boxed{1}\boxed{3}}$ and the initialization keeps this value at zero as required.

Note that the global element

$$S_{\boxed{2}\boxed{4}} = 0 + S_{23}^{(1)} + S_{13}^{(2)}. \tag{16.77}$$

The first value on the RHS is because of initialization, the second value comes from element (1), and the third from element (2). The main program and MATLAB CODE, for the function GLANT, are given in Appendix 16A.1.

16.3.3 Standard Area Integral

We will take a short digression from the two-dimensional FEM to discuss a standard integral useful in performing integration over an arbitrary triangle like that in Figure 16.6. The standard area integral obtained by integration by parts (see Appendix 16D) is given by

$$\iint_{\Delta^{(e)}} (\zeta_1)^\ell (\zeta_2)^m (\zeta_3)^n \, dx dy = \frac{\ell!m!n!}{(\ell+m+n+2)!} 2A^{(e)}, \tag{16.78}$$

where ζ_1, ζ_2, and ζ_3 are the area coordinates and ℓ, m, and n are the powers of these coordinates, respectively. We will illustrate how Equation 16.78 is useful in integrating a function $f(x, y)$ that is made up of polynomials in x and y over an arbitrary triangular area. Let

$$f(x,y) = x^2. \tag{16.79}$$

Since

$$\begin{aligned} x^2 &= \left(x_1\zeta_1 + x_2\zeta_2 + x_3\zeta_3 \right)^2 \\ &= x_1^2\zeta_1^2 + x_2^2\zeta_2^2 + x_3^2\zeta_3^2 + 2x_1x_2\zeta_1\zeta_2 + 2x_2x_3\zeta_2\zeta_3 + 2x_3x_1\zeta_1\zeta_3. \end{aligned}$$

We need to first study the integrals like

$$\iint\limits_{\Delta^{(e)}} \zeta_1^2 \, dx \, dy \tag{16.80}$$

and

$$\iint\limits_{\Delta^{(e)}} \zeta_1\zeta_2 \, dx \, dy. \tag{16.81}$$

The answer to Equation 16.80 can be written by inspection by using the standard integral Equation 16.78 and noting $\ell = 2$, $m = 0$, and $n = 0$. Thus, we get

$$\iint\limits_{\Delta^{(e)}} \zeta_1^2 \, dx \, dy = \frac{2!0!0!}{(2+0+0+2)!} 2A^{(e)} = \frac{A^{(e)}}{6}. \tag{16.82}$$

To evaluate Equation 16.81, we note $\ell = 1$, $m = 1$, and $n = 0$ in the standard integral and write

$$\iint\limits_{\Delta^{(e)}} \zeta_1\zeta_2 \, dx \, dy = \frac{1!1!0!}{(1+1+0+2)!} 2A^{(e)} = \frac{A^{(e)}}{12}. \tag{16.83}$$

Of course, we can generalize and write

$$\iint\limits_{\Delta^{(e)}} \zeta_k^2 \, dx \, dy = \frac{A^{(e)}}{6}, \tag{16.84}$$

$$\iint\limits_{\Delta^{(e)}} \zeta_j\zeta_k \, dx \, dy = \frac{A^{(e)}}{12}, \quad j \neq k. \tag{16.85}$$

$$\iint\limits_{\Delta^{(e)}} x^2 \, dx \, dy = \left(x_1^2 + x_2^2 + x_3^2 \right)\frac{A^{(e)}}{6} + \left(2x_1x_2 + 2x_2x_3 + 2x_3x_1 \right)\frac{A^{(e)}}{12}. \tag{16.86}$$

16.4 FEM for Poisson's Equation in Two Dimensions

The functional for Poisson's equation

$$\nabla_t^2 \Phi = -\frac{\rho}{\varepsilon} \tag{16.87}$$

is

$$I(\Phi) = \frac{1}{2} \iint \varepsilon \left| \overline{\nabla}_t \Phi \right|^2 ds - \iint \Phi \rho ds. \tag{16.88}$$

For a given element,

$$I^{(e)}(\Phi) = \frac{1}{2} \iint_{\Delta^{(e)}} \varepsilon^{(e)} \left| \overline{\nabla}_t \Phi \right|^2 ds - \iint_{\Delta^{(e)}} \Phi \rho ds. \tag{16.89}$$

We have already seen the contribution to the algebraic equation from the first term on the RHS of Equation 16.89. The second term

$$\iint_{\Delta^{(e)}} \Phi \rho ds \approx \iint_{\Delta^{(e)}} \sum_{k=1}^{3} \zeta_k \Phi_k \rho ds. \tag{16.90}$$

If we assume a constant value of $\rho^{(e)}$ in one element, then

$$\iint_{\Delta^{(e)}} \Phi \rho ds \approx \sum_{k=1}^{3} \rho^{(e)} \Phi_k \iint_{\Delta^{(e)}} \zeta_k ds. \tag{16.91}$$

From Equation 16.78,

$$\iint_{\Delta^{(e)}} \zeta_1 dx dy = \frac{1!0!0!}{(1+0+0+2)!} 2A^{(e)} = \frac{A^{(e)}}{3}. \tag{16.92}$$

Thus, we have

$$\iint_{\Delta^{(e)}} \Phi \rho ds = \frac{A^{(e)}}{3} \rho^{(e)} \left[\Phi_1 + \Phi_2 + \Phi_3 \right]. \tag{16.93}$$

The minimization of the term in Equation 16.93 with reference to Φ_1 gives the term

$$\frac{A^{(e)}}{3} \rho^{(e)}(1).$$

The algebraic equation obtained by minimizing the functional Equation 16.89 is

$$\varepsilon^{(e)} \left[S^{(e)} \right] \begin{bmatrix} \Phi_1 \\ \Phi_2 \\ \Phi_3 \end{bmatrix} = \begin{bmatrix} g \\ g \\ g \end{bmatrix}, \tag{16.94}$$

where $g = (A^{(e)}/3) \rho^{(e)}$.

Example

This problem is of interest in oil industry. Consider a typical container partly filled with charged liquid $\varepsilon = 2.5\varepsilon_0$ and $\rho = 10^{-5}$ c/m³. Determine the potential at $P(0.5, 1)$.

Let us divide the domain into six domains and assign the local and global nodes as shown in Figures 16.13 and 16.14.

Since the container is a PEC (grounded), all the nodes, except for $\boxed{1}$, are prescribed nodes with zero voltage. Hence, the equation we will be solving is $S_{11}\Phi_1 = g_1$, where S_{11} and g_1 are global values. We need to calculate $S_{11}^{(e)}$ for the elements shown in Figure 16.15; take, for example, the first element:

$$S_{11}^{(1)} = \frac{1}{4A^{(1)}}\left(b_1^2 + c_1^2\right),$$

$$A^{(1)} = \frac{1}{2}(0.5)(1) = \frac{1}{4},$$

$$b_1 = y_2 - y_3 = 1 - 0 = 1,$$

$$c_1 = x_3 - x_2 = 0 - 0 = 0,$$

$$S_{11}^{(1)} = \frac{1}{(4)1/4}\left(1^2 + 0^2\right) = 1.$$

Note that the elements (3), (4), and (6) are similar. Hence

$$S_{11}^{(1)} = S_{11}^{(3)} = S_{11}^{(4)} = S_{11}^{(6)}.$$

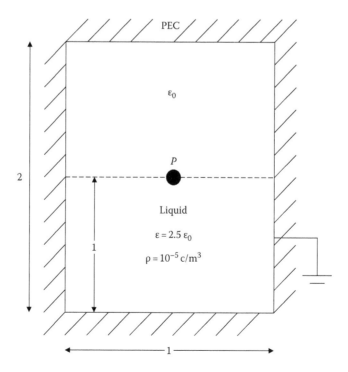

FIGURE 16.13
Container partly filled with charged liquid.

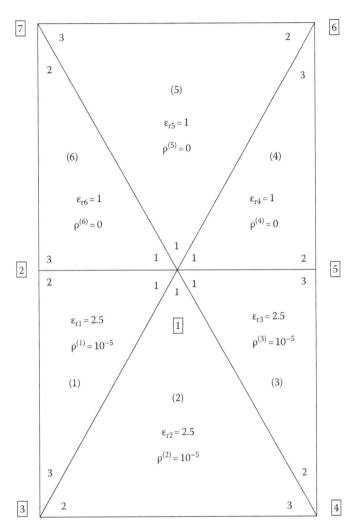

FIGURE 16.14
Local and global nodes for the six elements.

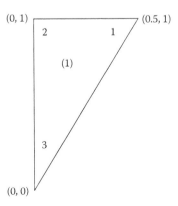

FIGURE 16.15
One of the typical elements of Figure 16.14.

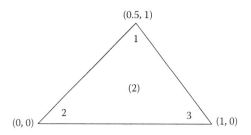

FIGURE 16.16
The second type of element of Figure 16.14.

For element (2) (Figure 16.16),

$$b_1 = y_2 - y_3 = 0,$$
$$c_1 = x_3 - x_2 = 1 - 0 = 1.$$
$$A^{(2)} = \frac{1}{2}(1)(1) = \frac{1}{2},$$
$$S_{11}^{(2)} = \frac{1}{4A^{(2)}}(b_1^2 + c_1^2) = \frac{1}{(4)1/2}(0^2 + 1^2) = \frac{1}{2}.$$

Since element (5) is similar to element (2),

$$S_{11}^{(5)} = \frac{1}{2}.$$

Let us now calculate $g_1^{(1)}$:

$$g_1^{(1)} = \rho^{(1)}\frac{A^{(1)}}{3} = \frac{10^{-5}}{12} = g_1^{(3)},$$
$$g_1^{(2)} = \rho^{(2)}\frac{A^{(2)}}{3} = \frac{10^{-5}}{6}.$$

Since

$$\rho^{(4)} = \rho^{(5)} = \rho^{(6)} = 0,$$
$$g_1^{(4)} = g_1^{(5)} = g_1^{(6)} = 0.$$

The global

$$S_{11} = \sum_{k=1}^{6} \varepsilon_k S_{11}^{(k)},$$
$$S_{11} = \varepsilon_0\left[(2.5)(1) + (2.5)\frac{1}{2} + (2.5)(1) + (1)(1) + (1)\frac{1}{2} + (1)(1)\right] = 8.75\varepsilon_0,$$
$$g_1 = \left[\sum_{k=1}^{6} g_1^{(k)}\right] = \frac{(2)10^{-5}}{12} + \frac{10^{-5}}{6} = \frac{10^{-5}}{3},$$
$$\therefore \Phi_1 = \frac{10^{-5}}{3}\frac{1}{8.75 \times 8.752 \times 10^{-12}} = 4.3026 \times 10^4 V = 43.026 \text{ kV}.$$

TABLE 16.6

Coordinate Table for the Example in Figure 16.14

n	x_n	y_n
1	0.5	1
2	0	1
3	0	0
4	1	0
5	1	1
6	1	2
7	0	2

TABLE 16.7

Element Parameters and Connectivity Table for the Example in Figure 16.14

Element #	n1L	n2L	n3L	ρ	ε_{PR}
(1)	1	2	3	10^{-5}	2.5
(2)	1	3	4	10^{-5}	2.5
(3)	1	4	5	10^{-5}	2.5
(4)	1	5	6	0	1
(5)	1	6	7	0	1
(6)	1	7	2	0	1

The function GLANT is modified as PGLANT2 to solve Poisson's equation function

$$\text{function}\left[S,T,g\right] = PGLANT2\left(Nn,Ne,n1L,n2L,n3L,Rho,Epr,xn,yn\right). \qquad (16.95)$$

This function requires additional inputs of the charge density ρ(Rho) and the relative permittivity ε_r(Epr) arrays. This function subprogram can also be used to solve the Laplace equation with different dielectrics in the domain, by inputting 0 values for Rho array but giving the specified values for *Epr*. The example is solved using the function.

The preparatory work (Tables 16.6 and 16.7) is given in this chapter, however, main program, function program PGLANT2 are given in Appendix 16A.2.

16.5 FEM for Homogeneous Waveguide Problem

The waveguide problem is an eigenvalue problem with the differential equation given by Equation 16.1, written here for convenience as

$$\nabla_t^2\Phi + k_t^2\Phi = 0, \qquad (16.96)$$

where $k_t^2 = \lambda$ is the eigenvalue. k_t is the cutoff wave number

$$k_t^2 = k^2 - \beta^2, \qquad (16.97)$$

where β^2 is the propagation constant in the z-direction. The functional for the problem is

$$I(\Phi) = \frac{1}{2} \iint_S \left| \bar{\nabla}_t \Phi \right|^2 ds - \iint_S k_t^2 \Phi^2 ds. \qquad (16.98)$$

For a given element, we have seen before, the contribution to the algebraic equation that comes from the first term on the RHS of Equation 16.98. Let us therefore discuss the second term:

$$\iint_{\Delta^{(e)}} \Phi \Phi ds = \iint_{\Delta^{(e)}} \sum_{i=1}^{3} \zeta_i \Phi_i \sum_{j=1}^{3} \zeta_j \Phi_j ds = \sum_{i=1}^{3} \sum_{j=1}^{3} \Phi_i \Phi_j \iint_{\Delta^{(e)}} \zeta_i \zeta_j ds. \qquad (16.99)$$

Using the standard integral, we have already shown that

$$\iint_{\Delta^{(e)}} \zeta_1 \zeta_2 \, dx dy = \frac{A^{(e)}}{12}, \qquad (16.100)$$

$$\iint_{\Delta^{(e)}} \zeta_1^2 \, dx dy = \frac{A^{(e)}}{6}. \qquad (16.101)$$

The algebraic equations that will be generated by minimizing the functional will be of the form

$$\left[S^{(e)} \right] \begin{bmatrix} \Phi_1 \\ \Phi_2 \\ \Phi_3 \end{bmatrix} - k_t^2 \left[T^{(e)} \right] \begin{bmatrix} \Phi_1 \\ \Phi_2 \\ \Phi_3 \end{bmatrix} = 0, \qquad (16.102)$$

where $S^{(e)}$ has been discussed before and

$$\left[T^{(e)} \right] = \frac{A^{(e)}}{12} \begin{bmatrix} 2 & 1 & 1 \\ 1 & 2 & 1 \\ 1 & 1 & 2 \end{bmatrix}. \qquad (16.103)$$

When the contributions from all the elements are taken into account, we get the global equation

$$[S][\Phi] - k_t^2 [T][\Phi] = 0. \qquad (16.104)$$

The eigenvalue problem in standard form is

$$[A][\Phi] = k_t^2[\Phi],$$ (16.105)

where

$$[A] = [T]^{-1}[S].$$ (16.106)

Example

Determine the cutoff wave number of a square waveguide shown in Figure 16.17. Consider the TM mode only.

Solution

For TM modes, the boundary condition of $E_z = 0$ on the walls translates to $\Phi = 0$. The only free node is $\boxed{1}$. Consider element (1). The coordinates of the local nodes are

Local node	Coordinates
1	$\left(\dfrac{a}{2}, \dfrac{a}{2}\right)$
2	$(0,0)$
3	$\left(\dfrac{a}{2}, 0\right)$

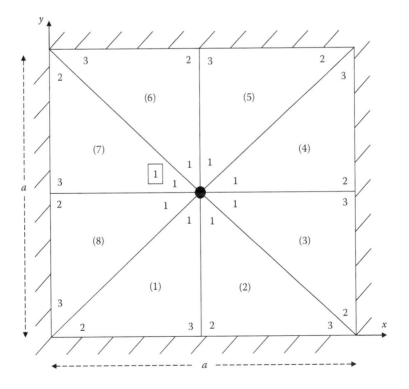

FIGURE 16.17
Square waveguide.

$$b_1 = y_2 - y_3 = 0,$$

$$c_1 = x_3 - x_2 = \frac{a}{2},$$

$$A^{(1)} = \frac{1}{2}\frac{a}{2}\frac{a}{2} = \frac{a^2}{8},$$

$$S_{11}^{(1)} = \frac{1}{4a^2/8}\left[0^2 + \left(\frac{a}{2}\right)^2\right] = \frac{1}{2},$$

$$T_{11}^{(1)} = \frac{A^{(1)}}{6} = \frac{a^2}{48}.$$

Note that $S_{11}^{(e)}$ are all same and $T_{11}^{(e)}$ are all same:

$$S_{11} = \sum_{e=1}^{8} S_{11}^{(e)} = \left(\frac{1}{2}\right)8 = 4,$$

$$T_{11} = \sum_{e=1}^{8} T_{11}^{(e)} = \frac{a^2}{48}8 = \frac{a^2}{6}.$$

Thus, the equation for one eigenvalue is

$$\left[4 - k_t^2\frac{a^2}{6}\right]\Phi_1 = 0,$$

$$\lambda = k_t^2 = \frac{(4)(6)}{a^2} = \frac{24}{a^2}.$$

The cutoff wave number is

$$k_t = \frac{\sqrt{24}}{a} = \frac{4.899}{a}.$$

The exact answer is

$$(k_c)_{TM11} = \sqrt{\left(\frac{\pi}{a}\right)^2 + \left(\frac{\pi}{a}\right)^2} = \sqrt{2}\frac{\pi}{a} = \frac{4.443}{a}.$$

The error is about 10%.

The function GLANT is actually written for a homogeneous waveguide problem and the project problem we will try is a problem of a ridged waveguide. In preparation for this, the problem specified in Appendix 16A.3, a regular rectangular waveguide problem, will be solved first. The exact answers can be obtained and are

$$(k_c)_{TE_{mn}} = \left[\left(\frac{m\pi}{a}\right)^2 + \left(\frac{n\pi}{b}\right)^2\right]^{1/2} \quad \begin{array}{l} m = 0,1,2,\ldots,\infty, \\ n = 0,1,2,\ldots,\infty, \\ \text{but } m = n = 0, \text{ is excluded.} \end{array}$$

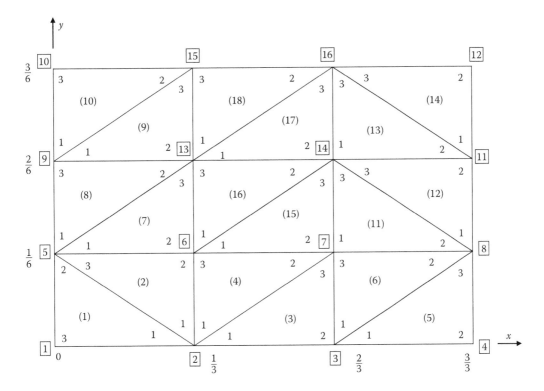

FIGURE 16.18
Elements of a standard rectangular waveguide.

$$\left(k_c\right)_{\text{TM}_{mn}} = \left[\left(\frac{m\pi}{a}\right)^2 + \left(\frac{n\pi}{b}\right)^2\right]^{1/2}, \quad m = 1,2,\dots,\infty, \, n = 1,2,\dots,\infty.$$

The dominant mode is TE_{10} mode, for $a > b$, and

$$\left(k_c\right)_{\text{TE}_{10}} = \left(\frac{\pi}{a}\right) = \pi, \quad \text{for } a = 1.$$

Let us use GLANT to determine the TE and TM modes. Figure 16.18 shows 18 elements, 16 nodes. Table 16.8 lists global node coordinates and Table 16.9 gives connectivity matrix. Note that for TE modes, since the boundary conditions are Neumann type, all the nodes are free nodes, whereas for TM modes the boundary conditions are homogeneous Dirichlet type, and hence the nodes on the boundary are prescribed nodes of zero potential.

For TM modes,

$$prn = \left[1,2,3,4,5,8,9,10,11,12,15,16\right],$$

$$frn = \left[6,7,13,14\right].$$

For TE modes, all are free modes.

TABLE 16.8

Coordinate Table for the Rectangular Waveguide Problem

Global Node #	X	Y
1	0	0
2	1/3	0
3	2/3	0
4	3/3	0
5	0	1/6
6	1/3	1/6
7	2/3	1/6
8	3/3	1/6
9	0	2/6
10	0	3/6
11	3/3	2/6
12	3/3	3/6
13	1/3	2/6
14	2/3	2/6
15	1/3	3/6
16	2/3	3/6

TABLE 16.9

Connectivity Table for the Rectangular Waveguide

Element #	(1)	(2)	(3)	(4)	(5)	(6)	(7)	(8)	(9)
$n1L$	2	2	2	2	3	3	5	5	9
$n2L$	5	6	3	7	4	8	6	13	13
$n3L$	1	5	7	6	8	7	13	9	15
Element #	(10)	(11)	(12)	(13)	(14)	(15)	(16)	(17)	(18)
$n1L$	9	7	8	14	11	6	6	13	13
$n2L$	15	8	11	11	12	7	14	14	16
$n3L$	10	14	14	16	16	14	13	16	15

The MATLAB statements that will be helpful in getting the eigenvalues and the cutoff wave number k_c are illustrated for TE modes. After getting S and T matrices, since all the nodes are free nodes, we can add the following statements in the main program:

```
ATE = inv(T) * S;
    [EVTE,kcsqTE] = eig(ATE);
kc TE = sqrt(diag(kcsqTE));
kcTES = sort(kcTE);
```

The first statement gives the A matrix. The second statement makes use of the built in function statement eig in MATLAB whose input is the square matrix and the outputs are eigenvectors (EVTE) and eigenvalues ($kcsq$TE). Similar steps can be taken for TM modes by keeping only the rows and columns of A matrix corresponding to free nodes.

Appendix 16A.4 gives the details of the MATLAB program and it can be seen that the first-order FEM we used gives us the lowest value

$$\left(k_c\right)_{TE} = 3.2795,$$

where as the exact answer is $(k_c)_{TE10} = \pi = 3.14$. The answer for TM modes that involved only four free modes give the lowest value of 8.2014, whereas the exact answer for TM_{11} mode is

$$(k_c)_{TM_{11}} = \pi\sqrt{5} = 7.0248.$$

Stiffness matrix in terms of Q-matrices: Another way of writing the formulas for the stiffness matrix $[S^{(e)}]$ in terms of the vertex angles [4] is

$$\left[S^{(e)}\right] = \sum_{i=1}^{3} \cot\theta_i Q_i, \tag{16.107}$$

where

$$Q_1 = \frac{1}{2}\begin{bmatrix} 0 & 0 & 0 \\ 0 & 1 & -1 \\ 0 & -1 & 1 \end{bmatrix}, \tag{16.108}$$

$$Q_2 = \frac{1}{2}\begin{bmatrix} 1 & 0 & -1 \\ 0 & 0 & 0 \\ -1 & 0 & 1 \end{bmatrix}, \tag{16.109}$$

$$Q_3 = \frac{1}{2}\begin{bmatrix} 1 & -1 & 0 \\ -1 & 1 & 0 \\ 0 & 0 & 10 \end{bmatrix}. \tag{16.110}$$

16.5.1 Second-Order Node-Based Method

Till now, we used first-order triangular elements with the shape functions

$$\alpha(x,y) = \varsigma(x,y) = a + bx + cy. \tag{16.111}$$

This leads to linear interpolation in two dimensions. A typical solution using first-order triangular elements is shown in Figure 16.19. The piecewise linear behavior of contour lines is clearly seen in Figure 16.19. A better approximation can be obtained if we use quadratic interpolation; we shall write the shape function as

$$\alpha(x,y) = a + bx + cy + dx^2 + exy + fy^2. \tag{16.112}$$

Note that we have used all the terms up to second order. Since there are six unknowns, we have six nodes (whose coordinates are known) to determine the constants $a, b, c, d, e,$ and f.

In addition to corner nodes 1, 4, and 6, let us introduce three more middle nodes, 2, 3, and 5 shown in Figure 16.20c. It can be shown that the second-order shape functions can be expressed in terms of the first-order shape functions $\varsigma_①, \varsigma_②, \varsigma_③$; the corner nodes for

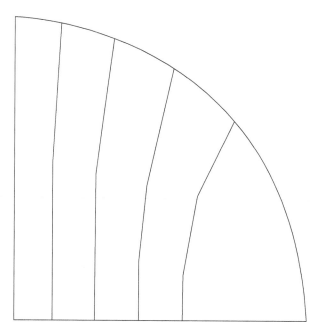

FIGURE 16.19
Solution due to linear interpolation in two dimensions.

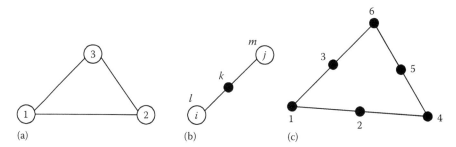

FIGURE 16.20
Second-order shape functions in terms of first-order shape functions. (a) First order shape functions, (b) nomenclature for the end nodes and the middle node, and (c) numbering of local nodes for the second order shape functions.

the first-order shape functions (area coordinates) shown in Figure 16.20a are with $\odot$ circle around the numbers:

$$\alpha_1 = (2\varsigma_{\odot} - 1)\varsigma_{\odot}, \tag{16.113}$$

$$\alpha_2 = 4\varsigma_{\odot}\varsigma_{\odot}, \tag{16.114}$$

$$\alpha_3 = 4\varsigma_{\odot}\varsigma_{\odot}, \tag{16.115}$$

$$\alpha_4 = (2\varsigma_{\odot} - 1)\varsigma_{\odot}, \tag{16.116}$$

$$\alpha_5 = 4\varsigma_{\odot}\varsigma_{\odot}, \tag{16.117}$$

$$\alpha_6 = (2\varsigma_{\odot} - 1)\varsigma_{\odot}. \tag{16.118}$$

Note that the formulas for the corner nodes (α_1, α_4, α_6) leave a pattern given by Equation 16.119 and the middle node have another pattern given by Equation 16.120 (see Figure 16.20b):

$$\alpha_l = (2\varsigma_① - 1)\varsigma_①, \tag{16.119}$$

$$\alpha_k = 4\varsigma_①\varsigma_①. \tag{16.120}$$

It may appear at first sight that the notation used is cumbersome using different numbers for the same corner nodes for the first- and the second-order methods. However, the formulas developed in [4] have this notation and we will use the same notation. Note that $\varsigma_①$ if needed can be calculated from Equations 16.38 through 16.44 based on coordinates of ①, ②, and ③ of Figure 16.20a. The second-order shape functions α_i must satisfy the usual requirement of

$$\alpha_i = \begin{array}{ll} 1 & \text{on node i,} \\ 0 & \text{on other node} \end{array}.$$

Let us check this for α_1: α_1 at node 1 is

$$\alpha_1 = (2\varsigma_① - 1)\varsigma_①, \quad \varsigma_① = 1 \text{ at node 2,}$$
$$\alpha_1 = (2 - 1)1 = 1.$$

α_1 at node 4 is

$$\alpha_1\big|_{node4} = (2\varsigma_① - 1)\varsigma_①, \quad \varsigma_① = 0 \text{ at node 4,}$$
$$= (0 - 1)0 = 0.$$

α_1 at node 2 is

$$\alpha_1\big|_{node2} = (2\varsigma_① - 1)\varsigma_①, \quad \varsigma_① = 0.5 \text{ at node 2,}$$
$$= (2 \times 0.5 - 1)0.5 = 0.$$

α_1 at nodes 5 and 6 is equal to zero since $\varsigma_① = 0$ at nodes 5 and 6. Thus, the shape function α_1 at node 1 is 1 and is zero at all other nodes. Also, it is of second order, since it is the product of $\varsigma_①$ and $2\varsigma_① - 1$, each of which is of first order.

Let us check α_2: from Equation 16.114,

$$\alpha_2 = 4\varsigma_①\varsigma_②, \quad \varsigma_① = 0.5, \quad \varsigma_② = 0.5; \text{ at node 2,}$$
$$\alpha_2 = 4(0.5)(0.5) = 1.$$

See Figure 16.21 to get a better understanding of the values of $\varsigma_①$ at various points. The dotted line parallel to the line BC has $\varsigma_① = $ constant. If this line goes through the middle point of AB, then $\varsigma_①$ at the middle point will be 0.5. Sketches of α_1 and α_3 as functions of (x, y) are shown in Figure 16.22. It can be shown that the $S^{(e)}$ matrix is given by Equation 16.107,

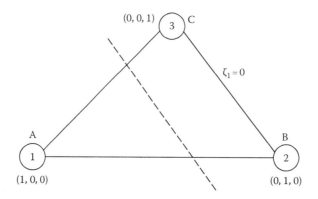

FIGURE 16.21
Sketch showing the value of ς_0 at various points.

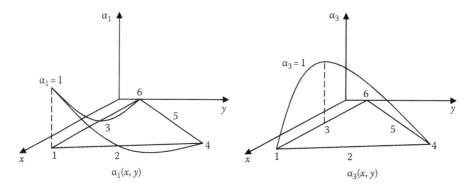

FIGURE 16.22
Sketch of second-order shape functions.

but the Q_i-matrices in this case are (6×6) matrices, and given by Equations 16.121 through 16.123. The $T^{(e)}$-matrix is given by Equation 16.124:

$$Q_1 = \frac{1}{6}\begin{bmatrix} 0 & 0 & 0 & 0 & 0 & 0 \\ 0 & 8 & -8 & 0 & 0 & 0 \\ 0 & -8 & 8 & 0 & 0 & 0 \\ 0 & 0 & 0 & 3 & -4 & 1 \\ 0 & 0 & 0 & -4 & 8 & -4 \\ 0 & 0 & 0 & 1 & -4 & 3 \end{bmatrix}, \tag{16.121}$$

$$Q_2 = \frac{1}{6}\begin{bmatrix} 3 & 0 & -4 & 0 & 0 & 1 \\ 0 & 8 & 0 & 0 & -8 & 0 \\ -4 & 0 & 8 & 0 & 0 & -4 \\ 0 & 0 & 0 & 0 & 0 & 0 \\ 0 & -8 & 0 & 0 & 8 & 0 \\ 1 & 0 & -4 & 0 & 0 & 3 \end{bmatrix}, \tag{16.122}$$

$$Q_3 = \frac{1}{6}\begin{bmatrix} 3 & -4 & 0 & 1 & 0 & 0 \\ -4 & 8 & 0 & -4 & 0 & 0 \\ 0 & 0 & 8 & 0 & -8 & 0 \\ 1 & -4 & 0 & 3 & 0 & 0 \\ 0 & 0 & -8 & 0 & 8 & 0 \\ 0 & 0 & 0 & 0 & 0 & 0 \end{bmatrix}, \tag{16.123}$$

$$T^{(e)} = \frac{A^{(e)}}{180}\begin{bmatrix} 6 & 0 & 0 & -1 & -4 & -1 \\ 0 & 32 & 16 & 0 & 16 & -4 \\ 0 & 16 & 32 & -4 & 16 & 0 \\ -1 & 0 & -4 & 6 & 0 & -1 \\ -4 & 16 & 16 & 0 & 32 & 0 \\ -1 & -4 & 0 & -1 & 0 & 6 \end{bmatrix}. \tag{16.124}$$

We can verify that $S^{(e)}$ and $T^{(e)}$ given in Equation 16.107 and Equations 16.121 through 16.124 are correct by evaluating them from their definitions: since

$$S_{ij}^{(e)} = \iint_{\Delta^{(e)}} \left(\bar{\nabla}\alpha_i \cdot \bar{\nabla}\alpha_j \right) dx\,dy, \tag{16.125}$$

$$S_{ij}^{(e)} = \iint_{\Delta^{(e)}} \left(\frac{\partial \alpha_i}{\partial x}\frac{\partial \alpha_j}{\partial x} + \frac{\partial \alpha_i}{\partial y}\frac{\partial \alpha_j}{\partial y} \right) dx\,dy. \tag{16.126}$$

Let us compute $S_{14}^{(e)}$:

$$\frac{\partial \alpha_1}{\partial x} = \frac{\partial}{\partial x}\left[(2\varsigma_{①}-1)\varsigma_{①} \right] = (2\varsigma_{①}-1)\frac{\partial \varsigma_{①}}{\partial x} + \varsigma_{①}\frac{\partial}{\partial x}(2\varsigma_{①}-1)$$

$$= (2\varsigma_{①}-1)\frac{\partial \varsigma_{①}}{\partial x} + \varsigma_{①}\frac{\partial}{\partial \varsigma_{①}}(2\varsigma_{①}-1)\frac{\partial \varsigma_{①}}{\partial x} = \frac{\partial \varsigma_{①}}{\partial x}[2\varsigma_{①}-1+2\varsigma_{①}]$$

$$= \frac{b_1}{2A^{(e)}}[4\varsigma_{①}-1],$$

$$\frac{\partial \alpha_4}{\partial x} = \frac{b_2}{2A^{(e)}}[4\varsigma_{②}-1],$$

$$\frac{\partial \alpha_1}{\partial y} = \frac{c_1}{2A^{(e)}}[4\varsigma_{①}-1],$$

$$\frac{\partial \alpha_4}{\partial y} = \frac{c_2}{2A^{(e)}}[4\varsigma_{②}-1],$$

$$S_{14}^{(e)} = \iint\limits_{\Delta^{(e)}} [(4\zeta_{\textcircled{1}} - 1)(4\zeta_{\textcircled{2}} - 1)] \frac{[b_1 b_2 + c_1 c_2]}{4\left[A^{(e)}\right]^2} dx\, dy$$

$$= \frac{[b_1 b_2 + c_1 c_2]}{4\left[A^{(e)}\right]^2} \iint\limits_{\Delta^{(e)}} [16\zeta_{\textcircled{1}}\zeta_{\textcircled{2}} - 4\zeta_{\textcircled{2}} - 4\zeta_{\textcircled{1}} + 1] dx\, dy$$

$$= \frac{b_1 b_2 + c_1 c_2}{4A^{(e)}} \left(\frac{16}{12} - \frac{4}{3} - \frac{4}{3} + 1 \right), \tag{16.127}$$

$$S_{14}^{(e)} = -\frac{1}{12 A^{(e)}} \left(b_1 b_2 + c_1 c_2 \right). \tag{16.128}$$

It can be shown that (see Appendix 16B)

$$b_i b_j + c_i c_j = -2 A^{(e)} \cot \theta_k, \quad i \neq j, \tag{16.129}$$

$$b_i^2 + c_i^2 = 2 A^{(e)} \left(\cot \theta_j + \cot \theta_k \right). \tag{16.130}$$

Thus, from Equation 16.128, we get

$$S_{14}^{(e)} = \frac{1}{6} \cot \theta_3. \tag{16.131}$$

From Equations 16.107 and 16.121 through 16.123, we get the same value for $S_{14}^{(e)}$. Let us verify the T-matrix elements by calculating one of the elements, say $T_{11}^{(e)}$:

$$T_{11}^{(e)} = \iint\limits_{\Delta^{(e)}} \alpha_1^2 dx\, dy, \tag{16.132}$$

$$\alpha_1^2 = \left[(2\zeta_{\textcircled{1}} - 1)\zeta_{\textcircled{1}} \right]^2 = \zeta_{\textcircled{1}}^2 \left(4\zeta_{\textcircled{1}}^2 + 1 - 4\zeta_{\textcircled{1}} \right),$$
$$\alpha_1^2 = 4\zeta_{\textcircled{1}}^4 + \zeta_{\textcircled{1}}^2 - 4\zeta_{\textcircled{1}}^3, \tag{16.133}$$

Using the standard integral Equation 16.78,

$$\iint\limits_{\Delta^{(e)}} \zeta_{\textcircled{1}}^4 dx\, dy = \frac{A^{(e)}}{15},$$

$$\iint\limits_{\Delta^{(e)}} \zeta_{\textcircled{1}}^3 dx\, dy = \frac{A^{(e)}}{10},$$

$$\therefore T_{11}^{(e)} = A^{(e)} \left(\frac{4}{15} + \frac{1}{6} - \frac{4}{10} \right) = \frac{A^{(e)}}{30}. \tag{16.134}$$

Appendix 16A.4 contains an example of using the node-based second-order method. The example is to determine the lowest cutoff wave number of an isosceles triangle

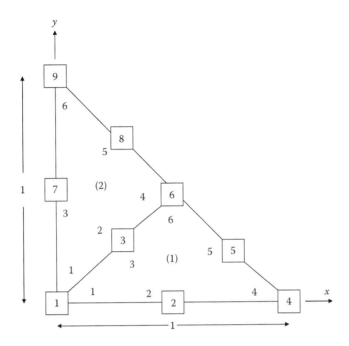

FIGURE 16.23
Waveguide with right-angled isosceles triangular cross section.

waveguide shown in Figure 16.23. We divide the waveguide into two elements, resulting in nine global nodes.

The function program GLAN2T and the main program are given. Table 16.10 gives the local node arrays n1L to n6L. Table 16.11 gives the coordinates of the global nodes 1, 4, 6, and 9 (corner nodes) of the two elements. The lowest value of k_c obtained is 3.25, whereas the exact answer is 3.14.

TABLE 16.10

Connectivity Table for the Two Elements

Element #	n1L $n(1, e)$	n2L $n(2, e)$	n3L $n(3, e)$	n4L $n(4, e)$	n5L $n(5, e)$	n6L $n(6, e)$
1	1	2	3	4	5	6
2	1	3	7	6	8	9

TABLE 16.11

The Coordinates of the Global Nodes of the Two Elements

$$x([1,4,6,9]) = \begin{bmatrix} 0 & 1 & 0.5 & 0 \end{bmatrix}$$
$$y([1,4,6,9]) = \begin{bmatrix} 0 & 0 & 0.5 & 1 \end{bmatrix}$$

16.5.2 Vector Finite Elements

Let us briefly review the type of problems we solved using the FEM method. They were all scalar problems.

a. Laplace equation (16.25):

$$\nabla_t^2 \Phi = 0.$$

We solved for Φ by minimizing the functional (Equation 16.26):

$$I(\Phi) = \iint \frac{1}{2}\varepsilon \left|\bar{\nabla}_t \Phi\right|^2 ds.$$

The FEM method applied to the minimization of Equation 16.26 results in the algebraic equations whose solution is given by Equation 16.75:

$$\left[\Phi_f\right] = -\left[S_{ff}\right]^{-1}\left[S_{fp}\right]\left[\Phi_p\right].$$

We obtained the global [S] after assembling the element [$S^{(e)}$] given by Equation 16.125:

$$S_{ij}^{(e)} = \iint_{\Delta^{(e)}} \left(\bar{\nabla}\alpha_i \cdot \bar{\nabla}\alpha_j\right)ds,$$

where α_i is the shape function.

b. In the homogeneous waveguide problem, the longitudinal electric field (TM modes) or the longitudinal magnetic field (TE modes) served as the potential and the associated problem was an eigenvalue problem. For the TM problem, Equation 15.3 can be written as

$$\tilde{E}_z(x,y,z) = F(x,y)e^{-j\beta z}$$

and $F(x, y) = \Phi$ satisfied Equations 16.96 and 16.97:

$$\nabla_t^2 \Phi + k_t^2 \Phi = 0,$$

$$k_t^2 = k^2 - \beta^2.$$

The functional for the problem is Equation 16.98:

$$I(\Phi) = \frac{1}{2}\iint_s \left|\bar{\nabla}_t \Phi\right|^2 ds - \iint_s k_t^2 \Phi^2 ds$$

and the discretization based on FEM lead to the algebraic equation 16.102

$$[S][\Phi] - k_t^2[T][\Phi] = 0.$$

The global [T] is obtained by assembling the elemental [$T^{(e)}$], where $T_{ij}^{(e)} = \iint_{\Delta^{(e)}} \alpha_i \alpha_j \, ds$.

For the first-order node-based method, the shape function α_i is the same as the area coordinate ζ_i. For the second-order method, α's are defined in terms of ζ's through Equations 16.119 and 16.120. The continuity of the scalar potential Φ at the common nodes and the common edges (sides of the triangular elements) of the finite elements is automatically satisfied in both the problems.

For the waveguide problem, we will encounter a difficulty if the medium inside the waveguide is not homogeneous. Suppose the waveguide is partially filled with a dielectric, then the modes cannot be completely separated as TE^z or TM^z modes. Some of the modes are hybrid $\left(\tilde{E}_z \neq 0,\ \tilde{H}_z \neq 0\right)$ and hence we do not have a suitable scalar potential to describe the problem. In such cases, it is best to solve for the vector electric field or vector magnetic field directly. The starting point is to write down the Maxwell equations for a sourceless region, however, containing inhomogeneous media, that is,

$$\varepsilon_r = \varepsilon_r\left(\bar{r}\right), \tag{16.135}$$

$$\mu_r = \mu_r\left(\bar{r}\right), \tag{16.136}$$

$$\bar{\nabla} \times \tilde{E} = -j\omega\mu_0\mu_r\left(\bar{r}\right)\tilde{H}, \tag{16.137}$$

$$\bar{\nabla} \times \tilde{H} = j\omega\varepsilon_0\varepsilon_r\left(\bar{r}\right)\tilde{E}. \tag{16.138}$$

By eliminating $\tilde{H}$, we can obtain the wave equation

$$\bar{\nabla} \times \left[\frac{\bar{\nabla} \times \tilde{E}}{\mu_r\left(\bar{r}\right)}\right] - k_0^2\varepsilon_r\left(\bar{r}\right)\tilde{E} = 0. \tag{16.139}$$

The functional for Equation 16.139 is given by

$$I\left(\tilde{E}\right) = \frac{1}{2}\iint_s \left[\frac{\left|\bar{\nabla} \times \tilde{E}\right|^2}{\mu_r} - k_0^2\varepsilon_r\left|\tilde{E}\right|^2\right]ds. \tag{16.140}$$

In a sourceless region, the divergence equations are

$$\bar{\nabla} \cdot \left[\varepsilon_r\tilde{E}\right] = 0, \tag{16.141}$$

$$\bar{\nabla} \cdot \left[\mu_r\tilde{H}\right] = 0. \tag{16.142}$$

To solve vector equations, one can use vector shape function $\bar{\alpha}_i$ and scalar field components:

$$\tilde{E}\left(\bar{r}\right) = \sum_{i=1}^{3}\bar{\alpha}_i\left(\bar{r}\right)E_i, \tag{16.143}$$

or one can use scalar shape functions and vector fields

$$\tilde{\bar{E}}(\bar{r}) = \sum_{i=1}^{3} \alpha_i(\bar{r}) \bar{E}_i. \tag{16.144}$$

Equation 16.144 is not the way to go since it would require continuity of the vector fields wherever the finite elements are connected. When there are discontinuous material properties in the problem domain, say at a dielectric interface, normal component of $\tilde{\bar{E}}$ is discontinuous, representation (Equation 16.144) is inappropriate. Representation (Equation 16.143) is appropriate for FEM solution, provided E_i is the tangential component. When two finite elements are connected having common edges, the boundary condition of the continuity of tangential E is satisfied.

In any given element, if ε_r is a constant, we can satisfy Equation 16.141 by imposing a requirement on the vector shape function $\bar{\alpha}_i$, that its divergence is zero:

$$\bar{\nabla} \cdot \bar{\alpha}_i = 0. \tag{16.145}$$

Note that the boundary condition

$$D_{n1} = D_{n2} \tag{16.146}$$

is automatically satisfied when FEM via variation principle is used. The boundary condition

$$\varepsilon_{r1} E_{n1} = \varepsilon_{r2} E_{n2} \tag{16.147}$$

turns out to be a natural boundary condition and hence not requiring specifically to be enforced. The vector shape functions need to satisfy the usual requirement of

$$\bar{\alpha}_i \cdot \hat{e}_j \Big|_{\text{edge } j} = 0, \quad i \neq j, \tag{16.148}$$

$$\bar{\alpha}_i \cdot \hat{e}_i \Big|_{\text{edge } i} = 1, \quad i = j, \tag{16.149}$$

where $\hat{e}_j$ is a unit vector denoting the direction of jth edge. Such vector shape functions can be constructed from the scalar shape functions α_i. For first-order shape functions,

$$\bar{\alpha}_1 = \ell_1 \left(\alpha_1 \bar{\nabla} \alpha_2 - \alpha_2 \bar{\nabla} \alpha_1 \right), \tag{16.150}$$

$$\alpha_1 = \frac{1}{2A} \left(a_1 + b_1 x + c_1 y \right), \tag{16.151}$$

$$\alpha_2 = \frac{1}{2A} \left(a_2 + b_2 x + c_2 y \right), \tag{16.152}$$

and ℓ_1 is the length of edge 1. Figure 16.24 shows the nodes and edges of a triangular element. The edge j is denoted #j. The beginning node of edge 1 (#1) is node 1 and the end node of #1 is node 2.

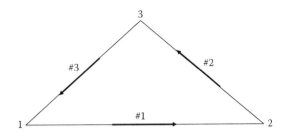

FIGURE 16.24
Nodes and edges of a triangle.

A few of the properties of the vector shape functions are derived and some of them are only stated and given as homework.

PROPERTY 1

$$\bar{\nabla} \cdot \bar{\alpha}_1 = 0. \tag{16.153}$$

From Equation 16.150,

$$\bar{\nabla} \cdot \bar{\alpha}_1 = \ell_1 \bar{\nabla} \cdot \left(\alpha_1 \bar{\nabla} \alpha_2 - \alpha_2 \bar{\nabla} \alpha_1 \right), \tag{16.154}$$

$$\bar{\nabla} \cdot \left(\alpha_1 \bar{\nabla} \alpha_2 \right) = \alpha_1 \nabla^2 \alpha_2 + \bar{\nabla} \alpha_1 \cdot \bar{\nabla} \alpha_2,$$

$$\nabla^2 \alpha_2 = \left(\frac{\partial^2}{\partial x^2} + \frac{\partial^2}{\partial y^2} \right) \frac{\left(a_2 + b_2 x + c_2 y \right)}{2A} = 0,$$

$$\therefore \bar{\nabla} \cdot \left(\alpha_1 \bar{\nabla} \alpha_2 \right) = \bar{\nabla} \alpha_1 \cdot \bar{\nabla} \alpha_2. \tag{16.155}$$

Similarly,

$$\bar{\nabla} \cdot \left(\alpha_2 \bar{\nabla} \alpha_1 \right) = \bar{\nabla} \alpha_2 \cdot \bar{\nabla} \alpha_1 = \bar{\nabla} \alpha_1 \cdot \bar{\nabla} \alpha_2, \tag{16.156}$$

$$\therefore \bar{\nabla} \cdot \bar{\alpha}_1 = \ell_1 \left(\bar{\nabla} \alpha_1 \cdot \bar{\nabla} \alpha_2 - \bar{\nabla} \alpha_1 \cdot \bar{\nabla} \alpha_2 \right) = 0.$$

PROPERTY 2

$$\bar{\nabla} \times \bar{\alpha}_1 = 2\ell_1 \bar{\nabla} \alpha_1 \times \bar{\nabla} \alpha_2. \tag{16.157}$$

PROPERTY 3

$$\bar{\alpha}_1 \cdot \hat{e}_1 = 1. \tag{16.158}$$

PROPERTY 4

$$\bar{\alpha}_1 \cdot \hat{e}_2 = \bar{\alpha}_1 \cdot \hat{e}_3 = 0. \tag{16.159}$$

We can generalize these properties:

$$\bar{\nabla} \cdot \bar{\alpha}_i = 0, \tag{16.160}$$

$$\bar{\alpha}_i \cdot \hat{e}_j = 0, \quad i \neq j, \tag{16.161}$$

$$\bar{\alpha}_i \cdot \hat{e}_i = 1, \quad i = j, \tag{16.162}$$

where $\hat{e}_j$ is the unit vector along the edge j. Equations 16.161 and 16.162 can be stated as follows: The tangential component of the shape function $\bar{\alpha}_i = 1$ on ith edge and zero on all other edges. The shape function $\bar{\alpha}_i$ is normal to the edges other than the ith edge. $\bar{\alpha}_i$ has both tangential and normal components on edge i and its tangential component is equal to 1.

Figure 16.25a shows a sketch of $\bar{\alpha}_1$ in the triangle [6]. The scalar electric field E_i in Equation 16.143 can be shown to be the tangential component of the electric field along edge i, by taking the dot product with $\hat{e}_i$. For example,

$$\tilde{\bar{E}} \cdot \hat{e}_1 = \sum_{i=\#1}^{\#3} \left[\bar{\alpha}_i(x,y) \cdot \hat{e}_1 \right] E_i = \left[\bar{\alpha}_1(x,y) \cdot \hat{e}_1 \right] E_i \Big|_{i=\#1}, \tag{16.163}$$

$$\tilde{\bar{E}} \cdot \hat{e}_1 = E_{\#1}.$$

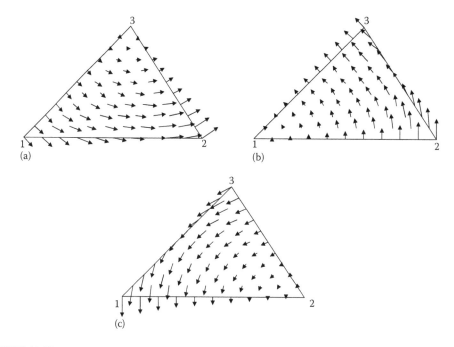

FIGURE 16.25
Sketch of vector shape function (a) $\bar{\alpha}_1$, (b) $\bar{\alpha}_2$, (c) $\bar{\alpha}_3$.

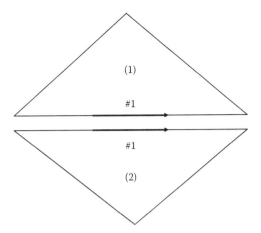

FIGURE 16.26
Continuity of tangential components at the common edge.

If edge i is a common edge to two triangles, the above property ensures the continuity of E_{tan} along the edge (Figure 16.26).

16.5.3 Fundamental Matrices for Vector Finite Elements

The stiffness matrix $[S^{(e)}]$ is obtained from formula 16.125 for the scalar problem. For the vector problem, the corresponding matrix for an element is $[E^{(e)}]$, whose element is

$$E_{ij}^{(e)} = \iint_{\Delta^{(e)}} \left(\bar{\nabla} \times \bar{\alpha}_i\right) \cdot \left(\bar{\nabla} \times \bar{\alpha}_j\right) ds. \tag{16.164}$$

By substituting for $\bar{\alpha}_i$ and $\bar{\alpha}_j$ and doing the integration, one obtains

$$E_{ij}^{(e)} = \frac{\ell_i \ell_j}{A^{(e)}}. \tag{16.165}$$

In Equation 16.164, ℓ_i is the length of the ith edge. For the homogeneous waveguide problem, $[T^{(e)}]$ is obtained from Equation 16.132. For the vector problem, the corresponding matrix for an element denoted by $[F^{(e)}]$ can be obtained by computing its ij element from

$$F_{ij}^{(e)} = \iint_{\Delta^{(e)}} \bar{\alpha}_i \cdot \bar{\alpha}_j \, ds. \tag{16.166}$$

For the purpose of easy coding, the result is written [4] in the following form:

$$F_{ij}^{(e)} = \frac{\ell_i \ell_j}{48 A^{(e)}} G_{ij}, \tag{16.167}$$

where the G-matrix is given by

$$[G] = \begin{bmatrix} 2(f_{22} + f_{11} - f_{12}) & (f_{23} - f_{22} - 2f_{13} + f_{12}) & (f_{21} - 2f_{23} - f_{11} + f_{13}) \\ G_{12} & 2(f_{22} + f_{33} - f_{23}) & (f_{31} - f_{33} - 2f_{21} + f_{23}) \\ G_{13} & G_{23} & 2(f_{33} + f_{11} - f_{13}) \end{bmatrix}, \quad (16.168)$$

and

$$f_{ij} = b_i b_j + c_i c_j. \quad (16.169)$$

Example

Calculate the E and F matrices for an isosceles triangle shown in Figure 16.27.
From formulas 16.165 and 16.167 through 16.169, we get the following results:

$$[f] = \begin{bmatrix} 2 & -1 & -1 \\ -1 & 1 & 0 \\ -1 & 0 & 1 \end{bmatrix},$$

$$[G] = \begin{bmatrix} 8 & 0 & -4 \\ 0 & 4 & 0 \\ -4 & 0 & 8 \end{bmatrix},$$

$$[E] = \begin{bmatrix} 2.0000 & 2.8284 & 2.0000 \\ 2.8284 & 4.0000 & 2.8284 \\ 2.0000 & 2.8284 & 2.0000 \end{bmatrix}, \quad (16.170a)$$

$$[F] = \begin{bmatrix} 0.3333 & 0 & -0.1667 \\ 0 & 0.3333 & 0 \\ -0.1667 & 0 & 0.3333 \end{bmatrix}. \quad (16.170b)$$

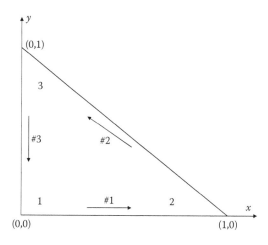

FIGURE 16.27
Sketch for calculating E and F metrics for an isosceles right-angled triangle.

We have not proved that Equation 16.167 is obtained from Equation 16.166 nor did we prove that Equation 16.165 is obtained from Equation 16.164. We will verify the results for a few elements of these matrices and leave it at that. Let us obtain $F_{11}^{(e)}$ and $F_{12}^{(e)}$:

$$\bar{\alpha}_1 = \ell_1\left(\alpha_1\bar{\nabla}\alpha_2 - \alpha_2\bar{\nabla}\alpha_1\right),$$
$$\bar{\alpha}_2 = \ell_2\left(\alpha_2\bar{\nabla}\alpha_3 - \alpha_3\bar{\nabla}\alpha_2\right),$$
$$\bar{\alpha}_3 = \ell_3\left(\alpha_3\bar{\nabla}\alpha_1 - \alpha_1\bar{\nabla}\alpha_3\right),$$

$$\alpha_1 = \frac{1}{2A}\left(a_1 + b_1 x + c_1 y\right),$$
$$\bar{\nabla}\alpha_1 = \frac{1}{2A}\left(b_1\hat{x} + c_1\hat{y}\right),$$
$$\bar{\nabla}\alpha_2 = \frac{1}{2A}\left(b_2\hat{x} + c_2\hat{y}\right),$$

$$\bar{\alpha}_1 = \frac{\ell_1}{2A}\left[\alpha_1\left(b_2\hat{x} + c_2\hat{y}\right) - \alpha_2\left(b_1\hat{x} + c_1\hat{y}\right)\right],$$
$$\bar{\alpha}_1 = \frac{\ell_1}{2A}\left[\hat{x}\left(\alpha_1 b_2 - \alpha_2 b_1\right) + \hat{y}\left(\alpha_1 c_2 - \alpha_2 c_1\right)\right],$$
$$\bar{\alpha}_2 = \frac{\ell_2}{2A}\left[\hat{x}\left(\alpha_2 b_3 + \alpha_3 b_2\right) - \hat{y}\left(\alpha_2 c_3 - \alpha_3 c_2\right)\right],$$

$$F_{11}^{(e)} = \iint_{\Delta^{(e)}} \bar{\alpha}_1\cdot\bar{\alpha}_1\,ds = \frac{\ell_1^2}{4A^2}\iint_{\Delta^{(e)}}\left[\left(\alpha_1 b_2 - \alpha_2 b_1\right)^2 + \left(\alpha_1 c_2 - \alpha_2 c_1\right)^2\right]ds.$$

The integrand is $\left(b_2^2 + c_2^2\right)\alpha_1^2 + \left(b_2^2 + c_2^2\right)\alpha_2^2 - 2\left(b_1 b_2 + c_1 c_2\right)\alpha_1\alpha_2$.
From the standard integral (for first-order shape functions $\alpha_1 = \zeta_1$, $\alpha_2 = \zeta_2$, etc.), we have

$$\iint_{\Delta^{(e)}}\alpha_1^2\,ds = \iint_{\Delta^{(e)}}\alpha_2^2\,ds = \frac{A}{6}, \quad \iint_{\Delta^{(e)}}\alpha_1\alpha_2\,ds = \frac{A}{12},$$

$$F_{11}^{(e)} = \frac{\ell_1^2}{4A^2}\left[\left(b_2^2 + c_2^2\right)\frac{A}{6} + \left(b_1^2 + c_1^2\right)\frac{A}{6} - 2\left(b_1 b_2 + c_1 c_2\right)\frac{A}{12}\right]$$
$$= \frac{\ell_1^2}{48A}\left[2\left(b_2^2 + c_2^2\right) + 2\left(b_1^2 + c_1^2\right) - 2\left(b_1 b_2 + c_1 c_2\right)\right].$$

Denoting $f_{ij} = b_i b_j + c_i c_j$, the above may be written as

$$F_{11}^{(e)} = \frac{\ell_1^2}{48A}\left[2f_{11} + 2f_{22} - 2f_{12}\right] = \frac{\ell_1^2}{48A}G_{11}.$$

$$F_{12}^{(e)} = \iint_{\Delta^{(e)}} \bar{\alpha}_1 \times \bar{\alpha}_2 \, ds$$

$$= \frac{\ell_1 \ell_2}{4A^2} \iint_{\Delta^{(e)}} \left[(\alpha_1 b_2 - \alpha_2 b_1)(\alpha_2 b_3 - \alpha_3 b_2) + (\alpha_1 c_2 - \alpha_2 c_1)(\alpha_2 c_3 - \alpha_3 c_2) \right] ds.$$

The integrand is $(b_2 b_3 + c_2 c_3)\alpha_1 \alpha_2 + (b_1 b_2 + c_1 c_2)\alpha_2 \alpha_3 - (b_2^2 + c_2^2)\alpha_1 \alpha_3 - (b_1 b_3 + c_1 c_3)\alpha_2^2$.
From the standard integral

$$\iint_{\Delta^{(e)}} \alpha_1 \alpha_2 \, ds = \iint_{\Delta^{(e)}} \alpha_2 \alpha_3 \, ds = \iint_{\Delta^{(e)}} \alpha_1 \alpha_3 \, ds = \frac{A}{12},$$

$$\iint_{\Delta^{(e)}} \alpha_2^2 \, ds = \frac{A}{6}.$$

Therefore,

$$F_{12}^{(e)} = \frac{\ell_1 \ell_2}{4A^2} \left[(b_2 b_3 + c_2 c_3)\frac{A}{12} + (b_1 b_2 + c_1 c_2)\frac{A}{12} - (b_2^2 + c_2^2)\frac{A}{12} - (b_1 b_3 + c_1 c_3)\frac{A}{6} \right]$$

$$= \frac{\ell_1 \ell_2}{48A} \left[f_{23} + f_{12} - f_{22} - 2f_{13} \right] = \frac{\ell_1 \ell_2}{48A} G_{12}.$$

Let us now calculate $E_{11}^{(e)} = \iint_{\Delta^{(e)}} \left| \bar{\nabla} \times \bar{\alpha}_1 \right|^2 ds$.
From property 2 of Equation 16.157,

$$\bar{\nabla} \times \bar{\alpha}_1 = 2\ell_1 \bar{\nabla}\alpha_1 \times \bar{\nabla}\alpha_2 = 2\ell_1 \frac{1}{2A}(\hat{x}b_1 + \hat{y}c_1) \times \frac{1}{2A}(\hat{x}b_2 + \hat{y}c_2) = \frac{2\ell_1}{4A^2}\left[\hat{z}(b_1 c_2 - b_2 c_1) \right],$$

$$\left| \bar{\nabla} \times \bar{\alpha}_1 \right|^2 = \left(\frac{2\ell_1}{4A^2} \right)^2 (b_1 c_2 - b_2 c_1)^2,$$

$$E_{11}^{(e)} = \left(\frac{2\ell_1}{4A^2} \right)^2 (b_1 c_2 - b_2 c_1)^2 \iint_{\Delta^{(e)}} ds.$$

Since $b_1 c_2 - b_2 c_1 = 2A$ (see Appendix 16B),

$$E_{11}^{(e)} = \frac{4\ell_1^2}{16A^4} 4A^2 A = \frac{\ell_1^2}{A}.$$

16.5.4 Application of Vector Finite Elements to Homogeneous Waveguide Problem

Let us consider the application of the edge elements to the homogeneous waveguide problem. Let $\varepsilon_r = \mu_r = 1$. For transverse electric modes, since the longitudinal component $\tilde{E}_z = 0$, by definition, Equation 16.139 becomes

$$\bar{\nabla} \times \bar{\nabla} \times \tilde{\bar{E}}_t - k_0^2 \tilde{\bar{E}}_t = 0, \tag{16.171}$$

where $\tilde{\bar{E}}_t$ is the transverse electric field.

Let

$$\bar{\nabla} = \bar{\nabla}_t + \hat{z}\frac{\partial}{\partial z}, \tag{16.172}$$

$$\bar{\nabla} \times \left(\bar{\nabla} \times \tilde{\bar{E}}_t \right) = \left(\bar{\nabla}_t + \hat{z}\frac{\partial}{\partial z} \right) \times \left[\left(\bar{\nabla}_t + \hat{z}\frac{\partial}{\partial z} \right) \times \tilde{\bar{E}}_t \right]$$

$$= \left(\bar{\nabla}_t + \hat{z}\frac{\partial}{\partial z} \right) \times \left(\bar{\nabla}_t \times \tilde{\bar{E}}_t + \hat{z}\frac{\partial}{\partial z} \times \tilde{\bar{E}}_t \right). \tag{16.173}$$

Let

$$\tilde{\bar{E}}_t\left(x,y,z\right) = \bar{E}_t\left(x,y\right)e^{-j\beta z}. \tag{16.174}$$

Substituting Equation 16.174 into 16.173, we get

$$\bar{\nabla}_t \times \left[\bar{\nabla}_t \times \bar{E}_t\left(x,y\right) \right] - k_c^2 \bar{E}_t = 0, \tag{16.175}$$

where

$$k_c^2 = k_0^2 - \beta^2,$$

$$k_0 = \omega\sqrt{\mu_0\varepsilon_0} = \frac{\omega}{c}. \tag{16.176}$$

The functional now is given by Equation 16.140 where $\bar{\nabla}$ is replaced by $\bar{\nabla}_t$. If we write for each element

$$\bar{E}_t^{(e)}\left(\bar{r}\right) = \sum_{i=1}^{3} \bar{\alpha}_i\left(\bar{r}\right)E_{\tan i}^{(e)}, \tag{16.177}$$

where $E_{\tan i}$ is the tangential component $\bar{E}_t$ along edge i, and after global assembly for all elements we get

$$\left[E_{ij}\right]\left[E_{\tan}\right] = k_c^2\left[F_{ij}\right]\left[E_{\tan}\right], \tag{16.178}$$

where $E_{\tan}$ is a column vector of tangential components along the edges. For TE modes, on PEC boundaries, $E_{\tan} = 0$, and hence all the edges on the PEC walls of the waveguide satisfy homogeneous Dirichlet boundary conditions and are thus prescribed edges. The rest of the edges (not on PEC walls) are free edges. Similar arguments for TM modes will give the following equations:

$$\bar{\nabla}_t \times \left[\bar{\nabla}_t \times \bar{H}_t\left(x,y\right) \right] - k_c^2 \bar{H}_t = 0, \tag{16.179}$$

$$\bar{H}_t^{(e)}(\bar{r}) = \sum_{i=1}^{3} \bar{\alpha}_i(\bar{r}) H_{\tan i}^{(e)},$$ (16.180)

$$\left[E_{ij}\right]\left[H_{\tan}\right] = k_c^2 \left[F_{ij}\right]\left[H_{\tan}\right].$$ (16.181)

The boundary condition on $H_{\tan}$ for PEC walls given by (2.26) is the Neumann type, $\partial H_{\tan}/\partial n = 0$, and in the FEM method, this boundary condition is a natural one and hence the edges on PEC walls are all free edges. Let us illustrate by a couple of examples.

Example 1

Calculate the dominant TM cutoff wave number of the PEC waveguide shown in Figure 16.27. Consider the entire triangle as one element.

For this waveguide, we have already obtained the [E] and [F] matrices given by Equations 16.170 and 16.171. For TM modes, the three edges are free edges. Hence, we have

$$[E][H_{\tan}] = k_c^2 [F][H_{\tan}],$$
$$[F]^{-1}[E][H_{\tan}] = k_c^2 [H_{\tan}].$$

So $k_c^2 = \lambda$ are the eigenvalues of the matrix:

$$[A] = [F]^{-1}[E].$$

For this problem,

$$[A] = \begin{bmatrix} 12.0000 & 16.9706 & 12.0000 \\ 8.4853 & 12.0000 & 8.4853 \\ 12.0000 & 16.9706 & 12.0000 \end{bmatrix}.$$

The nonzero eigenvalue $k_c^2 = 36$ and $k_c = 6$. The exact value of the lowest k_c is that of TM$_{21}$ mode and is equal to $\sqrt{5}\pi = 7.02$. We cannot find any significant answer for TE modes, since all the edges are prescribed edges.

Example 2

Consider that the waveguide of Example 1 is divided into two elements (Figure 16.28).

This has a total of five global edges, four global nodes, and two elements. For TM modes, all the five global edges are free edges and we get five eigenvalues. For TE modes, global edge $\boxed{3}$ is the only free edge. Hence, we get only one eigenvalue. The local numberings are shown inside the elements without a box enclosure. The global numbering is arbitrary. The local node numbering is arbitrary except that it has to maintain a counter-clockwise increase in numbering. When once the 1, 2, 3 numbering of the local nodes is decided, edge 1 (#1) has the beginning node 1 and ending node 2 and the direction of this edge, denoted by an arrow from 1 to 2. The assignment of a direction for global edge is arbitrary. For each element, the correspondence of the local edge numbering to that of the global edge numbering needs to be specified. Since it is possible that the direction of

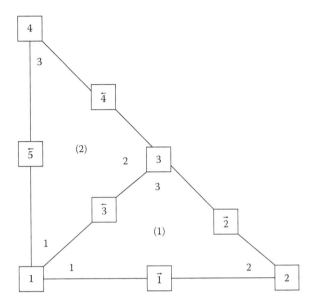

FIGURE 16.28
Two elements of Figure 16.27.

the local edge and the corresponding global edge have an opposite sense, we will append a negative sign to the global edge number when it has an opposite sense. These ideas are best illustrated through completing the example through a MATLAB code. The function is called GLAET (Global Assembly of two-dimensional Edge-based Triangular elements). The main program as well as the function program have many comment statements to explain the code and are given in Appendix 16A.5.

The comments should help the reader in correlating the algorithm with the MATLAB code.

Table 16.12 is generated based on the details of two elements shown in Figure 16.29.

Number of edges: $N_{eg} = 5$;
Number of nodes: $N_n = 4$;
Number of elements: $N_e = 2$;
$x_n = [0, 1, 0.5, 0]$;
$y_n = [0, 0, 0.5, 1]$.

TABLE 16.12

Connectivity Table for Nodes and Edges of Figure 16.29

	Local Node Array 1		
Element #	$n(1, e)$	$n(2, e)$	$n(3, e)$
①	1	2	3
②	1	3	4
	Local Edge Arrays		
Element #	$ne(1, e)$	$ne(2, e)$	$ne(3, e)$
①	1	2	3
②	−3	4	5

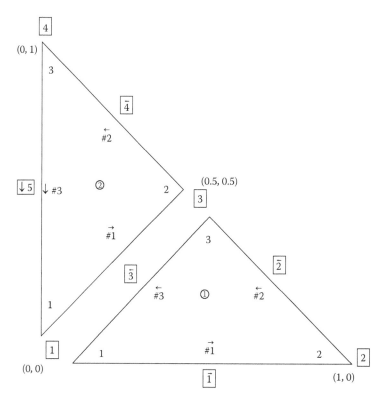

FIGURE 16.29
Details of the two elements.

16.6 Characteristic Impedance of a Transmission Line: FEM

A transmission line is a two-conductor system in which the electromagnetic wave propagates as a TEM wave. The wave does not have a longitudinal electric or magnetic field $\left(\tilde{E}_z = \tilde{H}_z = 0\right)$. The transmission line mode is a special case of the waveguide mode with zero-cutoff wave number ($k_c = 0$). Hence,

$$k = \beta. \tag{16.182}$$

The Helmholtz equation for the transverse component, say $\tilde{E}_x$, is given by

$$\nabla^2 \tilde{E}_x + k^2 \tilde{E}_x = 0,$$

$$\nabla_t^2 \tilde{E}_x + \frac{\partial^2}{\partial z^2}\tilde{E}_x + k^2 \tilde{E}_x = 0,$$

$$\nabla_t^2 \tilde{E}_x + \left(-j\beta\right)^2 \tilde{E}_x + k^2 \tilde{E}_x = 0,$$

$$\nabla_t^2 \tilde{E}_x + \left(k^2 - \beta^2\right)\tilde{E}_x = 0,$$

$$\nabla_t^2 \tilde{E}_x = 0, \tag{16.183}$$

$$\nabla_t^2 \Phi = 0. \tag{16.184}$$

The fields and the potentials satisfy the two-dimensional Laplace equation, even though the field quantities are harmonic. Hence, one can define the transmission line problem as one of solving the Laplace equation in the transverse plane. We can solve it by the FEM by minimizing the functional I, which is the electric energy:

$$I = W_e = \iint_s \frac{1}{2} \varepsilon \left| \bar{\nabla} \Phi \right|^2 ds. \tag{16.185}$$

We have earlier dealt with the Laplace equation and now we know how to solve it (through the stiffness matrix concept) and obtain the potentials. After obtaining the potentials of the nodes, one can obtain the electric energy by computing the functional for each element. We showed earlier that

$$W_e^{(e)} = \frac{1}{2} \varepsilon^{(e)} \sum_{j=1}^{3} \sum_{k=1}^{3} S_{jk} \Phi_j \Phi_k \tag{16.186}$$

and the total energy

$$W_e = \sum_{(e)=1}^{Ne} W_e^{(e)}. \tag{16.187}$$

Equation 16.187 gives us a way to calculate the capacitance C, since

$$\frac{1}{2} CV^2 = W_e, \tag{16.188a}$$

$$C = \frac{2W_e}{V^2}, \tag{16.188b}$$

where V is the potential difference between the two conductors. The characteristic impedance of a lossless transmission line is given by

$$Z_0 = \sqrt{\frac{L}{C}}, \tag{16.189}$$

where L is the inductance/meter length and C the capacitance per meter length. The inductance does not depend on the permittivity of the medium. Thus, the inductance is the same if the dielectric (even if inhomogeneous) is replaced by air. For an airtransmission line, the velocity of propagation is $c = 3 \times 10^8$ m/s, and is given by

$$c = \frac{1}{\sqrt{LC^{\mathrm{air}}}}. \tag{16.190}$$

From Equations 16.189 and 16.190, eliminating L in Equation 16.189:

$$Z_0 = \frac{1}{c\sqrt{C^{\mathrm{air}}C}}. \tag{16.191}$$

One can compute the characteristic impedance of a transmission line with multiple dielectrics by computing C^{air} and C. The FEM method can be used to compute these from Equation 16.188b. The wave number $\beta = k$ can be computed from Equation 16.192:

$$\beta = \omega\sqrt{LC} = \frac{\omega}{c}\frac{\sqrt{C}}{\sqrt{C^{air}}}. \tag{16.192}$$

16.7 Moment Method: Two-Dimensional Problems [7]

In Chapter 1, we solved an ordinary differential equation using the moment method. In this section, we illustrate its application in two-dimensional problems [7]: the first one being the calculation of capacitance of a parallel plate capacitor including the fringe effect. Let us first consider one plate of size $(2a \times 2a)$. If this plate is a PEC at V_0 volts, we have shown that the electrostatic problem can be formulated as an integral equation problem given by Equation 15.12 and Figure 15.3 defined the terms in Equation 15.12. The figure and the equation are repeated here as Figure 16.30 and Equation 16.193, respectively.

$$V_p = V_0 = \int_{-a}^{a}\int_{-a}^{a} \frac{\rho_s\,dx'dy'}{4\pi\varepsilon\left[(x-x')^2 + (y-y')^2\right]^{1/2}} \quad (P \text{ on the plate}). \tag{16.193}$$

The unknown in Equation 16.193 is ρ_s and the voltage V_0 is the known quantity and is a constant, since the plate is a PEC. In the language of the moment method established in Section 15.3.1 ($Lf = g$), the operator L is an integral operator:

$$L = \int_{-a}^{a}dx'\int_{-a}^{a} \frac{dy'}{4\pi\varepsilon\left[(x-x')^2 + (y-y')^2\right]^{1/2}} \tag{16.194}$$

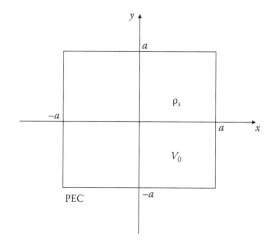

FIGURE 16.30
A square PEC plate.

and

$$f(x,y) = \rho_s(x,y), \tag{16.195}$$

$$g(x,y) = V_0. \tag{16.196}$$

We have to define an inner product for the problem:

$$\langle \psi_m, \psi_n \rangle = \int_{-a}^{a}\int_{-a}^{a} \psi_m \psi_n \, dx\, dy. \tag{16.197}$$

Let the unknown function be expressed as

$$\rho_s(x,y) = \sum_{n} \alpha_n f_n, \tag{16.198}$$

where the basis functions f_n can be chosen to be the two-dimensional pulse function

$$f_n = \begin{array}{ll} 1 & \text{on } \Delta S_n, \\ 0 & \text{on all other } \Delta S_m. \end{array} \tag{16.199}$$

In the above, ΔS_n and ΔS_m are the subareas of the plate. Let these subareas themselves be squares of dimensions $(2b \times 2b)$ as shown in Figure 16.31.

Let us choose the weight functions ω_m as the two-dimensional impulse functions:

$$\omega_m = \delta(x - x_m)\delta(y - y_m). \tag{16.200}$$

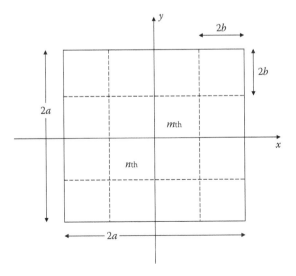

FIGURE 16.31
Square subareas of a square plate.

From Equation 15.75,

$$g_m = \langle \omega_m, g \rangle = \int_{-a}^{a} \int_{-a}^{a} V_0 \delta(x - x_m) \delta(y - y_m) dx dy = V_0. \qquad (16.201)$$

Note that

$$Lf_n = \int_{-a}^{a} dx' \int_{-a}^{a} \frac{f_n dy'}{4\pi\varepsilon \left[(x - x')^2 + (y - y')^2 \right]^{1/2}}. \qquad (16.202)$$

From Equation 15.74,

$$\ell_{mn} = \langle \omega_m, Lf_n \rangle = \int_{-a}^{a} \int_{-a}^{a} \delta(x - x_m) \delta(y - y_m) \int_{-a}^{a} dx' dy' \int_{-a}^{a} \frac{f_n dy'}{4\pi\varepsilon \left[(x - x')^2 + (y - y')^2 \right]^{1/2}} dx dy. \qquad (16.203)$$

Using the sampling property of the impulse functions,

$$\ell_{mn} = \int_{-a}^{a} \int_{-a}^{a} \frac{f_n}{4\pi\varepsilon \left[(x_m - x')^2 + (y_m - y')^2 \right]^{1/2}} dx' dy'. \qquad (16.204)$$

As per Equation 16.204, the integration extends over the entire plate. However, since f_n is zero on all subareas except for ΔS_n and is a pulse function ($f_n = 1$) on ΔS_n, Equation 16.204 reduces to

$$\ell_{mn} = \int_{\Delta S_n} \int \frac{dx' dy'}{4\pi\varepsilon \left[(x_m - x')^2 + (y_m - y')^2 \right]^{1/2}}. \qquad (16.205)$$

Let us give an interpretation of Equation 16.205.

From Equation 16.205 and Figure 16.32, we can give the following physical interpretation for ℓ_{mn}. It is the voltage at the center of the mth subarea due to a unit surface charge excitation of the nth subarea. The square root of the term in the square brackets in the denominator of the integrand in Equation 16.205 is R_{SP}, the distance between the source point and the field point. For $m \neq n$, one can approximate this distance by the distance between the centers of the two subareas, and can be treated as a constant, and taken outside the integral

$$R_{SP} \approx \left[(x_m - x_n)^2 + (y_m - y_n)^2 \right]^{1/2}, \quad m \neq n, \qquad (16.206)$$

$$\ell_{mn} \approx \frac{\Delta S_n}{4\pi\varepsilon R_{SP}}, \quad m \neq n, \qquad (16.207)$$

where $\Delta S_n = 4b^2$.

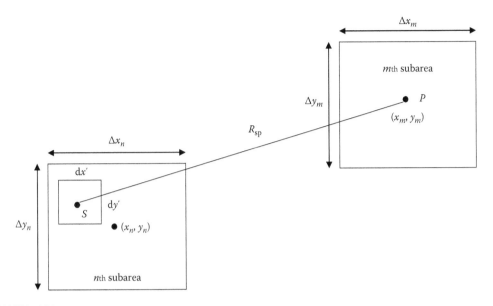

FIGURE 16.32
Subarea m and subarea n.

However, the approximation breaks down if $m = n$. In such a case, the point P is on the same subarea that has the unit charge density and is the center of the subarea. Figure 16.32 will now appear as Figure 16.33, and Equation 16.205 is reduced to

$$\ell_{nn} = \int\limits_{-b}^{b}\int\limits_{-b}^{b} \frac{dx'dy'}{4\pi\varepsilon\left[x'^2 + y'^2\right]^{1/2}}. \tag{16.208}$$

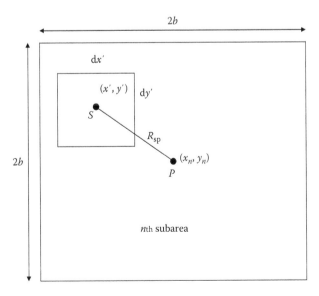

FIGURE 16.33
Geometry for calculating l_{nn}.

Integral in Equation 16.208 has an integrable singularity (when S coincides with P, $x' = 0$, $y' = 0$, and the denominator blows up). However, it can be integrated using the following known integrals:

$$\int_{-b}^{b} \frac{dy}{\left(x^2 + y^2\right)^{1/2}} = 2\ln\left(y + \sqrt{x^2 + y^2}\right)\Big|_0^b = 2\sinh^{-1}\frac{b}{|x|}. \tag{16.209}$$

Thus, we get

$$\ell_{nn} = \frac{2}{4\pi\varepsilon}\int_{-b}^{b} dx\left[\sinh^{-1}\frac{b}{|x|}\right] = \frac{2b}{\pi\varepsilon}\ln\left(1+\sqrt{2}\right) = \frac{2b}{\pi\varepsilon}0.8814. \tag{16.210}$$

Since $2b = \sqrt{A_n}$, one can write ℓ_{nn} in terms of the square root of the subarea:

$$\ell_{nn} = \frac{\sqrt{A_n}}{\pi\varepsilon}0.8814 = 0.2806\frac{\sqrt{A_n}}{\varepsilon}. \tag{16.211}$$

Thus, we can obtain the element of the $[\ell]$. On solving the algebraic equation

$$[\ell_{mn}][\alpha_n] = [g_m], \tag{16.212}$$

we get the surface charge density α_n on each of the subareas. The total charge Q on N subareas is given by

$$Q = \sum_{n=1}^{N} \alpha_n \Delta S_n \tag{16.213}$$

and the capacitance C of this PEC plate with reference to a sphere of infinite radius at zero voltage is given by

$$C = \frac{Q}{V_0}, \tag{16.214}$$

where Q is given by Equation 16.213.

We have illustrated the method using square subareas and a square PEC plate. A practical problem may have an odd-shaped plate and some of the subareas may not be squares. Do we need to modify Equations 16.207 and 16.211? Since ℓ_{mn} given by Equation 16.207 depends on the distance between the centers of the subareas and its area ΔS_n, the shape of ΔS_n is not terribly important. Since Equation 16.211 is obtained by doing integration with respect to x and y, it is not obvious as to how dependent it is on the shape of ΔS_n. To answer this question, let us determine ℓ_{nn} for a circular subarea as shown in Figure 16.34.

ℓ_{nn} is the potential at P due to a unit surface charge density excitation of the circular area:

$$\ell_{nn} = \Phi_P = \int_{\phi'=0}^{2\pi}\int_{\rho'=0}^{c} \frac{\rho_s ds'}{4\pi\varepsilon R_{SP}}$$

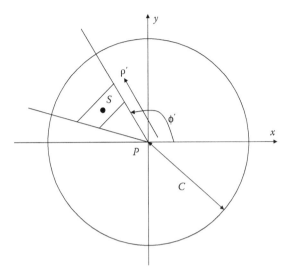

FIGURE 16.34
Geometry for calculating l_{nn} for a circular subarea.

and $\rho_s = 1$, $R_{SP} = \rho'$, and $ds' = d\rho'\rho'd\varphi'$:

$$\ell_{nn} = \Phi_P = \int_{\phi'=0}^{2\pi} \int_{\rho'=0}^{c} \frac{\rho'd\phi'd\rho'}{4\pi\epsilon\rho'} = \frac{2\pi}{4\pi\epsilon}\rho'\Big|_0^c = \frac{c}{2\epsilon}.$$

Since $\pi c^2 = A_n$, we can express ℓ_{nn} in terms of A_n as

$$\ell_{nn} = \frac{c}{2\epsilon} = \frac{1}{2\epsilon}\frac{\sqrt{A_n}}{\sqrt{\pi}} = 0.2821\frac{\sqrt{A_n}}{\epsilon} \quad \text{for a circular sub area.} \tag{16.215}$$

Equation 16.215 is approximately the same as Equation 16.211 from which we conclude that ℓ_{nn} does not depend much on the shape of the subarea. As long as the subarea does not have an oblong shape, we can use Equation 16.215 for ℓ_{nn}. The reason for this behavior is that the contribution to ℓ_{nn} comes mostly from the charges near the point P, since R_{SP} appears in the denominator. Another way of saying it is that the contribution comes mostly from the integrable singularity at the origin.

Let us now extend the method to the parallel plate capacitor shown in Figure 16.35.

Let

$$\rho_s^t(x,y) = \sum_n \alpha_n^t f_n \quad \text{for top plate,} \tag{16.216}$$

$$\rho_s^b(x,y) = \sum_n \alpha_n^b f_n \quad \text{for bottom plate.} \tag{16.217}$$

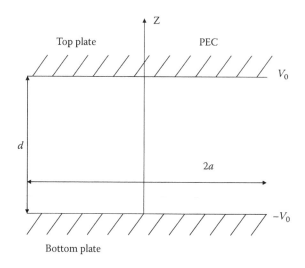

FIGURE 16.35
Parallel plate capacitor.

The algebraic equations can be written in the form of a partitioned matrix, whose matrix elements are

$$
\begin{bmatrix}
\begin{bmatrix} \ell^{tt} \end{bmatrix} & \begin{bmatrix} \ell^{tb} \end{bmatrix} \\
\begin{bmatrix} \ell^{bt} \end{bmatrix} & \begin{bmatrix} \ell^{bb} \end{bmatrix}
\end{bmatrix}
\begin{bmatrix} \alpha_n^t \\ \alpha_n^b \end{bmatrix}
=
\begin{bmatrix} g_m^t \\ g_m^b \end{bmatrix}.
\tag{16.218}
$$

The computation of the elements of $[\ell^{tt}]$ and $[\ell^{bb}]$ can be obtained as before. ℓ_{mn}^{tb} is the potential at the center of the element on the bottom plate due to a unit surface charge density excitation of an element on the top plate. A similar explanation holds good for the elements of $[\ell^{bt}]$. Figure 16.36 shows the geometry.

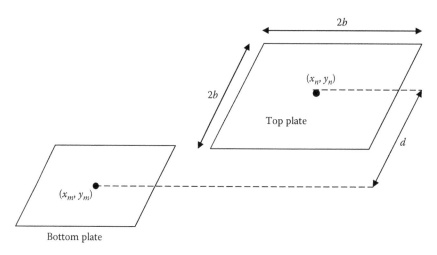

FIGURE 16.36
Subareas of bottom and top plates.

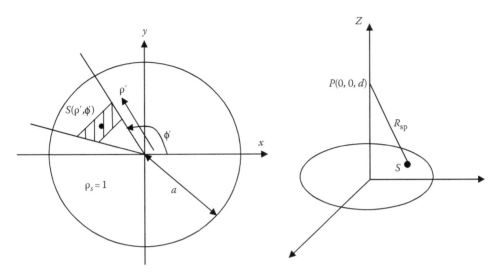

FIGURE 16.37
Improved calculations for aligned subareas of the top and bottom plates when the distance between the plates is small.

Even if $m = n$, we do not have a problem of singularity in the integrand. If $m = n$, the elements are aligned and their centers are separated by a distance d. Thus, we can write

$$\ell^{tb}_{mn} = \frac{A_n}{4\pi\varepsilon\sqrt{(x_m - x_n)^2 + (y_m - y_n)^2 + d^2}}. \tag{16.219}$$

However, the formula for ℓ^{tb}_{mn} can be improved. It becomes necessary to do so if d is small. Consider the geometry shown in Figure 16.37:

$$\Phi_P(0,0,d) = \int_{\phi'=0}^{2\pi}\int_{\rho'=0}^{c} \frac{\rho'd\rho'd\phi'}{4\pi\varepsilon(\rho'^2 + d^2)^{1/2}} = \frac{2\pi}{4\pi\varepsilon}\int_0^c \frac{\rho'd\rho'}{(\rho'^2 + d^2)^{1/2}}, \tag{16.220}$$

$$\ell^{tb}_{nn} = \Phi_P(0,0,d) = \frac{0.282(2c)}{\varepsilon}\left[\sqrt{1 + \frac{\pi}{4}\left(\frac{d}{c}\right)^2} - \frac{\sqrt{\pi}d}{2c}\right].$$

The above formula can be converted into one involving area rather than the radius by noting again $c = \sqrt{A_n / \pi}$. Thus, we have an improved formula for ℓ^{tb}_{nn}:

$$\ell^{tb}_{nn} = \frac{0.3182}{\varepsilon}\left[\sqrt{A_n + \frac{\pi^2 d^2}{4}} - \frac{\pi d}{2}\right]. \tag{16.221}$$

For a parallel plate capacitor with aligned identical plates, we can take advantage of symmetries and reduce the number of simultaneous equations. We note

$$\left[\ell^{tt}\right] = \left[\ell^{bb}\right], \tag{16.222}$$

$$\left[\ell^{tb} \right] = \left[\ell^{bt} \right].$$ (16.223)

If the top plate is at V_0 and the bottom plate at $-V_0$, then

$$g_m^t = -g_m^b.$$ (16.224)

From Equation 16.218,

$$\left[\ell^{tt} \right]\left[\alpha_n^t \right] + \left[\ell^{tb} \right]\left[\alpha_n^b \right] = g_m^t,$$ (16.225)

$$\left[\ell^{bt} \right]\left[\alpha_n^t \right] + \left[\ell^{bb} \right]\left[\alpha_n^b \right] = -g_m^t.$$ (16.226)

From Equations 16.222 through 16.226, it is obvious that

$$\left[\alpha_n^t \right] = -\left[\alpha_n^b \right].$$ (16.227)

Substituting Equation 16.227 into 16.225 and solving for $\left[\alpha_n^t \right]$, we obtain

$$\left[\alpha_n^t \right] = \left\{ \left[\ell^{tt} \right] - \left[\ell^{tb} \right] \right\}^{-1} \left[g_m^t \right].$$ (16.228)

The charge on the top plate is given by

$$Q = \sum_n \alpha_n \Delta S_n$$

and the potential difference between the two plates is $2V_0$. Hence,

$$C = \frac{Q}{2V_0}.$$ (16.229)

16.8 Moment Method: Scattering Problem [7]

16.8.1 Formulation

In Section 7.6, we considered the problem of scattering of electromagnetic waves by a conducting object. Specifically, we considered an incident plane wave propagating in the x-direction, with the electric field

$$\tilde{E}^i = \hat{z} E_0 e^{-j\beta x}.$$ (16.230)

A circular conducting cylinder infinitely long of radius a with z-axis as the axis of the cylinder creates a scattered wave and we showed, analytically, by solving the boundary value problem that the z-component of the total electric field along the x-axis is given by

$$E_z^t = E_z^i + E_z^{sc}$$

$$= E_0 \sum_{n=-\infty}^{n=\infty} \left(j^{-n}\right) \left[J_n\left(\beta x\right) - \frac{J_n\left(\beta a\right)}{H_n^{(2)}\left(\beta a\right)} H_n^{(2)}\left(\beta x\right) \right], \quad \begin{matrix} -\infty < x < -a, \\ a < x < \infty. \end{matrix} \tag{16.231}$$

The convergence of this series is quite slow for large values of βa [8]. An alternative to the analytical technique is the numerical method based on the moment method [7]. In this section, we describe this method.

We showed in Section 2.15 that the vector magnetic field $\tilde{A} = \hat{z}A(\rho)$ due to a harmonic current $I_0 e^{j\omega t}$ in an infinitely long filament along the z-axis in free space is given by

$$A\left(\rho\right) = -j\frac{\mu_0}{4\pi} I_0 H_0^{(2)}\left(\beta\rho\right) \tag{16.232}$$

and

$$\tilde{E}_z = -\frac{\beta\eta_0}{4} I_0 H_0^{(2)}\left(\beta\rho\right). \tag{16.233}$$

Suppose the wire is not along the z-axis but parallel to the z-axis passing through the point (x', y') or in cylinder coordinates (ρ', ϕ').

We obtain the fields for this case by replacing ρ in Equation 16.233 by $\left|\bar{\rho} - \bar{\rho}'\right|$:

$$\tilde{E}_z\left(\rho,\phi\right) = -\frac{\beta\eta_0}{4} I_0 H_0^{(2)}\left(\beta\left|\bar{\rho} - \bar{\rho}'\right|\right), \tag{16.234}$$

where

$$\left|\bar{\rho} - \bar{\rho}'\right| = \left[\rho^2 + \rho'^2 - 2\rho\rho'\cos\left(\phi - \phi'\right)\right]^{1/2}. \tag{16.235}$$

If the conductor is perfect, the harmonic currents can exist only on the surface as shown in Figure 16.38.

Let the surface current density be $\tilde{K}_z$ (A/m). Due to a differential length dl' on the curve c, the current $\tilde{I}_0 = \tilde{K}_z dl'$ and the total electric field $\tilde{E}_z$, from Equation 16.234, is given by

$$\tilde{E}_z\left(\rho,\phi\right) = -\frac{\beta\eta_0}{4} \oint \tilde{K}_z\left(\bar{\rho}'\right) H_0^{(2)}\left(\beta\left|\bar{\rho} - \bar{\rho}'\right|\right) dl'. \tag{16.236}$$

In Equation 16.236, $\tilde{K}_z$ is the surface current density induced on the surface due to the incident plane wave whose electric field $\tilde{E}_z^i$ is given. $\tilde{K}_z$ is not known and hence (Equation 16.236) is called the integral equation. We know that $\tilde{E}_z$ is the total electric field $\tilde{E}_z^t$, the sum of the incident and scattered fields and we also know the boundary condition

$$\tilde{E}_z\left(\rho,\phi\right) = 0 \quad \text{for a point on } C. \tag{16.237}$$

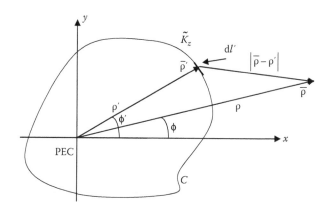

FIGURE 16.38
Surface current flow on a PEC cylinder.

Thus, for a point c, we get the integral equation

$$\frac{\beta\eta_0}{4}\oint_C \tilde{K}_z(\bar{\rho}')H_0^{(2)}\left(\beta\left|\bar{\rho}-\bar{\rho}'\right|\right)dl' = E_z^i \quad \text{for a point on } C. \tag{16.238}$$

Solving the integral equation, we obtain $\tilde{K}_z$. The fields due to $\tilde{K}_z$ give rise to the scattered field $\tilde{E}_z^{sc}$. Thus, the scattered fields are calculated from the equation

$$\tilde{E}_z^{sc}(\rho) = -\frac{\beta\eta_0}{4}\oint_C \tilde{K}_z(\bar{\rho}')H_0^{(2)}\left(\beta\left|\bar{\rho}-\bar{\rho}'\right|\right)dl' \quad \text{for any point } \bar{\rho}. \tag{16.239}$$

Thus, the moment method for scattering problem consists of two steps. Given $\tilde{E}_z^i$ solve the integral equation 16.238 by the moment method and obtain $\tilde{K}_z(\bar{\rho}')$. Determine $\tilde{E}_z^{sc}$ by doing the integration (Equation 16.239). Equation 16.238 is called electric field integral equation (EFIE).

16.8.2 Solution

See Figure 16.38. Divide C into N straight segments Δc_n. Let

$$\tilde{K}_z(\bar{\rho}) = \sum_{n=1}^{N}\alpha_n f_n, \tag{16.240}$$

when

$$f_n(\bar{\rho}) = \begin{cases} 1 & \text{on } \Delta c_n \\ 0 & \text{on all other } \Delta c_n \end{cases} \tag{16.241a}$$

that is, use pulse functions for expansion.

Choose impulses for testing (point watching)

$$\omega_m = \delta(\bar{\rho} - \bar{\rho}_m). \tag{16.242}$$

Define the inner product

$$\langle f, g \rangle = \int_C fg \, dl. \tag{16.243}$$

Note that the operator L for this problem is an integral operator given by

$$L = \frac{\beta \eta_0}{4} \int_C H_0^{(2)} \left(\beta |\bar{\rho} - \bar{\rho}'| \right) dl'. \tag{16.244}$$

The unknown function $f = \tilde{K}_z$ and the excitation function $g = E_z^i$.
We get the equations

$$[l_{mn}][\alpha_n] = [g_m], \tag{16.245}$$

where

$$l_{mn} = \int_C \delta(\bar{\rho} - \bar{\rho}_m) \left[\int_C \frac{\beta \eta_0}{4} H_0^{(2)} \left(\beta |\bar{\rho} - \bar{\rho}'| f_n \right) dl' \right] dl,$$

$$l_{mn} = \frac{\beta \eta_0}{4} \int_C H_0^{(2)} \left(\beta |\bar{\rho}_m - \bar{\rho}'| \right) f_n \, dl', \tag{16.246}$$

$$l_{mn} = \frac{\beta \eta_0}{4} \int_{\Delta c_n} H_0^{(2)} \left(\beta |\bar{\rho}_m - \bar{\rho}'| \right) dl', \tag{16.247}$$

$$l_{mn} = \frac{\beta \eta_0}{4} \int_{\Delta c_n} H_0^{(2)} \left\{ \left[\beta (x' - x_m)^2 + (y' - y_m)^2 \right]^{1/2} \right\} dl'. \tag{16.248}$$

It is clear that l_{mn} is the negative of the scattered field at (x_m, y_m) due to unit surface current density on Δc_n (see Figure 16.39).
If $m \neq n$, we can approximate Equation 16.248:

$$l_{mn} \approx \frac{\beta \eta_0}{4} \Delta c_n H_0^{(2)} \left\{ \left[\beta (x'_n - x_m)^2 + (y'_n - y_m)^2 \right]^{1/2} \right\}, \quad m \neq n. \tag{16.249}$$

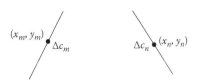

FIGURE 16.39
Interpretation of Equation 16.248.

When $m = n$, we have a singularity in the integrand of Equation 16.248. This can be evaluated by using a small argument formula for $H_0^{(2)}$:

$$H_0^{(2)}(z) = 1 - j\frac{2}{\pi}\ln\left(\frac{\gamma z}{2}\right), \tag{16.250}$$

where $|z| \ll 1$ and

$$\gamma = 1.781, \tag{16.251}$$

called Euler's constant.
    Thus,

$$l_{mn} = \frac{\beta\eta_0}{4}\int_{\Delta c_n}\left[1 - j\frac{2}{\pi}\ln\left(\frac{\gamma z}{2}\right)\right]dl', \quad m = n, \tag{16.252a}$$

where

$$z = \beta\left\{\left(x' - x_n\right)^2 + \left(y' - y_n\right)^2\right\}^{1/2}. \tag{16.252b}$$

The evaluation gives [7]

$$l_{mn} \approx \frac{\eta_0}{4}\beta\Delta c_n\left[1 - j\frac{2}{\pi}\ln\left(\frac{\gamma\beta\Delta c_n}{4e}\right)\right], \tag{16.253a}$$

where

$$e = 2.718. \tag{16.253b}$$

Since

$$
\begin{aligned}
g_m &= \langle\omega_m, g\rangle \\
&= \int\delta(\bar{\rho} - \rho_m)E_z^i \\
&= E_z^i(\bar{\rho}_m).
\end{aligned} \tag{16.254}
$$

Solving Equation 16.245 we get α_n, and hence we have solved for K_z numerically. The scattered field is now easily obtained by numerically evaluating Equation 16.239:

$$
\begin{aligned}
I &= \int_C \tilde{K}_z(\bar{\rho}')H_0^{(2)}\left[\beta|\bar{\rho} - \bar{\rho}'|\right]dl' \\
&= \int_C \sum \alpha_m f_m H_0^{(2)}\left[\beta|\bar{\rho} - \bar{\rho}'|\right]dl' \\
&= \int_{\Delta c_1}(\ )dl' + \int_{\Delta c_2}(\ )dl' + \cdots + \int_{\Delta c_N}(\ )dl' \\
&= \int_{\Delta c_1}\alpha_1 H_0^{(2)}(\cdot)dl' + \int_{\Delta c_2}\alpha_2 H_0^{(2)}(\cdot)dl' + \cdots + \int_{\Delta c_N}\alpha_N H_0^{(2)}(\cdot)dl'.
\end{aligned}
$$

For fields on axis, $(\cdot)$ can be taken as constant:

$$(\cdot) = \beta \left[\left(x_m - x \right)^2 + y_m^2 \right]^{1/2},$$

$$I = \sum_{m=1}^{N} \Delta c_m \alpha_m H_0^{(2)} \left\{ \beta \left[\left(x_m - x \right)^2 + y_m^2 \right]^{1/2} \right\}.$$

Thus,

$$E_z^{sc}(x) = -\frac{\beta \eta_0}{4} \sum_{m=1}^{N} \Delta c_m \alpha_m H_0^{(2)} \left\{ \beta \left[\left(x_m - x \right)^2 + y_m^2 \right]^{1/2} \right\}. \qquad (16.255)$$

Let us illustrate the algorithm by taking four segments of a circular cylinder as shown in Figure 16.40.

The coordinates of the centers of the four segments are

$$\left(x_1, y_1 \right) = \left(\frac{a}{2}, -\frac{a}{2} \right),$$

$$\left(x_2, y_2 \right) = \left(\frac{a}{2}, \frac{a}{2} \right),$$

$$\left(x_3, y_3 \right) = \left(-\frac{a}{2}, \frac{a}{2} \right),$$

$$\left(x_4, y_4 \right) = \left(-\frac{a}{2}, -\frac{a}{2} \right),$$

$$g_m = e^{-j\beta x_m},$$

$$g_1 = e^{-j\beta a/2}, \quad g_2 = e^{-j\beta a/2}, \quad g_3 = e^{j\beta a/2}, \quad g_4 = e^{j\beta a/2}.$$

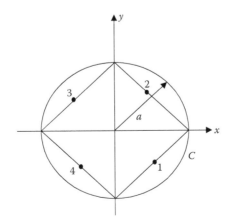

FIGURE 16.40
Circle of radius a approximated by four segments.

$$\Delta c_n = \sqrt{2}a, \quad n = 1, 2, 3, 4,$$

$$l_{mn} = \frac{\eta_0}{4} \beta\sqrt{2}a \left[1 - j\frac{2}{\pi}\ln\left\{ \frac{\gamma\beta\sqrt{2}a}{4e} \right\} \right], \quad n = 1, 2, 3, 4,$$

$$l_{mn} = \frac{\eta_0}{4} \beta\sqrt{2}a H_0^{(2)} \left\{ \left[\beta(x_n - x_m)^2 + (y_n - y_m)^2 \right]^{1/2} \right\}, \quad m \neq n.$$

Using

$$[\ell_{mn}][\alpha_n] = [g_m],$$

we can now determine α_1, α_2, α_3, and α_4.

The scattered field on the x-axis is given by

$$E_z^{sc}(x)\Big|_{y=0} = -\frac{\beta\eta_0}{4} \left[\sum_{m=1}^{4} \alpha_m \Delta c_m H_0^{(2)} \left\{ \beta\left[(x_m - x)^2 + y_m^2 \right]^{1/2} \right\} \right].$$

Problem P16.15 asks us to plot $\left| \left(E_z^i(x) + E_z^{sc}(x) \right)/E_z^i \right|$ as a function of x from $-20 < \beta x < 60$, for the data $a = 1$ and $\beta = 3.1$.

In part (a), use $N = 4$, and in (b) use $N = 72$.

References

1. Young, D., Iterative methods for solving partial differential equations of elliptic type, *Trans. Am. Math. Soc.*, 76, 92–111, 1954.
2. Frankel, S. P., Convergence rates of iterative treatments, *Math. Tables Other Aids Comput.*, 4, 65–75, 1950.
3. Kalluri, D., Potential gradients near rounded corners, MS thesis, University of Wisconsin, Madison, WI, 1959.
4. Pelosi, G., Selleri, S., and Coccioli, R., *Quick Finite Element Method for Electromagnetic Waves*, Artech House, Norwood, MA, 1998.
5. Volakis, J. L., Chafterjee, A., and Kempel, L., *Finite Element Method for Electromagnetics*, Wiley-Interscience, New York, 1998.
6. Jin, J. M., *The Finite Element Method in Electromagnetics*, Wiley, New York, 2002.
7. Harrington, R. F., *Field Computation by Moment Methods*, Macmillan, New York, 1968.
8. King, R. W. P. and Wu, T. T., *The Scattering and Diffraction of Waves*, Harvard University Press, Cambridge, MA, 1959.

17

Advanced Topics on Finite-Element Method*

In this Chapter, we consider briefly advanced topics on finite element method. References [1–5] can be consulted for a more detailed exposure to the topics.

17.1 Node- and Edge-Based FEM

Let us consider the example of a ridged waveguide. Point A is on a sharp conducting corner of 90° internal angle. A field point in the waveguide can be described by the cylindrical co-ordinates shown in Figure 17.1.

In the neighborhood of a point such as A, it can be shown that the longitudinal components E_z and H_z vary as $\rho^{1/3}$ and the transverse components E_ρ, E_ϕ, H_ρ, and H_ϕ have singularities at the point A (see Appendix 17A) and vary as $\rho^{-2/3}$. If the point A is on an edge, the transverse components of E or H at A becomes infinite. If we expand longitudinal components in terms of scalar shape functions and the transverse components in terms of vector shape functions, the problems of the spurious modes and the singularities are mitigated. The starting point of the finite-element method (FEM) for vector problems is the differential equation 16.139 and its functional (Equation 16.140).

For a waveguide problem, we will assume that

$$\tilde{\mathbf{E}}(x,y,z) = \mathbf{E}(x,y)e^{-\gamma z} = \left[\mathbf{E}_t(x,y) + \hat{z}E_z\right]e^{-\gamma z}. \tag{17.1}$$

The three-dimensional (3D) – operator can be expressed as

$$\nabla = \nabla_t - \gamma\hat{z}, \tag{17.2}$$

where ∇_t is the operator in the transverse plane and is given by

$$\nabla_t = \left[\hat{x}\frac{\partial}{\partial x} + \hat{y}\frac{\partial}{\partial y}\right]. \tag{17.3}$$

Then

$$\nabla \times \tilde{E} = \left[\nabla_t - \gamma\hat{z}\right] \times \left[\mathbf{E}_t + \hat{z}E_z\right]e^{-\gamma z}$$
$$= \left[\nabla_t \times \mathbf{E}_t - \gamma\hat{z} \times \mathbf{E}_t + \nabla_t \times \hat{z}E_z\right]e^{-\gamma z}. \tag{17.4}$$

Note that

$$\nabla_t \times \hat{z}E_z = \nabla_t E_z \times \hat{z} + E_z\nabla_t \times \hat{z} = \nabla_t E_z \times \hat{z}, \tag{17.5}$$

since $\nabla_t \times \hat{z} = \left[\hat{x}(\partial/\partial x) + \hat{y}(\partial/\partial y)\right] \times (1)\hat{z}$, $\partial/\partial x(1) = 0$, and $\partial/\partial y(1) = 0$.

---

* For chapter appendices, see 17A in the Appendices section.

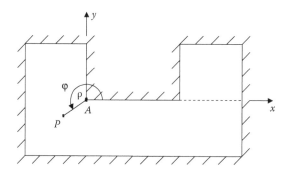

FIGURE 17.1
A sharp conducting corner.

Thus,

$$\nabla \times \tilde{E} = \left[\underbrace{\nabla_t \times E_t}_{\text{First term}} + \underbrace{\left(\nabla_t E_z + \gamma E_t\right) \times \hat{z}}_{\text{Second term}} \right] e^{-\gamma z}. \tag{17.6}$$

Note that the first term has only the z component and the second term lies in the transverse plane. Thus, one can write, for a loss-free system,

$$\left|\nabla \times \tilde{E}\right|^2 = \left|\nabla_t \times E_t\right|^2 + \left|\nabla_t E_z + \gamma E_t\right|^2$$

or alternatively

$$\left|\nabla \times \tilde{E}\right|^2 = \left(\nabla_t \times E_t\right) \cdot \left(\nabla_t \times E_t\right)^* + \left(\nabla_t E_z + \gamma E_t\right) \cdot \left(\nabla_t E_z + \gamma E_t\right)^*. \tag{17.7}$$

Substituting Equation 17.7 into Equation 16.140, we obtain

$$I\left(\tilde{E}\right) = \frac{1}{2} \iint \left[\frac{1}{\mu_r}\left(\nabla_t \times E_t\right) \cdot \left(\nabla_t \times E_t\right)^* - k_0^2 \varepsilon_r \tilde{E} \cdot \tilde{E}^* + \frac{1}{\mu_r}\left(\nabla_t E_z + \gamma E_t\right) \cdot \left(\nabla_t E_z + \gamma E_t\right)^* \right] ds. \tag{17.8}$$

Discretizing and minimizing the functional (Equation 17.8) will give equations in the form of an eigenvalue problem [1]. If γ is given, the eigenvalue $\lambda = k_0^2$ can be calculated. Usually, we wish to do it the other way round, that is, the operating frequency, and hence k_0^2 is given and one wishes to find out the propagation constant. The cutoff frequency is established at the frequency for which $\gamma = 0$ and γ changes from α to $j\beta$. A transformation given by

$$e_t = \gamma E_t$$

and

$$e_z = E_z$$

converts Equation 17.8 into

$$\gamma^2 I(e) = \frac{1}{2}\iint\left\{\frac{1}{\mu_r}(\nabla_t\times\overline{e}_t)\cdot(\nabla_t\times e_t)^* - k_0^2\varepsilon_r e_t\cdot e_t^*\right.$$
$$\left. + \gamma^2\left[\frac{1}{\mu_r}(\nabla_t e_z + e_t)\cdot(\nabla_t e_z + e_t)^* - k_0^2\varepsilon_r e_z e_z^*\right]\right\}ds. \tag{17.9}$$

Equation 17.9 can be descretized by expressing for each element

$$e_t^{(e)} = \sum_{i=1}^{3}\alpha_i e_{\tan i}^{(e)}, \tag{17.10}$$

$$e_z^{(e)} = \sum_{i=1}^{3}\alpha_i e_{zi}^{(e)}. \tag{17.11}$$

Substituting these into Equation 17.9, the functional on the LHS for each element can be written in the following form as

$$\frac{1}{2}\iint_{\Delta^{(e)}}\left[\frac{1}{\mu_r}(\nabla_t\times e_t)\cdot(\nabla_t\times e_t)^* - k_0^2\varepsilon_r^{(e)}e_t\cdot e_t^*\right]ds = \left\{e_{\tan}^{(e)}\right\}^{\mathrm{T}}\left[A_{tt}^{(e)}\right]\left\{e_{\tan}^{(e)}\right\}^*, \tag{17.12}$$

where

$$\left[A_{tt}^{(e)}\right] = \iint_{\Delta(e)}\left[\frac{1}{\mu_r^{(e)}}\left\{\nabla_t\times\alpha_i^{(e)}\right\}\cdot\left\{\nabla_t\times\alpha_j^{(e)}\right\}^{\mathrm{T}} - k_0^2\varepsilon_r^{(e)}\alpha_i^{(e)}\cdot\left(\alpha_j^{(e)}\right)^{\mathrm{T}}\right]ds$$
$$= \frac{1}{\mu_r^{(e)}}\left[E^{(e)}\right] - k_0^2\varepsilon_r^{(e)}\left[F^{(e)}\right]. \tag{17.13}$$

The second term on the RHS of Equation 17.9 needs more careful handling. Note that

$$(\nabla_t e_z\times e_t)\cdot(\nabla_t e_z\times e_t)^*$$
$$= e_t\cdot e_t^* + e_t^*\cdot(\nabla_t e_z) + (\nabla_t e_z)^*\cdot e_t + (\nabla_t e_z)\cdot(\nabla_t e_z)^*. \tag{17.14}$$

Thus,

$$\frac{1}{2}\gamma^2\iint_{\Delta(e)}\left[\frac{1}{\mu_r}(\nabla_t e_z\times e_t)\cdot(\nabla_t e_z\times e_t)^* - k_0^2\varepsilon_r e_z\cdot e_z^*\right]ds$$

can be written as

$$\gamma^2\left\{\begin{matrix}e_{\tan}^{(e)}\\ e_z^{(e)}\end{matrix}\right\}^{\mathrm{T}}\left[\begin{matrix}\left[B_{tt}^{(e)}\right] & \left[B_{tz}^{(e)}\right]\\ \left[B_{zt}^{(e)}\right] & \left[B_{zz}^{(e)}\right]\end{matrix}\right]\left[\begin{matrix}e_{\tan}^{(e)}\\ e_z^{(e)}\end{matrix}\right]^*, \tag{17.15}$$

where

$$B_{tt}^{(e)} = \iint_{\Delta(e)} \left[\frac{1}{\mu_r^{(e)}} \left\{\alpha_i^{(e)}\right\} \cdot \left\{\alpha_j^{(e)}\right\}^T \right] ds = \frac{1}{\mu_r^{(e)}} F^{(e)}, \tag{17.16}$$

$$B_{tz}^{(e)} = \iint_{\Delta(e)} \left[\frac{1}{\mu_r^{(e)}} \left\{\alpha_i^{(e)}\right\} \cdot \left\{\nabla_t \alpha_j^{(e)}\right\} \right] ds, \tag{17.17}$$

$$B_{zt}^{(e)} = \iint_{\Delta(e)} \left[\frac{1}{\mu_r^{(e)}} \left\{\nabla_t \alpha_i^{(e)}\right\} \cdot \left\{\alpha_j^{(e)}\right\} \right] ds, \tag{17.18}$$

$$B_{zz}^{(e)} = \iint_{\Delta(e)} \left[\frac{1}{\mu_r^{(e)}} \left\{\nabla_t \alpha_i^{(e)}\right\} \cdot \left\{\nabla_t \alpha_j^{(e)}\right\}^T - k_0^2 \varepsilon_r^{(e)} \left\{\alpha_i^{(e)}\right\} \left\{\alpha_j^{(e)}\right\}^T \right] ds$$

$$= \frac{1}{\mu_r^{(e)}} \left[S^{(e)} \right] - k_0^2 \varepsilon_r^{(e)} \left[T^{(e)} \right]. \tag{17.19}$$

The $A_{tt}^{(e)}$ matrix elements given by Equation 17.13 are easily evaluated, since $[E^{(e)}]$ and $[F^{(e)}]$ are given by Equations 16.165 and 16.167, respectively. Similarly, the $B_{zz}^{(e)}$ matrix elements given by Equation 17.19 are easily evaluated since $[S^{(e)}]$ and $[T^{(e)}]$ are given by Equations 16.52 and 16.103, respectively. It can be shown that

$$\left[B_{tz}^{(e)} \right] = \left[B_{zt}^{(e)} \right]^T = \frac{1}{\mu_r^{(e)}} \left[H_{tz}^{(e)} \right], \tag{17.20}$$

where

$$H_{11}^{(e)} = -K_1 \left[b_1 (b_1 - b_2) + c_1 (c_1 - c_2) \right], \tag{17.21a}$$

$$H_{12}^{(e)} = K_1 \left[b_2 (b_2 - b_1) + c_2 (c_2 - c_1) \right], \tag{17.21b}$$

$$H_{13}^{(e)} = -K_1 \left[-y_1 b_2 - y_2 b_1 - y_3 b_3 + x_1 c_2 + x_2 c_1 + x_3 c_3 \right], \tag{17.21c}$$

$$H_{21}^{(e)} = -K_2 \left[-y_1 b_1 - y_2 b_3 - y_3 b_2 + x_1 c_1 + x_2 c_3 + x_3 c_2 \right], \tag{17.21d}$$

$$H_{22}^{(e)} = -K_2 \left[b_2 (b_2 - b_3) + c_2 (c_2 - c_3) \right], \tag{17.21e}$$

$$H_{23}^{(e)} = K_2 \left[b_3 (b_3 - b_2) + c_3 (c_3 - c_2) \right], \tag{17.21f}$$

$$H_{31}^{(e)} = K_3 \left[b_1 (b_1 - b_3) + c_1 (c_1 - c_3) \right], \tag{17.21g}$$

$$H_{32}^{(e)} = -K_3 \left[-y_1 b_3 - y_2 b_2 - y_3 b_1 + x_1 c_3 + x_2 c_2 + x_3 c_1 \right], \tag{17.21h}$$

$$H_{33}^{(e)} = -K_3 \left[b_3 \left(b_3 - b_1 \right) + c_3 \left(c_3 - c_1 \right) \right], \tag{17.21i}$$

and

$$K_j = \frac{l_j}{12A}. \tag{17.22}$$

After global assembly for all elements, the equations can be written as [1]

$$\begin{bmatrix} A_{tt} & 0 \\ 0 & 0 \end{bmatrix} \begin{Bmatrix} e_{\text{tan\#}} \\ e_z \end{Bmatrix} = \gamma^2 \begin{bmatrix} B_{tt} & B_{tz} \\ (B_{tz})^{\text{T}} & B_{zz} \end{bmatrix} \begin{bmatrix} e_{\text{tan\#}} \\ e_z \end{bmatrix}. \tag{17.23}$$

The second equation in (17.23) is

$$\begin{aligned} \left[B_{tz} \right]^{\text{T}} \left[e_{\text{tan\#}} \right] + \left[B_{zz} \right] \left[e_z \right] &= 0, \\ \left[e_z \right] &= -\left[B_{zz} \right]^{-1} \left[B_{tz} \right]^{\text{T}} \left[e_{\text{tan\#}} \right]. \end{aligned} \tag{17.24a}$$

The first equation in (17.23) is

$$\left[A_{tt} \right] \left[e_{\text{tan\#}} \right] = \gamma^2 \left\{ \left[B_{tt} \right] \left[e_{\text{tan\#}} \right] + \left[B_{tz} \right] \left[e_z \right] \right\}. \tag{17.24b}$$

Substituting for $[e_z]$, from Equation 17.24a, we can formulate the above as the eigenvalue problem

$$\left[C \right] \left[e_{\text{tan\#}} \right] = \gamma^2 \left[e_{\text{tan\#}} \right], \tag{17.25a}$$

where

$$\left[C \right] = \left[\left[B_{tt} \right] - \left[B_{tz} \right] \left[B_{zz} \right]^{-1} \left[B_{tz} \right]^{\text{T}} \right]^{-1} \left[A_{tt} \right]. \tag{17.25b}$$

Before closing this section, let us show that the formulas given in Equations 17.21a through 17.21i are obtained by implementing Equation 17.18. We will illustrate for only two elements of $H^{(e)}$:

$$H_{11}^{(e)} = \iint_{\Delta^{(e)}} \alpha_1^{(e)} \cdot \nabla_t \alpha_1^{(e)} \, ds.$$

As shown before,

$$\nabla_t \alpha_1^{(e)} = \frac{1}{2A^{(e)}} \left[b_1 \hat{x} + c_1 \hat{y} \right],$$

$$\alpha_1^{(e)} = \frac{l_1^{(e)}}{2A^{(e)}}\left[\hat{x}\left(\alpha_1 b_2 - \alpha_2 b_1\right) + \hat{y}\left(\alpha_1 c_2 - \alpha_2 c_1\right)\right],$$

$$\alpha_1^{(e)}\cdot\nabla_t\alpha_1^{(e)} = \frac{l_1^{(e)}}{4\left(A^{(e)}\right)^2}\left[b_1\left(\alpha_1 b_2 - \alpha_2 b_1\right) + c_1\left(\alpha_1 c_2 - \alpha_2 c_1\right)\right]$$

$$= \frac{l_1^{(e)}}{4\left(A^{(e)}\right)^2}\left[\alpha_1\left(b_1 b_2 + c_1 c_2\right) + \alpha_2\left(-b_1^2 - c_1^2\right)\right],$$

$$\iint_{\Delta^{(e)}}\alpha_j\, \mathrm{d}x\mathrm{d}y = \frac{A^{(e)}}{3},$$

$$H_{11}^{(e)} = \frac{l_1^{(e)}}{12\left(A^{(e)}\right)^2}\left[b_1 b_2 + c_1 c_2 - b_1^2 - c_1^2\right]$$

$$= -K_1\left[b_1\left(b_1 - b_2\right) + c_1\left(c_1 - c_2\right)\right],$$

$$H_{12}^{(e)} = \iint_{\Delta^{(e)}}\alpha_1^{(e)}\cdot\nabla\alpha_2^{(e)}\mathrm{d}s,$$

$$\alpha_1^{(e)} = \frac{l_1^{(e)}}{2A^{(e)}}\left[\hat{x}\left(\alpha_1 b_2 - \alpha_2 b_1\right) + \hat{y}\left(\alpha_1 c_2 - \alpha_2 c_1\right)\right],$$

$$\alpha_1^{(e)}\cdot\nabla\alpha_2^{(e)} = \frac{l_1^{(e)}}{4\left(A^{(e)}\right)^2}\left[b_2\left(\alpha_1 b_2 - \alpha_2 b_1\right) + c_2\left(\alpha_1 c_2 - \alpha_2 c_1\right)\right]$$

$$= \frac{l_1^{(e)}}{4\left(A^{(e)}\right)^2}\left[\alpha_1\left(b_2^2 + c_2^2\right) + \alpha_2\left(-b_1 b_2 - c_1 c_2\right)\right],$$

$$H_{12}^{(e)} = \frac{l_1^{(e)}}{12\left(A^{(e)}\right)}\left[b_2\left(b_2 - b_1\right) + c_2\left(c_2 - c_1\right)\right]$$

$$= K_1\left[b_2\left(b_2 - b_1\right) + c_2\left(c_2 - c_1\right)\right].$$

17.2 Weak Formulation and Weighted Residual Method

In Section 15.3.2, we discussed the weighted residual method (WRM). Let us revisit this method by solving the 1D Helmholtz equation

$$\frac{\mathrm{d}^2 V}{\mathrm{d}z^2} + k^2 V = f\left(z\right). \tag{17.26}$$

In the WRM method, we define the residual function $R(z)$ as

$$R(z) = \frac{d^2V}{dz^2} + k^2V - f(z). \tag{17.27}$$

The domain Ω of the problem is subdivided into several subdomains $\Delta\Omega_m$, $m = 1,2,\ldots,N$, a weighted function W_m is defined for each subdomain, and the weighted residual for each subdomain is forced to be zero:

$$\int_{\Delta\Omega_m} W_m(z)R(z)dz = 0. \tag{17.28}$$

Substituting Equation 17.27 into Equation 17.28, we obtain

$$\int_{\Delta\Omega_m} W_m(z)\left[\frac{d^2V}{dz^2} + k^2V - f(z)\right]dz = 0. \tag{17.29}$$

In Section 15.3.2, we called the method as point matching method if we chose $W_m(z) = \delta(z - z_m)$ and Galerkin if we chose $W_m(z)$ the same as the basis function.

17.2.1 Weak Form of the Differential Equation

Integration by parts leads to the equation

$$\int W_m(z)\frac{d^2V}{dz^2}dz = W_m\frac{dV}{dz} - \int \frac{dW_m}{dz}\frac{dV}{dz}dz. \tag{17.30}$$

By substituting Equation 17.30 into Equation 17.29,

$$\int_{\Delta\Omega_m}\left[W_m(z)k^2V - W_mf - \frac{dW_m}{dz}\frac{dV}{dz}\right]dz - W_m\frac{dV}{dz}\bigg|_{\text{endpoints}} = 0. \tag{17.31}$$

Equation 17.31 is called a weak form of the differential equation [1–3] since it enforces the differential equation on the average (because of the integral), instead of satisfying the differential equation at every point.

17.2.2 Galerkin Formulation of the WRM Method: Homogeneous Waveguide Problem

In Section 14.4, we solved for the potential Φ in the homogeneous waveguide problem, the equation for which is given by

$$\nabla_t^2\Phi + k_t^2\Phi = 0, \tag{17.32}$$

subject to the boundary conditions by the FEM method. The key step was extremizing the functional. The derivation of the appropriate functional for practical problems with lossy and complex materials in the guide is often difficult. The WRM Galerkin formulation is a more flexible tool which side steps the derivation of the functional for the problem [2].

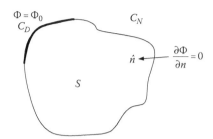

FIGURE 17.2
Helmholtz equation in 2D boundary by a closed curve C.

Let us formulate the problem in two dimensions for a domain S bounded by a closed curve C. Suppose C consists of C_N and C_D, where the Neumann boundary condition $\partial\Phi/\partial n = 0$ on C_N and the Dirichlet boundary condition $\Phi = \Phi_0$ is satisfied on C_D (see Figure 17.2).

Let W, W_D, and W_N be the weight functions defined on S, on C_D, on C_N, respectively. We can then write

$$\iint\limits_S W\left(\nabla_t^2\Phi + k_t^2\Phi\right)ds + \int\limits_{C_D} W_D\left(\Phi - \Phi_0\right)dl + \int\limits_{C_N} W_N\left(\nabla_t\Phi \cdot n\right)dl = 0. \tag{17.33}$$

The last two terms on the LHS are zero, since $\Phi = \Phi_0$ on C_D and $\partial\Phi/\partial n = 0$ on C_N.

Note that $\nabla_t\Phi \cdot n$ is the directional derivative of $\nabla_t\Phi$ in the direction of n and equal to $\partial\Phi/\partial n$.

First Green's theorem (Appendix 16C.5) for a surface is given by

$$\iint\limits_S W\nabla_t^2\Phi\, ds = \oint\limits_C W\frac{\partial\Phi}{\partial n}dc - \iint\limits_S \nabla_t W \cdot \nabla_t\Phi\, ds. \tag{17.34}$$

From Equations 17.33 and 17.34, we obtain

$$\iint\limits_S \nabla_t W \cdot \nabla_t\Phi\, ds - \iint\limits_S k_t^2 W\Phi\, ds - \oint\limits_C W\frac{\partial\Phi}{\partial n}dc - \int\limits_{C_D} W_D\left(\Phi - \Phi_0\right)dl - \int\limits_{C_N} W_N\frac{\partial\Phi}{\partial n}dl = 0. \tag{17.35}$$

We can reduce Equation 17.35 to Equation 17.36 is

$$\iint\limits_S \nabla_t W \cdot \nabla_t\Phi\, ds - \iint\limits_S k_t^2 W\Phi\, ds = 0, \tag{17.36}$$

by the following choices of the weight functions:

1. Choose $\Phi = \Phi_0$ on C_D. Irrespective of W_D, the fourth term on the LHS of Equation 17.35 is zero.

2. The third term on the LHS of Equation 17.35 is

$$-\int\limits_C W\frac{\partial\Phi}{\partial n}dc = -\left(\int\limits_{C_D} W\frac{\partial\Phi}{\partial n}dc + \int\limits_{C_N} W\frac{\partial\Phi}{\partial n}dc\right). \tag{17.37}$$

By choosing $W = 0$ on C_D, Equation 17.37 becomes

$$-\int_C W \frac{\partial \Phi}{\partial n} dc = -\int_{C_N} W \frac{\partial \Phi}{\partial n} dc. \tag{17.38}$$

3. The sum of the three line integrals in Equation 17.35 will given by

$$-\int_{C_N} W \frac{\partial \Phi}{\partial n} dc - \int_{C_N} W_N \frac{\partial \Phi}{\partial n} dc,$$

which can be made zero by choosing $W_N = -W$ on C_N.

The first choice $\Phi = \Phi_0$ on C_D is indeed the boundary condition.

Equation 17.36 is the weak form of Equation 16.96 and is an alternative to the variational formulation based on a functional. The steps in applying FEM in the framework of WRM are [2] as follows:

Step 1. Subdivide S into nonoverlapping elements.

Step 2. Approximate the unknown function on each element subject to the choice in (1).

Step 3. Define weight functions on each element subject to the choices of (2) and (3). Express the residue for each element.

Step 4. Sum contributions from all elements to obtain the residue on the whole domain.

Step 5. Solve the generalized eigenvalue problem formulated so as to annihilate the residue.

In step 2, we can express

$$\Phi^{(e)}(x,y) = \sum_j \Phi_j^{(e)} \alpha_j^{(e)}(x,y), \tag{17.39}$$

where $\alpha_j^{(e)}$ are the shape basis functions.

In step 3, if we use shape functions for the weight functions (Galerkin formulation), then Equation 17.36 becomes

$$\iint_{\Delta^{(e)}} \nabla_t \alpha_j^{(e)} \cdot \nabla_t \Phi^{(e)} ds - \iint_{\Delta^{(e)}} k_t^2 \alpha_j^{(e)} \Phi^{(e)} ds = R_i^{(e)}, \quad i = 1, 2, 3 \dots. \tag{17.40}$$

Substitution of Equation 17.39 in Equation 17.40 converts

$$\left[S^{(e)}\right]\left[\Phi^{(e)}\right] - k_t^2 \left[T^{(e)}\right]\left[\Phi^{(e)}\right] = \left[R_i^{(e)}\right], \tag{17.41}$$

where $[S^{(e)}]$ and $[T^{(e)}]$ are given by Equations 16.125 and 16.132, respectively. $[R^{(e)}]$ is approximated as zero.

These are exactly the same as those obtained from the variational approach. Note that if the vertex j of element (e) lies on C_D, then $\left[\Phi_j^{(e)}\right]$ is equal to prescribed Φ_0 and $W = 0$ on C_D. When once k_t^2 (eigenvalue) is obtained it is easy to compute β and α, since

$$(\beta - j\alpha)^2 = k_z^2 = k^2 - k_t^2.$$

For a lossy medium, $k^2 = \omega^2\mu\varepsilon$, where μ and ε may be complex. For a lossy medium, an exact functional is difficult to obtain but the method of weighted residuals overcomes this difficulty.

17.3 Inhomogeneous Waveguide Problem [2]

Equation 16.139 is the vector differential equation for the electric field in an inhomogeneous waveguide. If the dielectric is lossy, the relative permittivity is complex:

$$\hat{\varepsilon}_r = \frac{1}{\varepsilon_0}\left(\varepsilon + \frac{\sigma}{j\omega}\right), \tag{17.42}$$

and the electric field equation is given by

$$\tilde{\nabla}\times\left(\frac{1}{\mu_r}\nabla\times\tilde{E}\right) = k_0^2\hat{\varepsilon}_r\tilde{E}. \tag{17.43}$$

To solve the waveguide problem, we assume as usual

$$\tilde{E}(x,y,z) = E(x,y)e^{-\gamma z} = \left[E(x,y) + \hat{z}E_z(x,y)\right]e^{-\gamma z}. \tag{17.44}$$

Substituting Equation 17.44 into Equation 17.43, we can split Equation 17.43 into two equations:

$$\tilde{\nabla}_t\times\left(\frac{1}{\mu_r}\nabla_t\times\tilde{E}_t\right) - \frac{\gamma}{\mu_r}\left(\nabla_tE_z + \gamma\tilde{E}_t\right) = k_0^2\hat{\varepsilon}_r\tilde{E}_t, \tag{17.45}$$

$$\tilde{\nabla}_t\times\left[\frac{1}{\mu_r}\left(\nabla_tE_z + \gamma\tilde{E}_t\right)\times\hat{z}\right] = \hat{z}k_0^2\hat{\varepsilon}_rE_z. \tag{17.46}$$

Equation 17.45 has γ as well as γ^2 in it.

It is convenient to recast Equations 17.45 and 17.46 into more convenient form by using the transformation

$$e_t = \gamma E_t, \tag{17.47}$$

$$e_z = E_z. \tag{17.48}$$

Resulting in

$$\nabla_t \times \left(\frac{1}{\mu_r} \nabla_t \times e_t \right) - \frac{\gamma^2}{\mu_r} \left(\nabla_t e_z + e_t \right) = k_0^2 \hat{\varepsilon}_r e_t, \tag{17.49}$$

$$\gamma^2 \nabla_t \times \left[\frac{1}{\mu_r} \left(\nabla_t e_z + e_t \right) \times \hat{z} \right] = \gamma^2 \hat{z} k_0^2 \hat{\varepsilon}_r e_z. \tag{17.50}$$

The boundary conditions at the PEC wall are

$$\hat{n} \times e_t = 0, \tag{17.51}$$

$$e_z = 0, \tag{17.52}$$

where $\hat{n}$ is the unit vector perpendicular to the PEC walls.

It makes sense to use edge elements (vector elements) to approximate $e_t^{(e)}(x,y)$:

$$e_t^{(e)}(x,y) = \sum_{j=1}^{3} \alpha_j e_{\tan j}^{(e)}. \tag{17.53}$$

We will use nodes and scalar shape functions to approximate $e_z^{(e)}(x,y)$:

$$e_z^{(e)}(x,y) = \sum_{j=1}^{3} \alpha_j e_{zj}^{(e)}. \tag{17.54}$$

Using WRM in Galerkin formulation, Equations 17.49 and 17.50 yield Equation 17.25 whose solution gives the propagation constant $\gamma = \alpha + jb$.

17.3.1 Example of Inhomogeneous Waveguide Problem

A square waveguide with PEC walls has different dielectrics in the two elements as shown in Figure 17.3. Given $k_0 = \omega\sqrt{\mu_0\varepsilon_0} = 5$, determine γ.

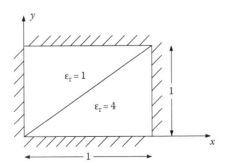

FIGURE 17.3
Cross section of a square waveguide with an inhomogeneous dielectric medium. A simple example for illustrating use of algorithm given by Equation 17.25.

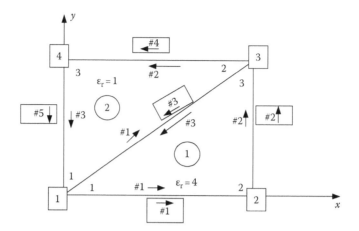

FIGURE 17.4
Global and local nodes and edges for the two elements of the loaded waveguide.

The function program INHWGD given in Appendix 18C.4 implements the algorithm given by Equation 17.25.

The student is encouraged to write the main program based on the local and global node and edge numberings given in Figure 17.4. The expected answer for γ is $0 + j6.5192$. The main program is given in Appendix 16A.6 (see Figure 17.4).

Examples with many more elements are given in Chapter 18.

17.4 Open Boundary, Absorbing Boundary, Conditions, and Scattering Problem [1–3]

We have so far considered Laplace's equations, Poisson's equation, and Helmholtz's equation, with Dirichlet or Neumann boundary conditions. A more general boundary condition is called mixed boundary condition on the potential, which, in the 1D case, has the general form

$$\left[\alpha_1 \frac{d\Phi}{dz} + \gamma\Phi\right]_{z=L} = q, \tag{17.55}$$

where α, γ, and q are known parameters. If $\gamma = 0$, then Equation 17.55 reduces to the Neumann BC and if $\alpha_1 = 0$, it can reduce to the Dirichlet BC; the boundary condition given in Equation 17.55 is also called boundary condition of the third kind. Moreover, the general form of the equations mentioned above (Laplace's, Poisson's, Helmholtz's) can also be written as one equation. For the 1D case (called the Sturm–Liouville equation), it is written as

$$-\frac{d}{dz}\left(\alpha \frac{d\Phi}{dz}\right) + \beta\Phi = f(z), \quad 0 < z < L. \tag{17.56}$$

The $\Phi(0)$ and $\Phi(L)$ are the specified boundary condition of first (Dirichlet), second (Neumann), or third (mixed) kind.

Equation 17.56 becomes the 1D Poisson equation if

$$\alpha = 1, \quad \beta = 0, \quad \text{and} \quad f(z) = \frac{\rho_v}{\varepsilon},$$

$$-\frac{d^2\Phi}{dz^2} = \frac{\rho_v}{\varepsilon}. \tag{17.57}$$

Equation 17.56 becomes the 1D homogeneous Helmholtz equation (applicable to transmission lines) if

$$\alpha = 1, \quad \beta(z) = -k^2, \quad \text{and} \quad f(z) = 0,$$

$$\frac{d^2\Phi}{dz^2} + k^2\Phi = 0. \tag{17.58}$$

Let us now formulate an FEM solution of Equation 17.56, subject to the boundary condition (Equation 17.55) at $z = L$ and a Dirichlet boundary condition at $z = 0$,

$$\Phi = p, \quad z = 0. \tag{17.59}$$

The functional for Equation 17.56 is given by

$$I(\Phi) = \frac{1}{2}\int_0^L \left[\alpha\left(\frac{d\Phi}{dz}\right)^2 + \beta\Phi^2\right]dz - \int_0^L f\Phi\,dz + \left[\frac{\gamma}{2}\Phi^2 - q\Phi\right]_{z=L}. \tag{17.60}$$

Note the last term on the RHS of Equation 17.60. It arises only because of the mixed boundary condition at $z = L$.

The discretization based on the FEM with variational formulation can be accomplished by using Equations 15.92 and 15.93:

$$N_i = \frac{z_j - z}{z_j - z_i}; \quad N_j = \frac{z - z_i}{z_j - z_i}, \tag{15.93}$$

$$\Phi^{(e)}(z) = N_i(z)\Phi_i^{(e)} + N_j(z)\Phi_j^{(e)}. \tag{15.92}$$

Then the integral terms in $I^{(e)}$ are

$$I^{(e)}\left(\Phi^{(e)}\right) = \frac{1}{2}\int_{z_i^{(e)}}^{z_j^{(e)}} \left[\alpha^{(e)}\left(\frac{d\Phi^{(e)}}{dz}\right)^2 + \beta^{(e)}\left[\Phi^{(e)}\right]^2\right]dz - \int_{z_i^{(e)}}^{z_j^{(e)}} f^{(e)}\Phi^{(e)}dz, \tag{17.61}$$

$$\frac{\partial I^{(e)}}{\partial \Phi_i^{(e)}} = \sum_{j=1}^{2}\Phi_j^{(e)}\int_{z_i^{(e)}}^{z_j^{(e)}} \left[\alpha^{(e)}\left(\frac{dN_i^{(e)}}{dz}\frac{dN_j^{(e)}}{dz}\right) + \beta^{(e)}N_i^{(e)}N_j^{(e)}\right]dz - \int_{z_i^{(e)}}^{z_j^{(e)}} N_i^{(e)}f^{(e)}dz. \tag{17.62}$$

Equation 17.62 in the matrix form can be written as

$$\frac{\partial I^{(e)}}{\partial \Phi_i^{(e)}} = \left[K^{(e)} \right] \left\{ \Phi^{(e)} \right\} - \left\{ b^{(e)} \right\},$$ (17.63)

where

$$\left\{ \Phi^{(e)} \right\} = \left\{ \begin{matrix} \Phi_i^{(e)} \\ \Phi_j^{(e)} \end{matrix} \right\}$$ (17.64a)

and

$$K_{ij}^{(e)} = \int_{z_i^{(e)}}^{z_j^{(e)}} \left[\alpha^{(e)} \underbrace{\left(\frac{dN_i^{(e)}}{dz} \frac{dN_j^{(e)}}{dz} \right)}_{\text{term 1}} + \underbrace{\beta^{(e)} N_i^{(e)} N_j^{(e)}}_{\text{term 2}} \right] dz,$$ (17.64b)

$$b_{ij}^{(e)} = \int_{z_i^{(e)}}^{z_j^{(e)}} \underbrace{N_i^{(e)} f^{(e)}}_{\text{term 3}} dz.$$ (17.64c)

Let us evaluate Terms 1, 2, and 3:

$$\frac{dN_i^{(e)}}{dz} = \frac{d}{dz} \left[\frac{z_j - z}{z_j - z_i} \right] = -\frac{1}{l^{(e)}},$$

where

$$l^{(e)} = z_j - z_i,$$

$$\frac{dN_j^{(e)}}{dz} = \frac{d}{dz} \left[\frac{z - z_i}{z_j - z_i} \right] = \frac{1}{l^{(e)}}.$$

Thus, we obtain

$$S_{11}^{(e)} = \int_{z_i^{(e)}}^{z_j^{(e)}} \frac{dN_j^{(e)}}{dz} \frac{dN_i^{(e)}}{dz} dz = \left(-\frac{1}{l^{(e)}} \right) \left(-\frac{1}{l^{(e)}} \right) l^{(e)} = \frac{1}{l^{(e)}},$$

$$S_{12}^{(e)} = \left(-\frac{1}{l^{(e)}} \right) \left(\frac{1}{l^{(e)}} \right) l^{(e)} = -\frac{1}{l^{(e)}}.$$

Term 1 gives the familiar S matrix for the 1D case

$$\left[S^{(e)}\right]_{1D} = \begin{bmatrix} \dfrac{1}{l^{(e)}} & -\dfrac{1}{l^{(e)}} \\ -\dfrac{1}{l^{(e)}} & \dfrac{1}{l^{(e)}} \end{bmatrix}. \tag{17.65}$$

It is obvious that Term 2 gives the familiar T matrix for the 1D case:

$$T_{11}^{(e)} = \int_{z_i^{(e)}}^{z_j^{(e)}} \left[N_i^{(e)}\right]^2 dz = \frac{1}{\left[l^{(e)}\right]^2} \int_{z_i^{(e)}}^{z_j^{(e)}} \left(z_j - z\right)^2 dz = \frac{1}{\left[l^{(e)}\right]^2} \int_{z_i^{(e)}}^{z_j^{(e)}} \left(z_j^2 + z^2 - 2z_j z\right) dz$$

or

$$\int_{z_i^{(e)}}^{z_j^{(e)}} \left(z_j - z\right)^2 dz = \int_{z_i^{(e)}}^{z_j^{(e)}} -\left(z_j - z\right)^2 d\left(z_j - z\right)$$

$$= -\frac{\left(z_j - z\right)^3}{3} \Bigg|_{z_i}^{z_j} = \frac{\left(z_j - z_i\right)^3}{3} = \frac{\left[l^{(e)}\right]^3}{3}.$$

Thus

$$T_{11}^{(e)} = \frac{l^{(e)}}{3}.$$

Similarly, it can be shown that $T_{12}^{(e)} = l^{(e)}/6$.
Thus, we have

$$\left[T^{(e)}\right]_{1D} = \begin{bmatrix} \dfrac{l^{(e)}}{3} & \dfrac{l^{(e)}}{6} \\ \dfrac{l^{(e)}}{6} & \dfrac{l^{(e)}}{3} \end{bmatrix}. \tag{17.66}$$

Now the K matrix which is a linear combination of $[S^{(e)}]_{1D}$ and $[T^{(e)}]_{1D}$,

$$\left[K^{(e)}\right] = \alpha^{(e)} \left[S^{(e)}\right]_{1D} + \beta^{(e)} \left[T^{(e)}\right]_{1D}, \tag{17.67a}$$

$$\left[K^{(e)}\right]_{1D} = \begin{bmatrix} \dfrac{\alpha^{(e)}}{l^{(e)}} + \dfrac{\beta^{(e)} l^{(e)}}{3} & -\dfrac{\alpha^{(e)}}{l^{(e)}} + \dfrac{\beta^{(e)} l^{(e)}}{3} \\ K_{21}^{(e)} & K_{22}^{(e)} \end{bmatrix}, \tag{17.67b}$$

where

$$K_{21}^{(e)} = K_{12}^{(e)} \quad \text{and} \quad K_{22}^{(e)} = K_{11}^{(e)}.$$

Since

$$\int\limits_{z_i^{(e)}}^{z_j^{(e)}} N_i^{(e)} \, dz = \frac{1}{l^{(e)}} \int\limits_{z_i^{(e)}}^{z_j^{(e)}} \left(z_j - z \right) dz$$

$$= \frac{1}{l^{(e)}} \int\limits_{z_i^{(e)}}^{z_j^{(e)}} -\left(z_j - z \right) d\left(z_j - z \right) = \frac{-\left(z_j - z_i \right)^2}{2l^{(e)}},$$

$$b_i^{(e)} = f^{(e)} \left[\frac{-\left(z_j - z \right)^2}{2l^{(e)}} \right]_{z_i}^{z_j} = \frac{f^{(e)} l^{(e)}}{2}. \tag{17.67c}$$

When we assemble globally, we get

$$[K][\Phi] = \{b\}. \tag{17.68}$$

Equation 17.68 is all we need if we have to deal with the boundary conditions of the first and second kind. The first kind gives us a prescribed node and the second kind is a free node.

17.4.1 Boundary Condition of the Third Kind

Let us now discuss as to how to incorporate the boundary condition of the third kind. The functional in Equation 17.60 has the extra term [1]

$$I_{\text{extra}} = \left[\frac{\gamma}{2} \Phi^2 - q\Phi \right]_{x=L}. \tag{17.69}$$

Let the node number at $x = L = x_j^{(M)}$ be labeled as N, where (M) is the element, whose terminal node is N:

$$I_{\text{extra}} = \frac{\gamma}{2} \Phi_N^2 - q\Phi_N,$$
$$\frac{\partial I_{\text{extra}}}{\partial \Phi_N} = \gamma \Phi_N - q. \tag{17.70}$$

Because of the extra terms in Equation 17.70, K_{NN} needs to be modified by adding γ. After getting the matrix K of Equation 17.68, MATLAB® statements

$$K_{NN} = K_{NN} + \gamma, \tag{17.71}$$

$$b_N = b_N + q, \tag{17.72}$$

will incorporate the mixed BC.

Steps (Equations 17.71 and 17.72) can be done in the main program, after getting K and b from a function program. The function program GLANL written on the same lines as GLANT is given in Appendix 18C.5. The output gives the $[K]$ and $[b]$ matrices. The implementation of the boundary condition of the third kind can be accomplished by modifying the $[K]$ and $[b]$ matrices. Let us illustrate by choosing a simple problem for which we have an exact solution.

17.4.1.1 A Simple Example

Solve

$$-\frac{d^2\Phi}{dx^2} = 1, \quad \Phi(0) = 0, \quad \left[\frac{d\Phi}{dx} + \frac{1}{2}\Phi\right]_{x=1} = -1.$$

The analytical solution is easily obtained by writing the general solution of the differential equation:

$$-\frac{d\Phi}{dx} = x + c_1,$$

$$-\Phi = \frac{x^2}{2} + c_1 x + c_2.$$

From $\Phi(0) = 0$, we have $c_2 = 0$; from the mixed BC at $x = 1$, we have

$$-x - c_1 + \frac{1}{2}\left[-\frac{x^2}{2} - c_1 x\right]\Bigg|_{x=1} = -1,$$

$$-1 - c_1 + \frac{1}{2}\left[-\frac{1}{2} - c_1\right] = -1,$$

$$-\frac{5}{4} - c_1\left[1 + \frac{1}{2}\right] = -1,$$

$$-c_1\left[\frac{3}{2}\right] = \frac{1}{4}, \quad c_1 = -\frac{2}{12} = -\frac{1}{6},$$

$$\therefore \Phi = -\frac{x^2}{2} + \frac{1}{6}x = \frac{1}{6}x(1 - 3x).$$

To illustrate the FEM solution, let us divide the domain into four elements ($N_e = 4$), then we have five global nodes ($N_n = 5$) (see Figure 17.5).

Thus, we construct Tables 17.1 and 17.2.

Comparing the specific differential equation with the general form

$$-\frac{d}{dz}\left[\alpha\frac{d\Phi}{dz}\right] + \beta\Phi = f(z),$$

$$z = x, \quad \alpha = 1, \quad \beta = 0, \quad f(z) = 1.$$

Thus, we have the element parameter (Table 17.2).

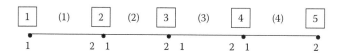

FIGURE 17.5
Four elements and five nodes of the 1D problem.

TABLE 17.1

Global Coordinates and Connectivity Table for the Simple Example

Global Coordinates		Connectivity Table		
		Element #	N1L	N2L
n	x_n	(1)	1	2
1	0	(2)	2	3
2	0.25	(3)	3	4
3	0.50	(4)	4	5
4	0.75			
5	1.00			

TABLE 17.2

α, β, f for the Elements in the Simple Example

Element #	$\alpha^{(e)}$	$\beta^{(e)}$	$f^{(e)}$
(1)	1	0	1
(2)	1	0	1
(3)	1	0	1
(4)	1	0	1

These data given in the main program can be passed onto the function program GLANL and obtain as output K and b matrices. Now we have to implement the boundary conditions. Let the prescribed node where the Dirichlet condition is satisfied be called Prn = [1]. The potential at Prn be called V_{Prn}. The free nodes f_{rn} are

$$f_{rn} = [2,3,4,5],$$

whose potential Φ's are to be found after implementing the mixed boundary condition at node 5.

For the problem at hand, the general mixed boundary condition.

$$\alpha_1 \frac{d\Phi}{dz} + \gamma\Phi \bigg|_{z=L} = q$$

translate to

$$\alpha_1 = 1, \quad \gamma = \frac{1}{2}, \quad q = -1, \quad L = 1.$$

This boundary condition may be implemented as follows:

Let the global node number array where mixed boundary condition are satisfied be denoted by *mxn*. In this case

$$mxn = [5].$$

Let α, γ, and *q* at *mxn* be denoted by

$$al\ mxn = [1],$$

$$ga\ mxn = [0.5],$$

$$q\ mxn = [-1].$$

Let us modified the elements in the *K* matrix as

$$K(mxn, mxn) = K(mxn, mxn) + g\,mxn.$$

Let us modify the *b* array element

$$b(mxn) = b(mxn) + q\,mxn.$$

The MATLAB statements to obtain and solve the system of equations are

$$Kff = K(frn, frn),$$

$$Kfp = K(frn, prn),$$

$$Vprn = [0],$$

$$bf = b(frn),$$

$$Vf = inv(Kff) + ((bf)' - Kfp * (Vprn)').$$

Thus, we obtain the potentials of the free nodes. The student is encouraged to write the main program and run it using the function GNANL given in Appendix 18C.5.

17.4.2 Example of Electromagnetic Problems with Mixed BC

Example 17.1: Reflection Problem

Formulation for FEM-1D: p-wave reflection from a PEC-backed dielectric [1] leads us to a mixed boundary value problem, which can be solved using 1D FEM using the function program given in the previous section (see Figure 17.6).

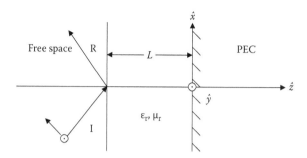

FIGURE 17.6
Reflection problem and the boundary condition of the third kind.

Let the magnetic field of the incident wave be written as

$$\tilde{\bar{H}}^I = \hat{y}\tilde{H}_y,$$ (17.73a)

where

$$\tilde{H}_y = H_0 e^{-jk_0(Sx+Cz)},$$ (17.73b)

$$S = \sin\theta,$$
$$k_0 = \omega\sqrt{\mu_0\varepsilon_0},$$ (17.73c)
$$C = \cos\theta.$$

Let ε_r and μ_r be functions of position in the domain, $-L < z < 0$.
The differential equation satisfied by $\tilde{H}_y$ can be obtained from Maxwell's equations

$$\bar{\nabla}\times\tilde{E} = -j\omega\mu_0\mu_r(z)\tilde{H},$$ (17.74)

$$\bar{\nabla}\times\tilde{H} = j\omega\varepsilon_0\varepsilon_r(z)\tilde{E}.$$ (17.75)

In the lossy dielectric $\varepsilon_r(z)$ can be complex.
From Equations 17.74 and 17.75,

$$\bar{\nabla}\times\left[\frac{1}{\varepsilon_r(z)}\bar{\nabla}\times\tilde{H}\right] = j\omega\varepsilon_0\bar{\nabla}\times\tilde{E}$$

$$= j\omega\varepsilon_0\left(-j\omega\mu_0\mu_r(z)\right)\tilde{H}$$

$$\bar{\nabla}\times\left[\frac{1}{\varepsilon_r}\bar{\nabla}\times\tilde{H}\right] = k_0^2\mu_r\tilde{H}.$$ (17.76)

It is obvious that the magnetic field in the domain $-L < z < 0$ will have the form

$$\tilde{H} = \hat{y}e^{-jk_0Sx}H_y(z)$$ (17.77)

and

$$\bar{\nabla} \times \tilde{H} = \begin{vmatrix} \hat{x} & \hat{y} & \hat{z} \\ -jk_0 S & 0 & \dfrac{\partial}{\partial z} \\ 0 & e^{-jk_0 Sx} H_y & 0 \end{vmatrix}$$

$$= -\hat{x} e^{-jk_0 Sx} \frac{\partial H_y(z)}{\partial z} + \hat{z}(-jk_0 S) e^{-jk_0 Sx} H_y,$$

$$\frac{1}{\varepsilon_r} \bar{\nabla} \times \tilde{H} = \underbrace{-e^{-jk_0 Sx} \frac{1}{\varepsilon_r}}_{C_1} \left[\frac{\partial H_y \hat{x}}{\partial z} + jk_0 S H_y \hat{z} \right],$$

$$\bar{\nabla} \times \left[\frac{1}{\varepsilon_r} \bar{\nabla} \times \tilde{H} \right] = \begin{vmatrix} \hat{x} & \hat{y} & \hat{z} \\ -jk_0 S & 0 & \dfrac{\partial}{\partial z} \\ C_1 \dfrac{\partial H_y}{\partial z} & 0 & jk_0 S C_1 H_y \end{vmatrix}$$

$$= \hat{x}(0-0) - \hat{y}(-jk_0 S jk_0 sc_1 H_y) + \hat{z}(0-0)$$

$$= -\hat{y} \left[k_0^2 S^2 C_1 H_y - \frac{\partial}{\partial z} \left(C_1 \frac{\partial H_y}{\partial z} \right) \right]$$

Thus, Equation 17.76 becomes

$$-\hat{y} \left[k_0^2 S^2 c_1 H_y - \frac{\partial}{\partial z} \left(c_1 \frac{\partial H_y}{\partial z} \right) \right] = k_0^2 \mu_r \hat{y} H_y e^{-jk_0 Sx},$$

$$\frac{\partial}{\partial z} \left[\frac{-e^{-jk_0 Sx}}{\varepsilon_r(z)} \frac{\partial H_y}{\partial z} \right] - k_0^2 S^2 \left[\frac{-e^{-jk_0 Sx}}{\varepsilon_r(z)} \right] H_y - k_0^2 \mu_r H_y = 0.$$

Cancelling $e^{-jk_0 Sx}$ and realizing that there is only one independent variable z,

$$-\frac{d}{dz} \left[\frac{1}{\varepsilon_r(z)} \frac{dH_y}{dz} \right] + k_0^2 S^2 \left[\frac{1}{\varepsilon_r(z)} \right] H_y - K_0^2 \mu_r H_y = 0,$$

$$-\frac{d}{dz} \left[\frac{1}{\varepsilon_r(z)} \frac{dH_y}{dz} \right] - k_0^2 \left[\mu_r - \frac{S^2}{\varepsilon_r(z)} \right] H_y = 0, \tag{17.78}$$

$$\frac{d}{dz} \left[\frac{1}{\varepsilon_r(z)} \frac{dH_y}{dz} \right] + k_0^2 \left[\mu_r - \frac{S^2}{\varepsilon_r(z)} \right] H_y = 0 \quad -L < z < 0.$$

The boundary condition at $z = 0$ is a PEC boundary condition

$$\frac{\partial H_y}{\partial z} \bigg|_{z=0} = 0. \tag{17.79}$$

The boundary condition at $z = -L$ is that

$$\tilde{H}_y\left(z = -L^+\right) = \tilde{H}_y\left(z = -L^-\right), \tag{17.80}$$

$$\tilde{E}_x\left(z = -L^+\right) = \tilde{E}_x\left(z = -L^-\right). \tag{17.81}$$

The RHS of Equation 17.80 is the sum of the $\tilde{H}_y$ of the incident wave plus the reflected wave at $z = -L^-$. Suppose we write the magnetic field of the reflected wave as

$$\tilde{\vec{H}}^R = \hat{y} R H_0 e^{-jk_0(Sx - Cz)}, \tag{17.82}$$

then

$$\tilde{H}_y\left(z < -L^-\right) = H_0 e^{-jk_0(Sx + Cz)} + R H_0 e^{-jk_0(Sx - Cz)}, \tag{17.83}$$

$$\frac{d\tilde{H}_y}{dz} = -jk_0 c H_0 \left[e^{-jk_0(Sx + Cz)} - R e^{-jk_0(Sx - Cz)} \right]. \tag{17.84}$$

The reflection coefficient R can be eliminated by multiplying Equation 17.83 by the factor $(-jk_0 C)$ and adding it to Equation 17.84:

$$\frac{d\tilde{H}_y}{dz} - jk_0 c \tilde{H}_y = -j2k_0 c H_0 e^{-jk_0(Sx + Cz)}. \tag{17.85}$$

The boundary condition (Equation 17.81) translates to a boundary condition on the derivative of $\tilde{H}_y$ from Maxwell's equation:

$$\nabla \times \tilde{\vec{H}} = j\omega\varepsilon_0\varepsilon_r\left(z\right)\tilde{\vec{E}},$$

$$\therefore \tilde{E}_x = \frac{1}{j\omega\varepsilon_0\varepsilon_r\left(z\right)}\left[-e^{-jk_0 Sx} \frac{\partial H_y\left(z\right)}{\partial z} \right].$$

Thus, Equation 17.81 is the same as

$$\left.\frac{1}{\varepsilon_r}\frac{dH_y}{dz}\right|_{z=-L^+} = \left.\frac{dH_y}{dz}\right|_{z=-L^-}. \tag{17.86}$$

The result of the continuity condition on H_y and E_x, when applied to Equation 17.85 at $z = -L^-$, gives

$$\left.\frac{d\tilde{H}_y}{dz} - jk_0 C \tilde{H}_y\right|_{z=-L^-} = -j2k_0 c H_0 e^{-jk_0\left[Sx + C(-L)\right]}$$

$$= \left.\frac{1}{\varepsilon_r}\frac{d\tilde{H}_y}{dz} - jk_0 C \tilde{H}_y\right|_{z=-L^+}.$$

The above may be rewritten as a mixed boundary condition at $z = -L^+$:

$$\frac{1}{\varepsilon_r}\frac{dH_y}{dz} - jk_0CH_y\bigg|_{z=-L^+} = -j2k_0CH_0e^{jk_0CL}. \qquad (17.87)$$

We can now solve Equation 17.78 subject to the boundary condition (Equation 17.79) at $z = 0$ and the mixed boundary condition Equation 17.87 at $z = -L^+$ and obtain H_y at $z = -L^+$. From this value, we can compute the reflection coefficient R using Equation 17.80 and Equation 17.83:

$$R = \frac{H_y\big|_{z=L} - H_0e^{jk_0CL}}{H_0e^{-jk_0CL}}. \qquad (17.88)$$

The FEM formulation developed in Section 17.4 can be applied to this problem by noting, from Equations 17.56 and 17.78.

$$\Phi = H_y, \quad \alpha = \frac{1}{\varepsilon_r},$$

$$\beta = -k_0^2\left(\mu_r - \frac{S^2}{\varepsilon_r}\right), \qquad (17.89)$$

$$f = 0.$$

Similarly comparing Equation 17.55 with Equation 17.87, we obtain

$$\alpha_1 = 1,$$

$$\gamma = -jk_0C\varepsilon_{RL}, \qquad (17.90)$$

$$q = -2jk_0\varepsilon_{RL}CH_0e^{jk_0CL}.$$

where $\varepsilon_{RL} = \varepsilon_R\big|_{-L}^+$.

Example 17.2[1]

Equation 17.79 is the boundary condition of the second kind valid for a PEC boundary. If we wish to consider the case of a good conductor as opposed to a PEC, then the boundary condition, called the impedance boundary condition [1–3], is again of the third kind given by

$$\frac{\partial H_y}{\partial z} = -jk_0\varepsilon_{r0}\eta H_y \qquad (17.91a)$$

$$\eta = \sqrt{\frac{\mu_{r2}}{\varepsilon_{r2}}}, \qquad (17.91b)$$

where ε_{r0} is the relative permittivity of the dielectric medium at $z = 0$, μ_{r2} and ε_{r2} are the relative permeability and relative permittivity of the good conductor. For a conductor such as copper $\mu_{r2} = 1$ and $\varepsilon_{r2} = 1 - j\sigma/\omega\varepsilon_0$, σ being the conductivity of copper ($\sigma \approx 5.8 \times 10^7$ s/m). Thus, we have to replace Equation 17.55 by Equation 17.91 that is a mixed boundary condition with

$$\Phi = H_y, \qquad (17.92a)$$

$$\alpha = \frac{1}{\varepsilon_r},$$ (17.92b)

$$\gamma = jk_0\eta,$$ (17.92c)

$$q = 0.$$ (17.92d)

In arriving at Equation 17.92, we wrote Equation 17.91a as

$$\frac{1}{\varepsilon_{r0}}\frac{\partial H_y}{\partial z} = -jk_0\eta H_y.$$ (17.93)

PEC WITH THIN DIELECTRIC COATING

Suppose L is very small. The problem can be solved by treating the dielectric layer as a dielectric coating. Then η in Equation 17.92c will be replaced by the input impedance of a short-circuited transmission line:

$$\eta = j\sqrt{\frac{\mu_{r2}}{\varepsilon_{r2}}}\tan\left(k_0\sqrt{\mu_{r2}\varepsilon_{r2}}\,\Delta\right)$$ (17.94)

and ε_{r0} in Equation 17.92b by 1 (see Figure 17.7).

Example 17.3: Radiation Boundary Condition

Figure 17.8 shows a fictitious boundary beyond which we consider fields are only those of an outgoing wave. For a 2D problem, the Sommerfield radiation condition can be written as [1]

$$\lim_{\rho\to\infty}\sqrt{\rho}\left[\frac{\partial}{\partial\rho}\begin{pmatrix}E_z\\H_z\end{pmatrix} + jk_0\begin{pmatrix}E_z\\H_z\end{pmatrix}\right] = 0,$$ (17.95)

where $\rho = \sqrt{x^2 + y^2}$ is the cylindrical radial coordinate. This can be written as

$$B_{1/2}\begin{pmatrix}E_z\\H_z\end{pmatrix} = O\left(\rho^{-3/2}\right),$$ (17.96)

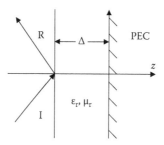

FIGURE 17.7
Impedance boundary condition and the boundary condition of the third kind.

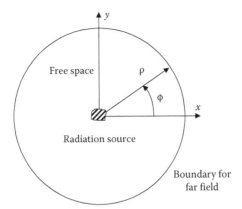

FIGURE 17.8
Radiation boundary condition.

where $B_{1/2}$ is an operator given by

$$B_{1/2} = \frac{\partial}{\partial \rho} + jk_0. \tag{17.97}$$

The term on the RHS shows the order of approximation, that is, the terms $\rho^{-3/2}$ and smaller are neglected. For this order of the approximation, ρ needs to be sufficiently large (the computational domain for the problem need to be sufficiently large).

The computational domain can be reduced by increasing the order of the approximation. Higher-order radiation conditions are discussed in the literature. Examples are

$$B_1 = \frac{\partial}{\partial \rho} + jk_0 + \frac{1}{2\rho} + o\left(\rho^{-5/2}\right), \tag{17.98}$$

$$B_2 = \frac{\partial}{\partial \rho} + jk_0 + \frac{1}{2\rho} - \frac{1}{8\rho(1+jk_0\rho)} - \frac{1}{2\rho(1+jk_0\rho)} \frac{\partial^2}{\partial \phi^2} + o\left(\rho^{-9/2}\right). \tag{17.99}$$

These equations are called absorbing boundary conditions, in the sense that only a small fraction of the incident energy is reflected by the fictitious boundary. The boundary conditions are of mixed type, for example, the BC $(\partial E_z/\partial \rho) + jk_0 E_z = 0$ is of the type (Equation 17.55).

17.5 The 3D Problem [1,3,4]

Three-dimensional analog of a 2D triangle is a tetrahedron (four-faced elements) (see Figure 17.9). Let the vertices 1, 2, 3, and 4, have the Cartesian coordinates, x_i, y_i, z_i, $i = 1,2,3,4$.
The triangle faces of the tetrahedral are

$$
\begin{array}{ccc}
1 & 2 & 3 \\
1 & 2 & 4 \\
2 & 3 & 4 \\
1 & 4 & 4 \\
\end{array}
$$

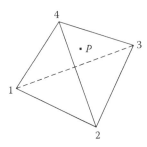

FIGURE 17.9
Tetrahedral element.

17.5.1 Volume Coordinates

A point P inside the tetrahedron has the coordinates x, y, and z. We define the volume coordinates:

$$\zeta_1 = \frac{\text{Volume } p234}{\text{Volume } 1234}, \tag{17.100}$$

$$\zeta_2 = \frac{\text{Volume } p341}{\text{Volume } 1234}, \tag{17.101}$$

$$\zeta_3 = \frac{\text{Volume } p412}{\text{Volume } 1234}, \tag{17.102}$$

$$\zeta_4 = \frac{\text{Volume } p123}{\text{Volume } 1234}. \tag{17.103}$$

The volume coordinates and the (x, y, z) coordinates are related by

$$\begin{bmatrix} 1 \\ x \\ y \\ z \end{bmatrix} = \begin{bmatrix} 1 & 1 & 1 & 1 \\ x_1 & x_2 & x_3 & x_4 \\ y_1 & y_2 & y_3 & y_4 \\ z_1 & z_2 & z_3 & z_4 \end{bmatrix} \begin{bmatrix} \zeta_1 \\ \zeta_2 \\ \zeta_3 \\ \zeta_4 \end{bmatrix}. \tag{17.104}$$

The potential at P can be expressed in terms of the volume coordinates and the potentials at the vertices:

$$\Phi^e(x,y,z) = \sum_{i=1}^{4} \zeta_i \Phi_i. \tag{17.105}$$

Also, in the first-order (linear) approximation in three dimensions $\Phi^e(x,y,z) = a + bx + cy + dz$. Applying this equation at the vertices, we obtain Equations 17.106a through 17.106d:

$$\Phi_1^e = a + bx_1 + cy_1 + dz_1, \tag{17.106a}$$

$$\Phi_2^e = a + bx_2 + cy_2 + dz_2, \tag{17.106b}$$

$$\Phi_3^e = a + bx_3 + cy_3 + dz_3, \tag{17.106c}$$

$$\Phi_4^e = a + bx_4 + cy_4 + dz_4. \tag{17.106d}$$

Solving for the coefficients a, b, c, and d, we obtain

$$a = \frac{1}{6V} \begin{bmatrix} \Phi_1^e & \Phi_2^e & \Phi_3^e & \Phi_4^e \\ x_1 & x_2 & x_3 & x_4 \\ y_1 & y_2 & y_3 & y_4 \\ z_1 & z_2 & z_3 & z_4 \end{bmatrix}, \tag{17.107}$$

$$b = \frac{1}{6V} \begin{bmatrix} 1 & 1 & 1 & 1 \\ \Phi_1^e & \Phi_2^e & \Phi_3^e & \Phi_4^e \\ y_1 & y_2 & y_3 & y_4 \\ z_1 & z_2 & z_3 & z_4 \end{bmatrix}, \tag{17.108}$$

$$c = \frac{1}{6V} \begin{bmatrix} 1 & 1 & 1 & 1 \\ x_1 & x_2 & x_3 & x_4 \\ \Phi_1^e & \Phi_2^e & \Phi_3^e & \Phi_4^e \\ z_1 & z_2 & z_3 & z_4 \end{bmatrix}, \tag{17.109}$$

$$d = \frac{1}{6V} \begin{bmatrix} 1 & 1 & 1 & 1 \\ x_1 & x_2 & x_3 & x_4 \\ y_1 & y_2 & y_3 & y_4 \\ \Phi_1^e & \Phi_2^e & \Phi_3^e & \Phi_4^e \end{bmatrix}, \tag{17.110}$$

where the determinant

$$\Delta = \begin{bmatrix} 1 & 1 & 1 & 1 \\ x_1 & x_2 & x_3 & x_4 \\ y_1 & y_2 & y_3 & y_4 \\ z_1 & z_2 & z_3 & z_4 \end{bmatrix} = 6V. \tag{17.111}$$

From Equation 17.104, we can show

$$\zeta_j = \frac{1}{6V}\left[a_j + b_j x + c_j y + d_j z \right]. \tag{17.112}$$

The coefficients a_j, b_j, c_j, and d_j may be obtained as follows.

By expanding Equation 17.107, we obtain

$$a = \frac{1}{6V} \left[a_1 \Phi_1^e + a_2 \Phi_2^e + a_3 \Phi_3^e + a_4 \Phi_4^e \right], \tag{17.113}$$

thus a_1, a_2, a_3, and a_4 are the coefficients of the potentials at the corresponding vertices when the determinant of matrix in Equation 17.107 is expanded.

Similarly, from Equations 17.108 through 17.110, we obtain

$$b_j, \quad j = 1, \ldots, 4, \tag{17.114}$$

$$c_j, \quad j = 1, \ldots, 4, \tag{17.115}$$

$$d_j, \quad j = 1, \ldots, 4, \tag{17.116}$$

respectively.

Note that the volume coordinates satisfy the requirements of a shape function

$$\zeta_i \left(x_j, y_j, z_j \right) = \delta_{ij} = \begin{cases} 1, & i = j, \\ 0, & i \neq j. \end{cases} \tag{17.117}$$

Thus, $\alpha_j = \zeta_j$, the first-order shape functions for the tetrahedra are determined.

17.5.2 Functional

The scalar wave equation in its most general form can be written as

$$\frac{\partial^2 \Phi}{\partial x^2} + \frac{\partial^2 \Phi}{\partial y^2} + \frac{\partial^2 \Phi}{\partial z^2} + k^2 \Phi = -f \tag{17.118}$$

in V.

The boundary conditions, in general, may include $\Phi = p$ on S_1 (Dirichlet)

$$\left(\hat{x} \frac{\partial \phi}{\partial x} + \hat{y} \frac{\partial \phi}{\partial y} + \hat{z} \frac{\partial \phi}{\partial z} \right) \cdot \hat{n} + \nu \Phi = q$$

on S_2 (mixed boundary condition, which include Neumann boundary condition), where $S_1 + S_2$ is the closed surface enclosing the volume V.

The functional for this case is

$$F(\Phi) = \frac{1}{2} \iiint\limits_V \left[\left(\frac{\partial \phi}{\partial x} \right)^2 + \left(\frac{\partial \phi}{\partial y} \right)^2 + \left(\frac{\partial \phi}{\partial z} \right)^2 - k^2 \Phi^2 \right] dV$$

$$+ \iint\limits_{S_2} \left(\frac{\nu}{2} \Phi^2 - q\Phi \right) ds - \iiint\limits_V f\Phi \, dV. \tag{17.119}$$

The relevant matrices are obtained just as in the case of two dimensions by using the 3D standard integral

$$\iiint\limits_{V^e} \left(\zeta_1\right)^k \left(\zeta_2\right)^l \left(\zeta_3\right)^m \left(\zeta_4\right)^n dV$$

$$= \frac{k!\,l!\,m!\,n!}{\left(k+l+m+n+3\right)!} 6V,$$ (17.120)

where V is the volume of the element.

17.5.3 S, T, and g Matrices

The elemental matrices

$$S_{ij}^e = \iiint\limits_{V^e} \left[\frac{\partial \zeta_i}{\partial x} \frac{\partial \zeta_j}{\partial x} + \frac{\partial \zeta_i}{\partial y} \frac{\partial \zeta_j}{\partial y} + \frac{\partial \zeta_i}{\partial x} \frac{\partial \zeta_j}{\partial x} \right] dV,$$ (17.121)

$$T_{ij}^e = \iiint\limits_{V^e} \zeta_i \zeta_j \, dV,$$ (17.122)

$$g_{ij}^e = \iiint\limits_{V^e} f \zeta_j dV,$$ (17.123)

when evaluated give

$$S_{ij}^e = \frac{1}{36V} \left(b_i b_j + c_i c_j + d_i d_j\right),$$ (17.124)

$$T_{ij}^e = \frac{V}{20} \left(1 + \zeta_{ij}\right),$$ (17.125)

$$g_i^e = \frac{V}{4} f^e.$$ (17.126)

To incorporate the boundary condition of the third kind, we have to deal additionally with the term in the functional (Equation 17.119)

$$F_b\left(\Phi\right) = \iint\limits_{S_2} \left(\frac{\nu}{2} \Phi^2 - q\Phi \right) ds,$$ (17.127)

$$F_b\left(\Phi\right) = \sum_{s=1}^{M_s} F_b^s\left(\Phi^s\right),$$

$$\Phi^s = \sum_{j=1}^{3}\left(\zeta_j^S \Phi_j^s\right),$$

$$\frac{\partial F_b^s}{\partial \Phi_i^s} = \sum_{j=1}^{3}\Phi_i^s \iint_{S^s} \nu \zeta_i \zeta_j \, ds - \iint_{S^s} q \zeta_i \, ds.$$

S^s denotes the surface of the Sth triangle.

Thus, we get

$$\frac{\partial F_b^s}{\partial \Phi^3} = \left[S^s\right]\{\Phi^2\} - k^2\left[T^s\right] - \{g^2\}, \tag{17.128}$$

where

$$T_{ij}^s = \iint_{S^s} \nu \zeta_i \zeta_j \, ds, \tag{17.129}$$

$$g_{ij}^s = \iint_{S^s} q \zeta_i \, ds. \tag{17.130}$$

If v and q are constant,

$$T_{ij}^s = v^s \frac{\Delta^s}{12}\left(1+\zeta_{ij}\right), \tag{17.131}$$

$$g_i^s = q^s \frac{\Delta^s}{3}. \tag{17.132}$$

To include the effect of mixed boundary condition, T_{ij}^s and g_i^s should be suitably added in assembling the global matrix. We then get the equation

$$[A][\Phi] = [g]. \tag{17.133}$$

17.5.4 3D Edge Elements [1,3]

In the 3D case, there are six edges as shown in Figure 17.10.

An edge i has a beginning node i_1 and ending node i_2 as shown in Table 17.3.

We can define vector shape functions as follows:

$$\alpha_i = \left(\alpha_{i_1}\nabla\alpha_{i_2} - \alpha_{i_2}\nabla\alpha_{i_1}\right)l_i.$$

17.5.5 Higher-Order Edge Elements

In two dimensions, higher-order edge elements are obtained as in Figure 17.11.

The tangential projection of the vector field along edge $\{i, j\}$ is determined by two unknowns E_i^j and E_j^i and two facet unknowns—F_1 and F_2—to provide quadratic approximation of the normal component along two of the three edges. Only two facet elements are

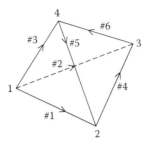

FIGURE 17.10
Edges of tetrahedral element.

TABLE 17.3

Edge Definitions for Tetrahedral Element

Edge # i	Node i_1	Node i_2
1	1	2
2	1	3
3	1	4
4	2	3
5	4	2
6	3	4

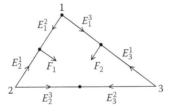

FIGURE 17.11
Higher-order edge elements.

required to make the range space of time curl operator complete to first order. These are eight degrees of freedom. The vector field $\mathbf{E}^e$ is expanded as

$$E^e = \sum_{i=1}^{3}\sum_{j=1}^{3} E_i^j \alpha_i \nabla \alpha_j + F_1 \alpha_i \left(\alpha_j \nabla \alpha_k - \alpha_k \nabla \alpha_j \right) + F_2 \alpha_j \left(\alpha_k \nabla \alpha_i - \alpha_i \nabla \alpha_k \right), \quad i \neq j. \quad (17.134)$$

For the 3D case, higher-order edge elements have

$$E^e = \sum_{i=1}^{4}\sum_{j=1}^{4} E_i^j \alpha_i \nabla \alpha_j + \sum_{i=1}^{4} F_1 \alpha_i \left(\alpha_j \nabla \alpha_k - \alpha_k \nabla \alpha_j \right) + F_2^i \alpha_j \left(\alpha_k \nabla \alpha_i - \alpha_i \nabla \alpha_k \right), \quad (17.135)$$

where F_1^i and F_2^i are facet variables. In total, there are (16) degrees of freedom in the cc first term and the second term has four degrees of freedom with a total of 20 degrees of freedom [3,5].

References

1. Jin, J., *The Finite Element Method in Electromagnetics*, 2nd Edition, Wiley, New York, 2002.
2. Pelosi, G., Coccioli, G., and Selleri, S., *Quick Finite Elements for Electromagnetic Waves*, Artech House, Norwood, MA, 1989.
3. Volakis, J. L., Chatterjee, A., and Kempel, L. C., *Finite Element Method for Electromagnetics*, IEEE Press, New York, 1998.
4. Silvester, P. P. and Ferrari, R. L., *Finite Elements for Electrical Engineers*, 3rd Edition, Cambridge University Press, Cambridge, U.K., 1996.
5. Ansoft Publication, Ansoft HFSS, Technical notes, Pittsburgh, PA, 2011.

18

Case Study Ridged Waveguide with Many Elements*

In Chapters 16 and 17, the theory of FEM was developed. The algorithms were obtained and illustrated using as few elements as possible, so as not to obscure the understanding of the algorithm as well as giving an opportunity to do hand computation or perform dry run of the code. Obviously with so few elements, we did not expect the answer to be accurate.

In this chapter, we illustrate the additional steps needed to get accurate results. Larger number of elements requires computer-generated input data for the various function programs used in Chapters 16 and 17. It is very tedious to hand-generate the coordinates of the nodes, connectivity tables, node and edge numbering, and so on. Downloadable free software is available on the Internet, and the details of using this software to computer-generate the input data are illustrated through a case study of the homogeneous as well as the dielectric-loaded ridged guide.

18.1 Homogenous Ridged Waveguide

The objective is to find the cutoff frequency for the ridged waveguide for both TM and TE modes. The mesh generation is done via GID software, which can be downloaded from http://www.gidhome.com/download/. The *evaluation version* of GID is completely functional but geometry is limited to 25 surfaces and meshes are limited to 1000 nodes. This can be downloaded at free of cost.

We can create our structure by selecting, *Straight Line* tool from the *Geometry* menu under *Create* submenu; draw the structure with lengths shown in Figure 18.1a, then convert the contour into surface using *NURBS surface by contour* tool under the same *Create* submenu; this process is depicted in Figure 18.1.

The mesh generation process can be done by selecting *Generate Mesh* command under *Mesh* menu. In order to create the Ridge, just delete the elements occupying that region ($12 \times 4 = 48$ elements); the GID will automatically renumber the nodes and the elements. In this case, there are 384 elements and 227 nodes. The mesh, element numbers, and node numbers are shown in Figures 18.2 and 18.3, respectively.

Node numbers, element numbers, and the coordinate dimensions of the nodes are all exported to a text file using this function in GID *Export to text data Report* from *File* menu, and this text file can be loaded into a MATLAB® function (subroutine) (see Appendix 18A). This file contains the node numbers with corresponding coordinates (x_n, y_n), the element numbers with corresponding global nodes. Also the boundary nodes can be selected from GID by using this function *View Boundary Mesh* under the *Mesh* submenu, those boundary

---

* For chapter appendices, see 18A through 18C in Appendices section.

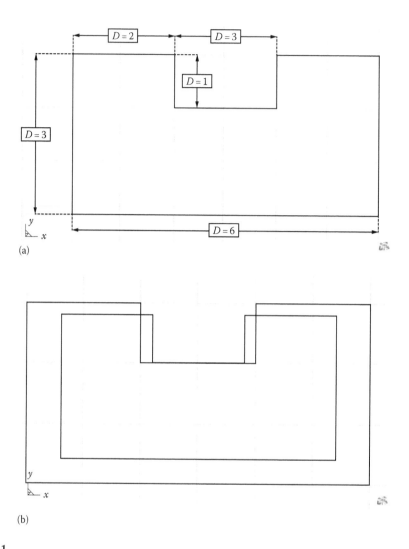

FIGURE 18.1
(a) The ridged WG contour lines and (b) converted surface.

nodes can be copied to a text file using the command *List Some Filtered Results*, the boundary nodes or the prescribed nodes are shown in Figure 18.4.

18.1.1 Node-Based FEM

In this part, the first-order node-based FEM is implemented to find the cutoff frequencies for TE and TM modes for the ridged WG, the Information Data in Appendix 18A that was exported from GID can be utilized in a MATLAB subroutine (see Appendix 18B.1), the inputs for this subroutine are:

N_n the total number of nodes

N_e the total number of triangular elements

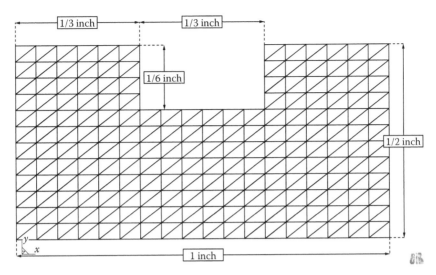

FIGURE 18.2
Mesh geometry with dimensions in inches.

Meshread This function reads nodal coordinates and connectivity matrix from an ASCII mesh file created by the GID preprocessor (see Appendix 18C.2)

GLANT This function finds the stiffness matrix from the global nodes assembly (Appendix 18C.1)

For the TE modes, all nodes are treated as free nodes, but for the TM case the nodes at the boundaries should not be taken into account, thus only free nodes are needed to calculate the cutoff modes in the TM case.

The first seven modes for both TE and TM modes are listed in Table 18.1.

18.1.2 Edge-Based FEM

In this part, the cutoff frequencies are found using the edge-based FEM, the function named Edges (Appendix 18C.4) was used to generate local edge arrays $n1EL$, $n2EL$, $n3EL$, and the number of edges which were later used in GLAET function to find the global stiffness matrix (see Appendix 18C.3 for the MATLAB code). Apparently, the function Edges works only for structured elements (structured: right-angled triangular element), because using unstructured elements gave inaccurate results.

The structured elements can be constructed in GID by going to *Mesh* menu and choosing *Assign* for *Surfaces* under *Structured* submenu, the popup window will appear, enter the step size in that window. Figure 18.5 shows the structured elements.

For TM modes, all edges are considered free edges, but for the TE modes, the boundary edges are treated as prescribed edges and only the inner edge should be treated as free. The common edges were extracted by intersecting $n1EL$, $n2EL$, and $n3EL$ as can be seen from the code in Appendix 18B.2.

The first seven modes for TE and TM are listed in Table 18.2.

FIGURE 18.3
(a) Global nodes.

(Continued)

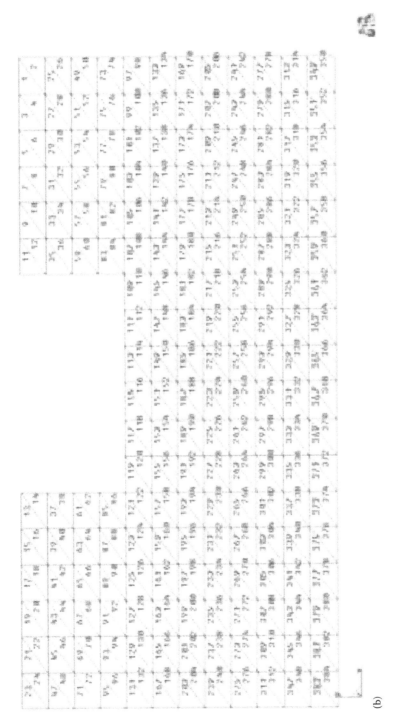

(b)

FIGURE 18.3 (*Continued*)
(b) Elements numbering.

(*Continued*)

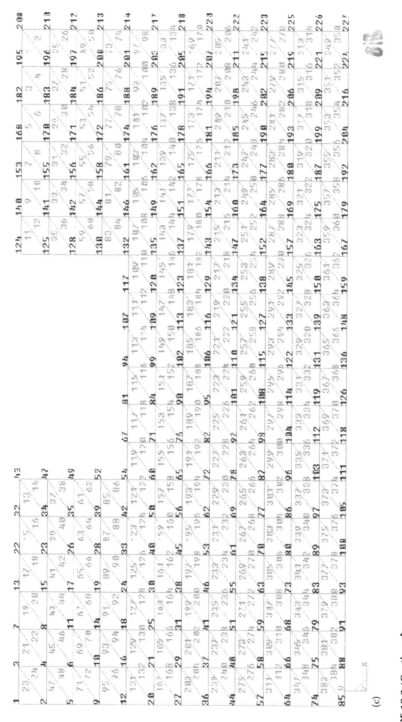

FIGURE 18.3 (*Continued*)
(c) Larger view of the first-order nodes (bold) and elements numbering.

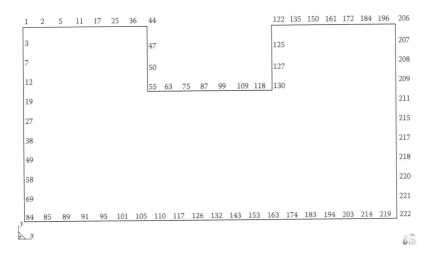

FIGURE 18.4
Prescribed nodes for TM modes.

TABLE 18.1

First-Order FEM: Cutoff Frequencies
of the First Seven Modes

TE Modes		TM Modes	
k_c (rad/m)	f (GHz)	k_c (rad/m)	f (GHz)
102.9	4.91	367.2	17.53
227.8	10.88	396.6	18.94
267.2	12.76	481.2	22.97
281.9	13.46	570.8	27.25
374.2	17.87	619.1	29.56
417.3	19.92	650.5	31.06
450.7	21.52	721.6	34.45

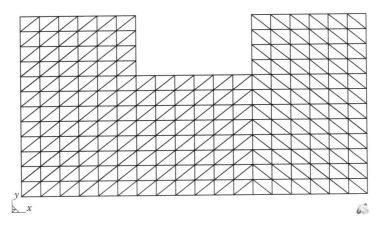

FIGURE 18.5
Element setup for edge-based method.

TABLE 18.2

Edge-Based FEM: Cutoff Frequencies
of the First Seven Modes

TE Modes		TM Modes	
k_c (rad/m)	f (GHz)	k_c (rad/m)	f (GHz)
102.4	4.89	365.7	17.46
210.3	10.04	398.3	19.02
261.8	12.50	479.2	22.88
298.2	14.24	553.3	26.42
383.2	18.30	603.6	28.82
403.0	19.24	651.9	31.13
440.5	21.03	718.1	34.29

18.1.3 Second-Order Node-Based FEM

Using *Quadratic9* from *Element Type* menu in GID, we can generate the second-order nodes for this geometry as shown in Figure 18.6. The function *meshread2T* is used to read data from GID-exported text file (see the MATLAB code in Appendix 18C.6). Then these data are used in *GLAN2T* subroutine in Appendix 18C.5 to find the stiffness matrix. The main code is listed in Appendix 18B.3.

The first seven modes are listed in Table 18.3.

18.1.4 HFSS Simulation

In this part, the ridged WG cutoff frequencies are obtained using the commercial EM simulator HFSS which is based on FEM. The WG structure shown in Figure 18.7 has the same dimensions used in the previous parts. The simulation in Figure 18.8 shows the first eight modes cutoff frequencies.

The first- and second-order results are very close to the HFSS results but the edge-based method is different (see Table 18.4).

18.2 Inhomogeneous Waveguide

18.2.1 Loaded Square Waveguide

The problem is composed of a square waveguide loaded with dielectric material with a dielectric constant of 1.5, and the dimensions of this WG is depicted in Figure 18.9 where the dielectric material occupies half of the WG.

The first step is to generate the structure using the commercial software GID, following the same steps that we used in ridged WG, the mesh generation was done using structured triangular elements as shown in Figure 18.10, and the geometry consists of 72 elements and 49 nodes. The data from GID can be exported to a text file that contains Local nodes

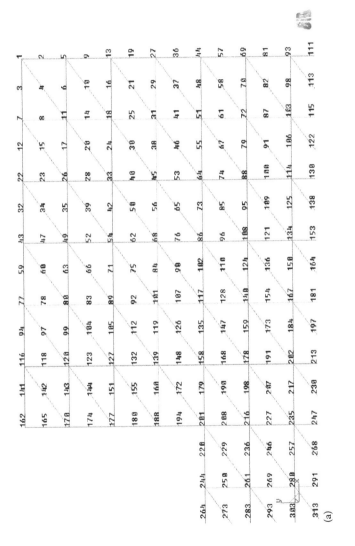

FIGURE 18.6
(a) Nodes assignment for the second-order node-based FEM.

(Continued)

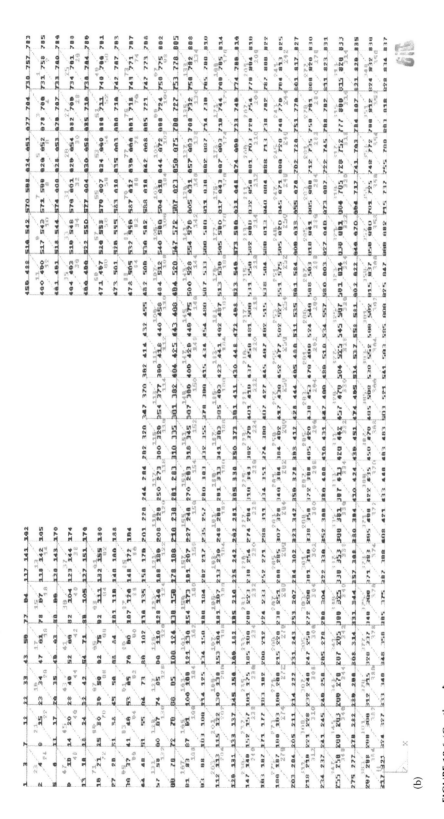

FIGURE 18.6 (*Continued*)
(b) Large view of the second-order (bold) nodes and element numbering.

TABLE 18.3

Second-Order FEM: Cutoff Frequencies
of the First Seven Modes

TE Modes		TM Modes	
k_c (rad/m)	f (GHz)	k_c (rad/m)	f (GHz)
102.3	4.89	339.0	16.19
225.6	10.77	369.5	17.64
265.6	12.68	451.1	21.54
280.6	13.40	533.3	25.46
371.1	17.72	569.4	27.19
412.4	19.69	598.8	28.59
442.4	21.12	668.7	31.93

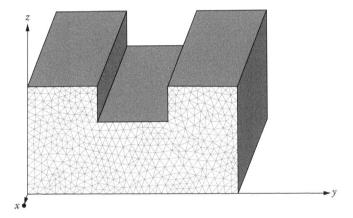

FIGURE 18.7
Constructing triangular elements in HFSS.

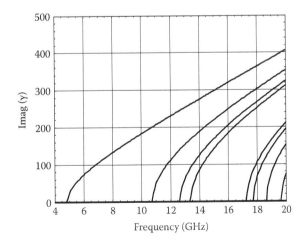

FIGURE 18.8
First eight modes in ridged WG using HFSS.

TABLE 18.4

Comparison of the First Seven Modes Cutoff Frequencies
of the Three Methods

	Cutoff Frequency (GHz)						
First-order	4.97	10.88	12.76	13.46	17.53	17.87	18.94
Second-order	4.89	10.77	12.68	13.40	16.19	17.64	17.72
Edge-based	4.89	10.04	12.50	14.24	17.46	18.30	19.02
HFSS simulation	4.8	10.7	12.6	13.3	17.2	17.7	18.5

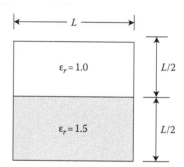

FIGURE 18.9
Loaded square wave guide.

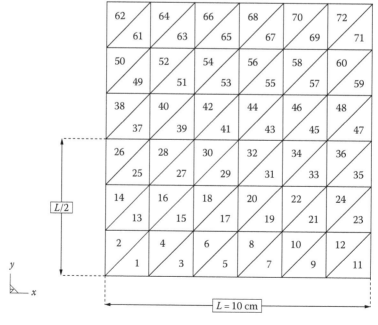

FIGURE 18.10
Elements generation using GID.

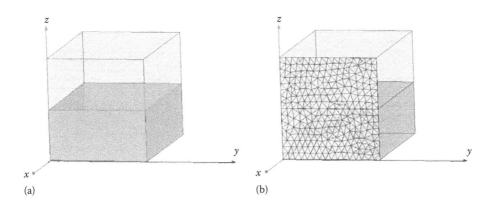

FIGURE 18.11
(a) HFSS structure and (b) mesh generation.

numbers for each element ($n1L$, $n2L$, and $n3L$) and each node coordinates (x_n and y_n). The dielectric elements can also be selected using the command *List Some Filtered Results* and stored in a separate text file to be used in the main code.

Meshread subroutine is used to read the data from the exported GID file while the *Edges* function is used to generate local edge numbering for each element ($n1EL$, $n2EL$, and $n3EL$) (see Appendices 18C.2 and 18C.4, respectively). These functions are fed to the *INHWGD* function (see Appendix 18B.4), which computes the matrix **C** whose eigenvalues are the wave numbers.

The simulation of the same waveguide was carried out using HFSS as shown in Figure 18.11.

For inhomogeneous square WG with length = 0.1 m filled with dielectric material that has a dielectric constant of 1.5, the dielectric occupies half the WG.

Figures 18.12 and 18.13 show the propagating modes in the WG for both HFFS simulation and numerical method using *INHWGD* code.

In the loaded waveguide, all the modes are not TE or TM but the modes in general are the hybrid modes.

18.2.2 Inhomogeneous Ridged WG

The ridged WG shown below is loaded with a dielectric material that occupies the central region (elements in the middle region), the dielectric constant of this material has been chosen to be $\varepsilon_r = 4.0$ and the rest of the WG is filled with air.

The MATLAB subroutine called INHWG.m is used to find γ at a given frequency for the structure depicted in Figure 18.14. The same procedure that was carried out in the first-order FEM mentioned earlier is used to generate nodes and element numbering in GID, and the data from GID was exported to the MATLAB subroutine as explained earlier. The main MATLAB program in Appendix 18B.4 is written to determine the propagation constant β at a specified wave number k_0.

A comparison between the HFSS simulation and the result obtained from applying the INHWG MATLAB subroutines (called numerical in Table 18.5) is shown in Figure 18.15.

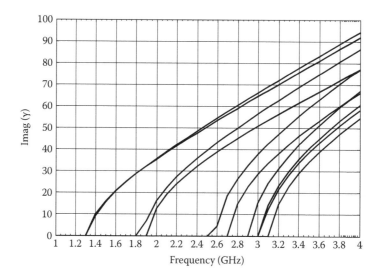

FIGURE 18.12
HFSS simulation of loaded square waveguide.

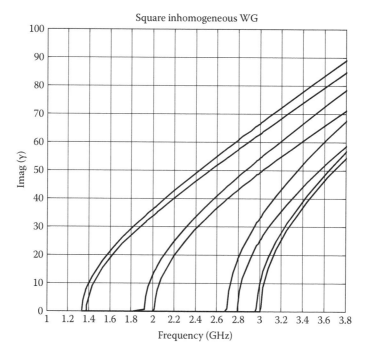

FIGURE 18.13
INHWG MATLAB code solution: loaded square waveguide.

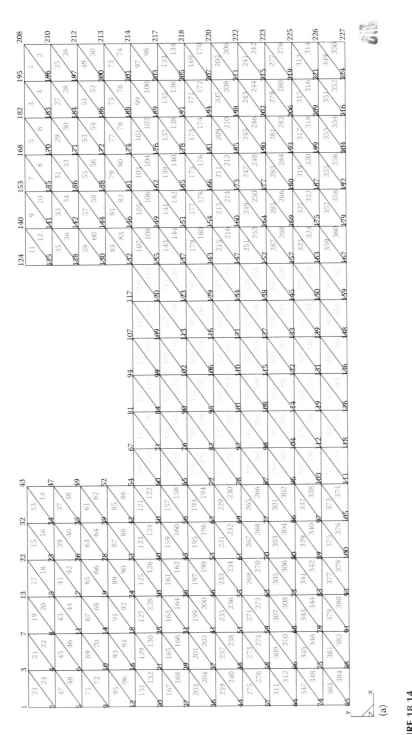

FIGURE 18.14

Ridged WG loaded with dielectric material: (a) nodes and elements.

(Continued)

144 Advanced Electromagnetic Computation

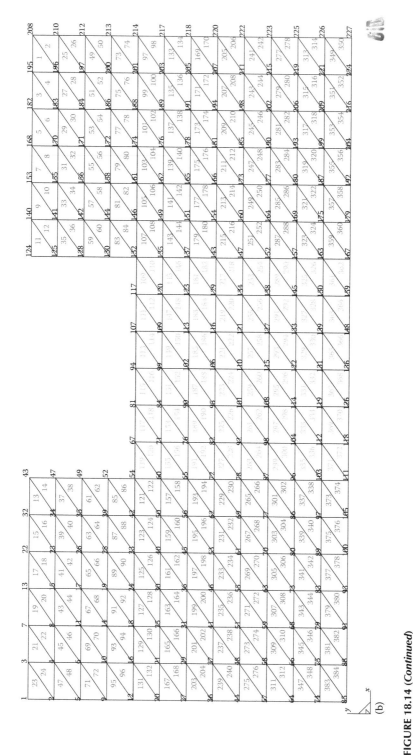

FIGURE 18.14 (*Continued*)
Ridged WG loaded with dielectric material: (b) large view of the nodes and element numbering: nodes (bold) and elements, loaded elements in middle.

TABLE 18.5

INHWG First Seven Cutoff Frequencies: HFSS Simulation Compared with the Numerical Code

Frequency (GHz)	f_0	f_1	f_2	f_3	f_4	f_5	f_6
HFSS	2.9	8.2	10.1	11.1	11.6	12.9	13.9
Numerical	2.86	8.188	10.26	11.08	11.55	12.84	14.32
%error	1.38	0.15	1.58	0.18	0.43	0.465	3.02

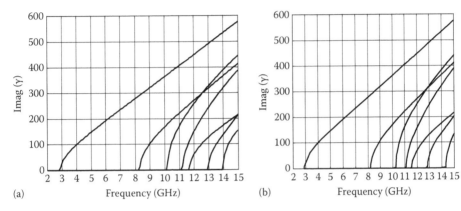

(a)   Frequency (GHz)   (b)   Frequency (GHz)

FIGURE 18.15
Loaded ridged WG modes: (a) HFSS simulation and (b) INHWG MATLAB code solution.

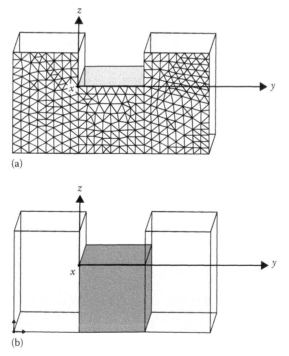

FIGURE 18.16
HFSS (a) mesh generation and (b) structure.

There is a very good agreement between the two graphs. Table 18.5 compares the cutoff frequencies f_n.
For the first cutoff, f_0,

$$\%\text{error} = \left| \frac{2.9 - 2.86}{2.9} \right| \times 100\% = 1.38\%.$$

The errors are consistent with the chosen numbers of elements and nodes in (Figure 18.16a). Even for the moderate number of elements involved, the input data as given in Appendix 18A are difficult to hand-generate.

19

Finite-Difference Time-Domain Method

In this Chapter we introduce the finite-difference time-domain (FDTD) method [1–5]. Section 19.3–19.7 follow the notation and subject development in Ref. [1–2]. Section 19.8–19.10 follow the subject development given in Ref. [3] and Ref. [5]. Maxwell's equations are PDEs in time domain. Their solution with appropriate initial and boundary conditions give results for many practical problems in classical electrodynamics. Maxwell's equations are supplemented by the "constitutive relations" to describe the properties of the medium. In previous chapters, we explored such solutions using numerical methods. Most of the examples involved static or time-harmonic solution. The corresponding PDEs are called elliptic type. We explored their solution using finite differences. The technique consisted of replacing the continuous domain by a discrete set of points, which are the intersection points of a grid superposed on the continuous domain. The differential equation is converted into a set of algebraic equations by expressing the derivatives in terms of the potential at the neighboring points of the grid. This technique resulted in a set of simultaneous algebraic equations. The boundary of the domain is a closed one and the boundary conditions are the values of the potential or its functions on the boundary. The difficulty is the slow convergence of the solution if iterative methods are used. If the simultaneous equations are solved by matrix inversion, then it could take a large amount of computer time. Moreover, due to floating point arithmetic used by the computers, the round off error involved in performing a large number of arithmetic operations can overwhelm the actual solution leading to large errors. The waveguide problem we solved in the previous chapters is based on the Helmholtz equation which is an elliptic type PDE. However, the problem involved determination of the eigenvalues.

Maxwell's equations in time domain are PDEs of hyperbolic type. FD solution of this type if properly formulated and discretized will lead to a step-by-step solution rather than requiring the solution of simultaneous equations. Step-by-step solutions are also called "marching" solutions in the sense that the solution at the present time step is expressed in terms of the solution at previous time steps, already obtained by a simple calculation. In that sense, when it works, the solution takes much less time than the solution of equilibrium problem. However, if it not properly done, numerical instability can set in and one can get absurd results. Let us explain the above-mentioned points through a simple example.

19.1 Air-Transmission Line

The simple wave equation for an air-transmission line is

$$\frac{\partial^2 V}{\partial x^2} - \frac{1}{c^2}\frac{\partial^2 V}{\partial t^2} = 0. \tag{19.1}$$

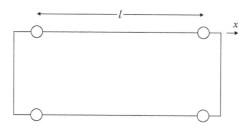

FIGURE 19.1
Air-transmission line.

The boundary conditions for the short-circuited line shown in Figure 19.1 are

$$V(0,t) = 0, \tag{19.2}$$

$$V(\ell,t) = 0. \tag{19.3}$$

Let us assume that the line is charged and has an initial voltage of $V_0 \sin(\neq x/\ell)$: such a static voltage distribution can arise due to an overhead cloud. The current I is zero. Suppose the cloud moves away at $t = 0$, and we wish to find the voltage variation on the line as a function of x and t. By separation of variable technique one can obtain an analytical solution for this simple problem. The initial conditions for the problem are

$$V(x,0) = V_0 \sin\frac{\pi x}{\ell}, \tag{19.4}$$

$$\frac{\partial V}{\partial t}(x,0) = 0. \tag{19.5}$$

Equation 19.5 is obtained from $I(x,0) = 0$, I being proportional to $(\partial V)/(\partial t)$. Equation 19.1 is a second-order differential equation, requiring two boundary conditions and two initial conditions. The analytical solution for this well-posed problem is given by

$$V(x,t) = V_0 \sin\frac{\pi x}{\ell}\cos\frac{\pi c t}{\ell}. \tag{19.6}$$

This solution will be useful in ascertaining the accuracy of the numerical solution, which we will obtain by using FDs.

19.2 Finite-Difference Time-Domain Solution

We shall obtain FD solution by using central difference approximations for the space and time second-order partial derivatives. Let

$$t = n\Delta t, \quad n = 0,1,2,\dots, \tag{19.7}$$

$$x = i\Delta x, \quad i = 0,1,2,\dots N. \tag{19.8}$$

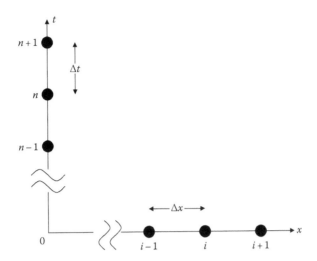

FIGURE 19.2
Notation for the grid points.

We can now abbreviate (Figure 19.2)

$$V(x_i, t_n) = V_i^n. \tag{19.9}$$

From central difference approximations

$$\left. \frac{\partial^2 V}{\partial x^2} \right|_{x_i, t_n} = \frac{V_{i+1}^n - 2V_i^n + V_{i-1}^n}{(\Delta x)^2} + 0(\Delta x)^2, \tag{19.10}$$

$$\left. \frac{\partial^2 V}{\partial t^2} \right|_{x_i, t_n} = \frac{V_i^{n+1} - 2V_i^n + V_i^{n-1}}{(\Delta t)^2} + 0(\Delta t)^2. \tag{19.11}$$

From Equations 19.1, 19.9, and 19.10 and treating V_i^{n+1} as the unknown to be calculated, we obtain the following:

$$V_i^{n+1} = r^2 \left(V_{i+1}^n - 2V_i^n + V_{i-1}^n \right) + 2V_i^n - V_i^{n-1},$$
$$= r^2 V_{i-1}^n + 2 \left(1 - r^2 \right) V_i^n + r^2 V_{i+1}^n - V_i^{n-1}, \tag{19.12}$$

where

$$r = c \frac{\Delta t}{\Delta x}. \tag{19.13}$$

The computational molecule for the problem can be written as a sketch given in Figure 19.3.

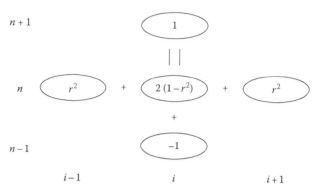

FIGURE 19.3
Computational molecule.

The values in the circles are the weights of multiplication of the potentials at various grid points. The choice of $r = 1$, results in a simple computational molecule given in Figure 19.4 and Equation 19.14.

$$V_i^{n+1} = V_{i-1}^n + V_{i+1}^n - V_i^{n-1}. \tag{19.14}$$

We can apply this computational molecule and calculate the voltage at any time and at any point on the transmission line. Let us illustrate this by hand computation to a simple example where the length of the line $\lambda = 1$. Let us prepare a table where the values of the voltage will be recorded as shown in Figure 19.5.

Note that the problem has a line of symmetry at $x = 0.5(i = 5)$. It then follows

$$V_4^n = V_6^n. \tag{19.15}$$

From the initial condition, Equation 19.4, $V_0^0 = 0$, $V_1^0 = \sin\pi(0.1) = 0.309$, and $V_2^0 = \sin\pi(0.2) = 0.58979$, and so on marked in Figure 19.5. The second initial condition given by Equation 19.5, when approximated by central difference formula gives

$$\frac{\partial V}{\partial t}\bigg|_{t=0} = 0 = \frac{V_i^1 - V_i^{-1}}{2\Delta t}$$

$$V_i^1 = V_i^{-1}. \tag{19.16}$$

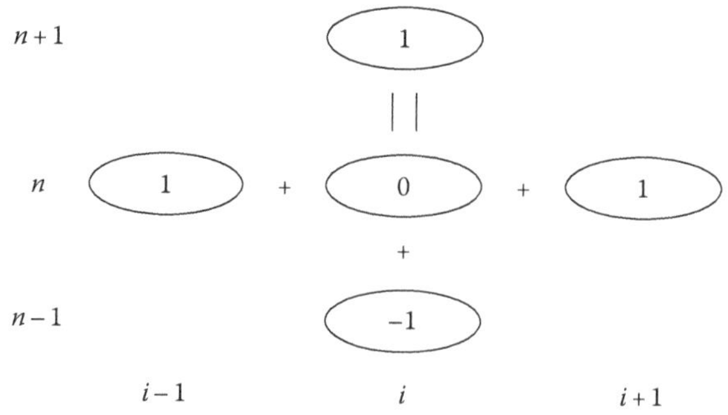

FIGURE 19.4
Computational molecule for $r = 1$.

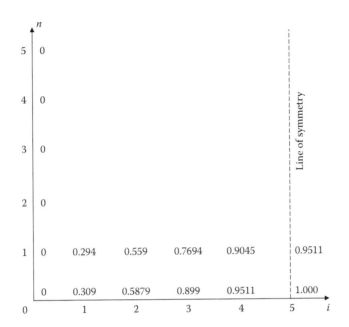

FIGURE 19.5
Values of the voltage variable at $n = 0$ and $n = 1$ time-step levels.

The first boundary condition Equation 19.2 gives

$$V_0^n = 0. \tag{19.17}$$

From Equation 19.14, when applied for $n = 0$, $i = 1$,

$$V_1^1 = V_0^0 + V_2^0 - V_1^{-1}. \tag{19.18}$$

From Equations 19.16 and 19.18

$$2V_1^1 = 0 + 0.5879, \quad V_1^1 = 0.294.$$

Applying Equation 19.14 at $n = 0$, $i = 2$ and using Equation 19.18

$$V_2^1 = V_1^0 + V_3^0 - V_2^{-1}, \quad 2V_2^1 = V_1^0 + V_3^0,$$

$$V_2^1 = 0.559.$$

Similar computations at $i = 3, 4$ give

$$V_3^1 = 0.7694, \quad V_4^1 = 0.9045.$$

For $n = 0$ and $i = 5$, Equation 19.14 gives

$$V_5^1 = V_4^0 + V_6^0 - V_5^{-1}.$$

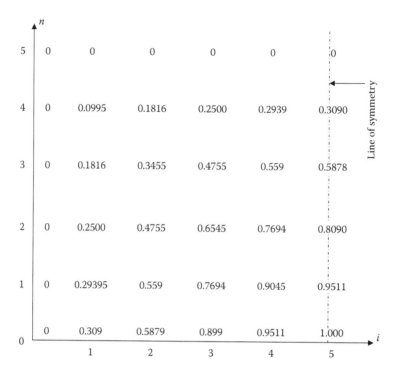

FIGURE 19.6
Values of the voltage variable up to $n = 5$.

From Equation 19.16, $V_5^1 = V_5^{-1}$, and from Equation 19.15, $V_4^0 = V_6^0$.

$$\therefore V_5^1 = V_4^0 = 0.9511.$$

Note that we obtained the voltage at the time level $t = \Delta t$ by a step-by-step calculation. Applying Equation 19.14 at interior points and applying Equation 19.15 for points on the line of symmetry, we can calculate the voltage at the next time step and march the solution to any time level. Figure 19.6 show the table of values up to $n = 5a$. Let us compare the exact solution with the value obtained by FDTD for V_3^1. The exact solution is given by substituting $x = 3\Delta x$ and $t = \Delta t$ in Equation 19.6 with $V_0 = 1$ and $\lambda = 1$;

$$V_3^1 = \sin 3\pi \Delta x \cos \pi c \Delta t,$$

$$\Delta x = 0.1, \quad c\frac{\Delta t}{\Delta x} = 1, \quad \Delta t = \frac{\Delta x}{c} = \frac{0.1}{c},$$

$$V_3^1 = \sin 0.3\pi \cos 0.1\pi = 0.7694,$$

which is the value obtained by FDTD. The result is surprising since the central difference approximation is only accurate up to second order, that is, of the order of $O[(\Delta x)^2, (\Delta t)^2]$. This result is not accidental, but it can be shown that the FDTD solution is exact for $r = 1$. For example, if we have chosen $\Delta x = 0.25$, then we choose Δt such that $r = c(\Delta t)/(\Delta x) = 1$, we again get the exact answer.

19.3 Numerical Dispersion

For the simple 1D wave equation given by Equation 19.1, the harmonic solution of a positive-going wave is

$$V = Ae^{j(\omega t - kx)}. \tag{19.19}$$

A negative-going wave is also a solution and the solution Equation 19.6 is a superposition of the above two waves satisfying the initial and boundary conditions. The relation between ω and k

$$\omega = ck, \tag{19.20}$$

where c is the velocity of light. The $\omega = ck$ diagram is a straight line. The phase velocity $v_p = (\omega)/(c)$ and the group velocity $v_g = (d\omega)/(dk)$ are both equal to c. When we approximate Equation 19.1 by using central difference formulas, we obtained the computational molecule given by Equation 19.12. Since the central difference formula we used has an error of the order of $(\Delta t)^2$ and $(\Delta x)^2$, we expect the wave solutions to be approximate. The error can be studied by studying a monotonic wave of frequency ω and an approximate corresponding wave number $\tilde{k}$ and the solution can be written as

$$\tilde{V} \approx Ae^{j(\omega t - \tilde{k}x)}. \tag{19.21}$$

The relation between ω and the numerical wave number $\tilde{k}$ is called numerical dispersion relation. We can obtain this relation by substituting Equation 19.21 into Equation 19.12. The constant A will cancel and hence we will concentrate on the exponential factors involved in Equation 19.12. The exponential factor ψ, $\exp[j(\omega t - kx)]$, in V will assume the form ψ_i^n in V_i^n:

$$\psi_i^n = e^{j(\omega n \Delta t - \tilde{k}i\Delta x)}. \tag{19.22}$$

Thus Equation 19.12 can be approximated as

$$e^{j\left[\omega(n+1)\Delta t - \tilde{k}i\Delta x\right]} = r^2 \left\{ e^{j\left[\omega n \Delta t - \tilde{k}(i+1)\Delta x\right]} - 2e^{j\left(\omega n \Delta t - \tilde{k}i\Delta x\right)} + e^{j\left[\omega n \Delta t - \tilde{k}(i-1)\Delta x\right]} \right\} + 2e^{j\left(\omega n \Delta t - \tilde{k}i\Delta x\right)} - e^{j\left[\omega(n-1)\Delta t - \tilde{k}i\Delta x\right]}.$$

Factoring out $e^{j\left(\omega n \Delta t - \tilde{k}in\Delta x\right)}$ one can obtain

$$e^{j\omega \Delta t} = r^2 \left(e^{-j\tilde{k}\Delta x} - 2 + e^{j\tilde{k}\Delta x} \right) + 2 - e^{-j\omega \Delta t}. \tag{19.23}$$

By grouping the terms Equation 19.23 can be written as [1]

$$\cos(\omega \Delta t) = r^2 \left[\cos(\tilde{k}\Delta x) - 1 \right] + 1. \tag{19.24}$$

Let us now examine two special cases [1]:

Case 1: r = 1.

Equation 19.24 gives

$$\cos\omega\Delta t = \cos\tilde{k}\Delta x, \quad \omega\Delta t = \tilde{k}\Delta x,$$

$$\tilde{v}_p = \frac{\omega}{\tilde{k}} = \frac{\Delta x}{\Delta t} = c. \tag{19.25}$$

The last equality in Equation 19.25 is because

$$r = c\frac{\Delta t}{\Delta x} = 1.$$

For this case the numerical wave has the same phase velocity as the physical wave, the numerical approximation has not introduced any dispersion. This explains the exact answer we obtained when we used $r = 1$.

Case 2: Very fine mesh $\Delta t \to 0$, $\Delta x \to 0$.

The arguments of the cosine functions in Equation 19.24 are small and we can use the approximations

$$\cos\theta = \left(1 - \sin^2\theta\right)^{1/2} \approx 1 - \frac{1}{2}\sin^2\theta \approx 1 - \frac{1}{2}\theta^2 \tag{19.26}$$

and obtain

$$1 - \frac{1}{2}(\omega\Delta t)^2 = r^2\left[1 - \frac{1}{2}\tilde{k}^2(\Delta x)^2 - 1\right] + 1,$$

$$-\frac{1}{2}(\omega\Delta t)^2 = c^2\frac{(\Delta t)^2}{(\Delta x)^2}\left[-\frac{1}{2}\tilde{k}^2(\Delta x)^2\right], \tag{19.27}$$

$$\omega^2 = c^2\tilde{k}^2.$$

As expected, the numerical wave and the physical wave are identical.

General case: Let us look at a general case of arbitrary r and R, where

$$R = \frac{\lambda_0}{\Delta x}. \tag{19.28}$$

R is the number of cells per wavelength. Equation 19.24 can be written as

$$\tilde{k} = \frac{1}{\Delta x}\cos^{-1}\left\{1 + \frac{1}{r^2}\left[\cos(\omega\Delta t) - 1\right]\right\} \tag{19.29}$$

and the normalized numerical phase velocity is given by

$$\frac{\tilde{v}_p}{c} = \frac{\omega}{\tilde{k}c} = \frac{(\omega\Delta x)/(c)}{\cos^{-1}\left\{1 + (\cos(\omega\Delta t) - 1)/(r^2)\right\}}$$

Noting

$$\frac{\omega \Delta x}{c} = \frac{2\pi f \Delta x}{c} = \frac{2\pi \Delta x}{\lambda_0} = \frac{2\pi}{R}$$

$$\omega \Delta t = 2\pi f \Delta t = \frac{2\pi f}{c} \frac{c \Delta t}{\Delta x} \Delta x = \frac{2\pi}{\lambda_0} r \Delta x = 2\pi \frac{r}{R},$$

we obtain

$$\frac{\tilde{v}_p}{c} = \frac{2\pi}{R \cos^{-1}\left\{1 + (1)/\left(r^2\right)\left[\cos 2\pi (r)/(R) - 1\right]\right\}}. \tag{19.30}$$

Let us look at a couple of numbers obtained from Equation 19.30 [1].

- *Case* 1: $r = 0.5$, $R = 10$.

  Let $r = 0.5$ and the number of cells per wavelength $R = 10$; the normalized numerical phase velocity

$$\frac{\tilde{v}_p}{c} = 0.9873. \tag{19.31}$$

  This is an error of 1.27%.

- *Case* 2: $r = 0.5$, $R = 20$.

$$\frac{\tilde{v}_p}{c} = 0.99689. \tag{19.32}$$

By doubling R, the phase velocity error is reduced to 0.31%, 0.25 of the previous value. Figure 19.7 show the normalized phase velocity versus R with r as a parameter. As we note from the figure, the decrease in accuracy by decreasing r can be compensated by increasing R, that is, by increasing the number of cells per wavelength.

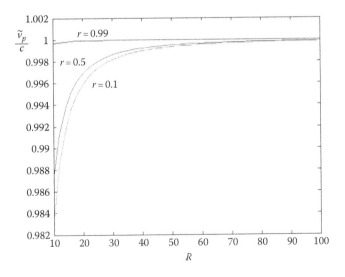

FIGURE 19.7
Normalized numerical phase velocity versus R with r as the parameter.

19.4 Waves in Inhomogeneous, Nondispersive Media: FDTD Solution

Let us apply FDTD method to wave propagation in an inhomogeneous, nondispersive dielectric medium. Consider the one-dimensional case, where the fields as well as the permittivity vary in one spatial dimension:

$$\varepsilon = \varepsilon(x), \tag{19.33}$$

$$\bar{D} = \hat{z}D_z(x), \tag{19.34a}$$

$$\bar{E} = \hat{z}E_z(x), \tag{19.34b}$$

$$\bar{H} = \hat{y}H_y(x), \tag{19.35}$$

$$D_z = \varepsilon(x)E_z. \tag{19.36}$$

A simple example of Equation 19.36 is a dielectric slab of width d and permittivity ε_2. The input medium has permittivity ε_1 and the output medium has permittivity ε_3 as shown in Figure 19.8.

The relevant Maxwell's equations for a nonmagnetic medium are

$$\bar{\nabla} \times \bar{E} = -\mu_0 \frac{\partial \bar{H}}{\partial t}, \tag{19.37}$$

$$\bar{\nabla} \times \bar{H} = \frac{\partial \bar{D}}{\partial t}. \tag{19.38}$$

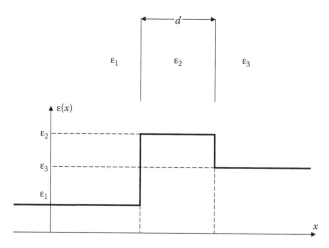

FIGURE 19.8
Example of an inhomogeneous medium.

From Equations 19.37, 19.34b, and 19.35, we have

$$\frac{\partial H_y}{\partial t} = \frac{1}{\mu_0} \frac{\partial E_z}{\partial x}. \tag{19.39}$$

From Equations 19.38, 19.35, and 19.34a, we have

$$\frac{\partial D_z}{\partial t} = \frac{\partial H_y}{\partial x}. \tag{19.40}$$

We can discretize Equation 19.39 using the central difference formula for the first derivative. Let $t = \Delta t$ and $x = i\Delta x$, and approximate Equation 19.39 at the spatial index $(i + (1/2))$ and time index n;

$$\left.\frac{\partial H_y}{\partial t}\right|_{i+(1/2)}^{n} = \frac{1}{\mu_0} \left.\frac{\partial E_z}{\partial x}\right|_{i+(1/2)}^{n}, \tag{19.41}$$

$$\frac{\left.H_y\right|_{i+(1/2)}^{n+(1/2)} - \left.H_y\right|_{i+(1/2)}^{n-(1/2)}}{\Delta t} = \frac{1}{\mu_0} \frac{\left.E_z\right|_{i+1}^{n} - \left.E_z\right|_{i}^{n}}{\Delta x}, \tag{19.42}$$

$$\left.H_y\right|_{i+(1/2)}^{n+(1/2)} = \left.H_y\right|_{i+(1/2)}^{n-(1/2)} + \frac{\Delta t}{\mu_0 \Delta x}\left[\left.E_z\right|_{i+1}^{n} - \left.E_z\right|_{i}^{n}\right]. \tag{19.43}$$

We can discretize Equation 19.40 using FDTD at the space index i and time index $n + (1/2)$;

$$\left.\frac{\partial D_z}{\partial t}\right|_{i}^{n+(1/2)} = \left.\frac{\partial H_Y}{\partial x}\right|_{i}^{n+(1/2)}, \tag{19.44}$$

$$\left.D_z\right|_{i}^{n+1} = \left.D_z\right|_{i}^{n} + \frac{\Delta t}{\Delta x}\left[\left.H_y\right|_{i+(1/2)}^{n+(1/2)} - \left.H_y\right|_{i-(1/2)}^{n+(1/2)}\right]. \tag{19.45}$$

The algebraic equation (19.36) can be discretized at the space index i and time index i:

$$\left.E_z\right|_{i}^{n+1} = \frac{1}{\varepsilon_i} \left.D_z\right|_{i}^{n+1}, \tag{19.46}$$

where

$$\varepsilon_i = \varepsilon(i\Delta x). \tag{19.47}$$

Equations 19.43, 19.45, and 19.46 give a leap-frog step-by-step computational algorithm [4] to compute and lead the solution to the time domain. Figure 19.9 helps us to understand the algorithm.

The left-hand side of Equation 19.43 can be computed from the value of $\left.H_y\right|_{i+(1/2)}^{n-(1/2)}$ and $\left.E_z\right|_{i}^{n}$ and $\left.E_z\right|_{i+1}^{n}$ on the right-hand side of Equation 19.43, which are the values at previous time

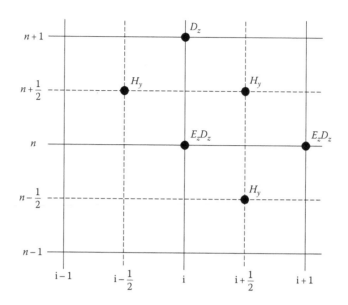

FIGURE 19.9
Leap-frog computational algorithm.

steps and are already computed. Thus H_y's at time step $(n + (1/2))$ are calculated. The left-hand side of Equation 19.45 is now computed from D_z at time step n and H_y at time step $(n + (1/2))$. Thus, D_z's at time step $(n + 1)$ are computed. Using Equation 19.46, from D_z, E_z can be computed. Repeating these three computational steps, the values of the fields can be computed at the next level of time step and the solution is thus marched in time. To start the process you have to supply the initial values: H_y at the time step $-(1/2)$ $(n = 0, t = -(\Delta t / 2)$ and E_z at the time step 0 $(n = 0, t = 0)$.

19.5 Waves in Inhomogeneous, Dispersive Media

For a dispersive media, where ε is a function of the frequency

$$\varepsilon = \varepsilon(\omega), \tag{19.48}$$

the algebraic equation (19.36) will be true only if a monochromatic wave propagation is considered. A pulse propagation problem requires the constitutive relation for the dielectric medium to be written as

$$D(s) = \varepsilon(s)E(s), \tag{19.49}$$

which in time domain

$$D(t) = L^{-1}\big[\varepsilon(s)E(s)\big] = \varepsilon(s) * E(t) = \int_{-\infty}^{\infty} \varepsilon(\tau)E(t-\tau)\mathrm{d}\tau. \tag{19.50}$$

In the above * denotes convolution. Thus, the constitutive relation

$$D(t) = \varepsilon(\omega)E(t) \tag{19.51}$$

will lead to errors, unless ε is nondispersive. The constitutive relation needs to be stated as a convolution integral given in Equation 19.50 or it can be stated as an auxiliary differential equation. Given below are three examples of a dispersion relation and the auxiliary differential equation and a sketch of relative permittivity versus frequency.

1. Drude dispersive relation (cold, isotropic, and lossy plasma):

$$\varepsilon_r' - j\varepsilon_r'' = \varepsilon_p(\omega) = 1 - \frac{\omega_p^2}{\omega(\omega - j\nu)}. \tag{19.52}$$

Constitutive relation as an auxiliary differential equation

$$\frac{d\overline{J}}{dt} + \nu\overline{J} = \varepsilon_0\omega_p^2\overline{E}. \tag{19.53}$$

Example values: plasma frequency

$$f_p = \frac{\omega_p}{2\pi} = 30\,\text{GHz}.$$

Collision frequency (Figure 19.10) $\nu = 2 \times 10^{10}$ rad/s.

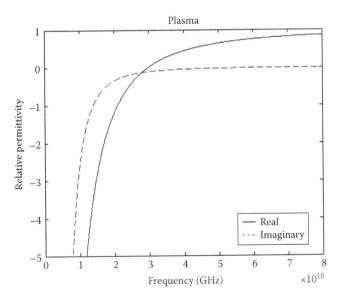

FIGURE 19.10
Relative permittivity versus frequency for a Drude type of dispersive material, like, cold, isotropic, and lossy plasma.

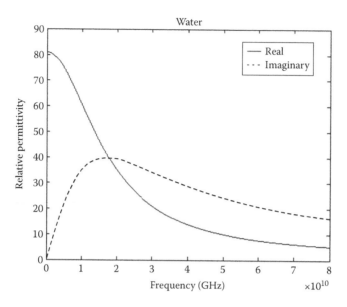

FIGURE 19.11
Relative permittivity versus frequency for a Debye type of dispersive material like water.

2. Debye dispersion relation (water)

$$\varepsilon_r(\omega) = \varepsilon_\infty + \frac{\varepsilon_s - \varepsilon_\infty}{1 + j\omega t_0}, \tag{19.54}$$

$$t_0 \frac{d\bar{D}}{dt} + \bar{D} = \varepsilon_s \varepsilon_0 \bar{E} + t_0 \varepsilon_\infty \varepsilon_0 \frac{d\bar{E}}{dt}. \tag{19.55}$$

Example values: Low-frequency relative permittivity $\varepsilon_s = 81$.
High-frequency relative permittivity $\varepsilon_s = 1.8$.
Relaxation time $t_0 = 9.4 \times 10^{-12}$ s (Figure 19.11).

3. Lorentz dispersive material (second-order, and optical material)

$$\varepsilon_r(\omega) = \varepsilon_\infty + \frac{(\varepsilon_s - \varepsilon_\infty)\omega_R^2}{\omega_R^2 + 2j\omega\delta - \omega^2}, \tag{19.56}$$

$$\omega_R^2 \bar{D} + 2\delta \frac{d\bar{D}}{dt} + \frac{d^2\bar{D}}{dt^2} = \omega_R^2 \varepsilon_s \varepsilon_0 \bar{E} + 2\delta\varepsilon_\infty \varepsilon_0 \frac{d\bar{E}}{dt} + \varepsilon_\infty \varepsilon_0 \frac{d^2\bar{E}}{dt^2}. \tag{19.57}$$

Example values: $\varepsilon_s = 2.25$, $\varepsilon_\infty = 1$.
Resonant frequency $\omega_R = 4 \times 10^{16}$ rad/s.
Damping constant $\delta = 0.28 \times 10^{16}$/s (Figure 19.12).

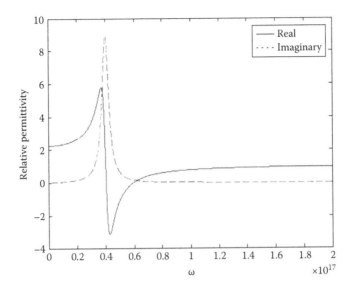

FIGURE 19.12
Relative permittivity versus frequency for a second-order Lorentz type of dispersive material like optical material.

The technique of obtaining the auxiliary differential equation from the dispersion relation is illustrated for the Debye material

$$\bar{D}(\omega) = \varepsilon_0 \varepsilon_r (\omega) \bar{E}(\omega) = \varepsilon_0 \left[\varepsilon_\infty + \frac{\varepsilon_s - \varepsilon_\infty}{1 + j\omega t_0} \right] \bar{E}(\omega).$$

$$\bar{D}(\omega)(1 + j\omega t_0) = \varepsilon_0 \varepsilon_\infty (1 + j\omega t_0) \bar{E}(\omega) + \varepsilon_0 (\varepsilon_s - \varepsilon_\infty) \bar{E}(\omega).$$

$$j\omega t_0 \bar{D}(j\omega) + \bar{D}(j\omega) = \varepsilon_0 (\varepsilon_\infty + \varepsilon_s - \varepsilon_\infty) \bar{E}(j\omega) + \varepsilon_0 \varepsilon_\infty j\omega t_0 \bar{E}(j\omega).$$

Using $(\partial)/(\partial t) = j\omega$

$$t_0 \frac{\partial \bar{D}}{\partial t} + \bar{D} = \varepsilon_0 \varepsilon_s t_0 \frac{\partial \bar{E}}{\partial t} + \varepsilon_0 \varepsilon_s \bar{E}.$$

Note, however, that in example (1), the constitutive relation in time domain relates the current density with the electric field. The medium is modeled as a conductor. The frequency domain constitutive relation relates $\bar{D}$ with $\bar{E}$.

19.6 Waves in Debye Material: FDTD Solution

The auxiliary differential equation (Equation 19.55) is the constitutive relation for the Debye material. The computational algorithm consists of Equations 19.43 and 19.45, but Equation 19.46 needs to be replaced. The replacement can be obtained by discretizing

Equation 19.55 through FDTD. Applying Equation 19.55 at the space step i and time step $(n + (1/2))$, we obtain

$$\varepsilon_0 \varepsilon_\infty t_0 \left.\frac{dE_z}{dt}\right|_i^{n+(1/2)} + \varepsilon_0 \varepsilon_s \left.E_z\right|_i^{n+(1/2)} = t_0 \left.\frac{dD_z}{dt}\right|_i^{n+(1/2)} + \left.D_z\right|_i^{n+(1/2)}. \tag{19.58}$$

Using central difference approximation in Equation 19.58,

$$\varepsilon_0 \varepsilon_\infty t_0 \frac{\left.E_z\right|_i^{n+1} - \left.E_z\right|_i^{n}}{\Delta t} + \varepsilon_0 \varepsilon_s \left.E_z\right|_i^{n+(1/2)} = t_0 \frac{\left.D_z\right|_i^{n+1} - \left.D_z\right|_i^{n}}{\Delta t} + \left.D_z\right|_i^{n+(1/2)}. \tag{19.59}$$

Since E_z and D_z are available at integral time steps (n integer), we cannot directly get values for $\left.E_z\right|_i^{n+(1/2)}$ and $\left.D_z\right|_i^{n+(1/2)}$ in Equation 19.59. However, we can use averaging operator on these (averaging operator is accurate to second order in Δt):

$$\left.E_z\right|_i^{n+(1/2)} = \frac{\left.E_z\right|_i^{n+1} + \left.E_z\right|_i^{n}}{2}, \tag{19.60}$$

$$\left.D_z\right|_i^{n+(1/2)} = \frac{\left.D_z\right|_i^{n+1} + \left.D_z\right|_i^{n}}{2}. \tag{19.61}$$

Substituting Equations 19.60 and 19.61 into Equation 19.59 and rearranging the unknown to be on the left-hand side of the equation, we obtain

$$\left.E_z\right|_i^{n+1} = \frac{\Delta t + 2t_0}{2t_0\varepsilon_\infty + \varepsilon_s\Delta t} \left.D_z\right|_i^{n+1} + \frac{\Delta t - 2t_0}{2t_0\varepsilon_\infty + \varepsilon_s\Delta t} \left.D_z\right|_i^{n} + \frac{2t_0\varepsilon_\infty - \varepsilon_s\Delta t}{2t_0\varepsilon_\infty + \varepsilon_s\Delta t} \left.E_z\right|_i^{n}. \tag{19.62}$$

Equations 19.45, 19.46, and 19.62 give the computational algorithm for the Debye material.

19.7 Stability Limit and Courant Condition

A plane wave propagating through a one-dimensional FDTD grid. In any one time step, a point on the wave cannot pass through more than one cell. If the medium is free space and if the point exactly passes through one cell, then $c(\Delta t)/(\Delta x) = r = 1$. This condition is called the courant condition. If $r > 1$, then the point passes through a distance more than that of one cell resulting in a velocity $(\Delta x)/(\Delta t) > c$ violating the principle that the wave velocity in the free space can not be greater than that of the light. Thus, the choice of $r > 1$, results in an instability of the algorithm, which cannot be corrected by choosing a larger value for R. For 2D and 3D problems, the stability criterion can be stated as

$$r = \frac{c\Delta t}{\sqrt{N}\Delta} \leq 1, \tag{19.63}$$

where

$$\Delta = \Delta x = \Delta y = \Delta z \qquad (19.64)$$

and N is the dimensionality.

19.8 Open Boundaries

The problem solved in Section 19.1 has bounded space $0 < x < l$ and the problem is solved by using the boundary conditions at the end points $x = 0$ and 1. The problem of finding the reflection and transmission coefficients involves open boundaries and the domain is $-\infty < x < \infty$. Unless we terminate the problem space we cannot solve the problem (Figure 19.13).

Suppose we decide to consider problem space as $-L_1 < x < L_2$. At $x = -L_1$, the reflected wave is to be an outgoing wave and should not be reflected back into the problem space. This can be ensured if the wave is absorbed by the boundary. Such a boundary condition is called absorbing boundary condition. A simple way of implementing the absorbing boundary condition is illustrated below. Let the left boundary $x = -L_1$ has the space index $i = 0$ and let us choose

$$r = c\frac{\Delta t}{\Delta x} = 0.5. \qquad (19.65)$$

The above equation tells us that it takes two time steps for a wave front to cross one cell, that is,

$$E_0^n = E_1^{n-2}. \qquad (19.66)$$

As you approach the left boundary, store the value of E_1^{n-2} in E_0^n. Similar arrangement at $x = L_2$ terminates the right boundary.

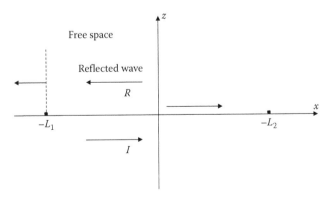

FIGURE 19.13
One-dimensional problem involving open boundaries.

19.9 Source Excitation

The problem solved in Section 19.1 has the initial value of the voltage of the charged line given. For a problem in Section 19.8, the incident wave is specified. If the incident wave is a monochromatic wave, one can simulated it by assigning a sinusoidal value at a fixed point. This is called a hard source. A propagating wave sees the hard source as a metal wall to FDTD and when the pulse passes through, it will be reflected. Suppose the fixed point is indexed by $i = F$, then

$$E_F^n = \sin\left(2 * pi * \text{freq} * dt * n\right) \tag{19.67}$$

generates a hard source at the space point $i = F$.
   On the other hand,

$$\text{pulse} = \sin\left(2 * pi * \text{freq} * dt * n\right), \tag{19.68}$$

$$E_F^n = E_F^n + \text{pulse} \tag{19.69}$$

generates a soft source. With a soft source, the propagating pulse will pass through.

19.10 Frequency Response

Suppose we wish to find the frequency response of a system. Suppose the output medium in Section 19.8 is a plasma or a lossy dielectric and we wish to find the reflection coefficient at various frequencies. One way is to run the problem at various frequencies one by one and calculate the reflection coefficient for each of the frequencies. A more efficient way is to have a pulse as the source and obtain the reflected pulse. The ratio of the Fourier transform of the reflected pulse to that of the incident pulse gives the frequency response data. If the source is an impulse, its Fourier transform is 1, that is, the frequency spectrum has an amplitude 1 at all frequencies. A more practical source is a Gaussian pulse given by

$$f(t) = e^{-(1/2)((t-t_0)/\sigma)2} \tag{19.70}$$

which is centered at t_0 and has a spread (standard deviation) of σ (Figure 19.14).
   As an example, suppose $t_0 = 40\Delta t$, $t = n\Delta t$ and $\sigma = 12\Delta t$, then [5]

$$f(n) = e^{-0.5((n-40)/12)^2}$$

$$f(0) = f(80) = e^{-0.5((10/3))^2} = e^{-5.55} \approx 0$$

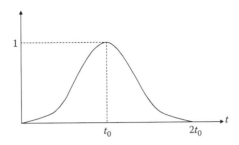

FIGURE 19.14
Gaussian pulse, $\Delta t = 1.875 \times 10^{-11}$, $\sigma = 5\sqrt{2}\Delta t$, $t_0 = 40\Delta t$.

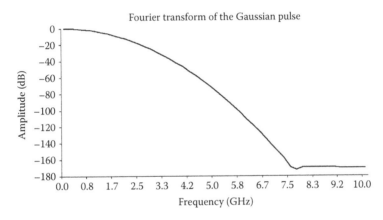

FIGURE 19.15
Fourier transform of the Gaussian pulse $\Delta t = 1.875 \times 10^{-11}$s, $\sigma = 5\sqrt{2}\Delta t$, $t_0 = 40\Delta t$. (Reprinted from Kunz, K. S. and Luebbers, R., *Finite Differences Time Domain Method for Electromagnetics*, CRC Press LLC, Boca Raton, FL, 1993, p. 36. With permission.)

The pulse can be made narrower by choosing a smaller value of σ. For example, if $\sigma = 5\sqrt{2}\Delta t$

$$f(0) = f(80) = e^{-0.5\left(\left(40/5\sqrt{2}\right)\right)^2} = e^{-0.5*64*0.5} = e^{-16} = -140 \text{ dB}.$$

The amplitude of the Fourier transform of Equation 19.70 is sketched in Figure 19.15. The truncation of the pulse at $t = 0$ or $t = 40\Delta t$ did not introduce unwanted high frequencies.

From the FFT it is clear that the amplitude of the signal is quite adequate, say up to 3 GHz. However, we will be concerned about noise if say 7 or 8 GHz is of interest, since the signal amplitude is 120 dB at this frequency.

After obtaining the reflected pulse, one can obtain the FFT of the reflected pulse and thus one can obtain the ratio of reflected pulse amplitude to the incident pulse amplitude at various frequencies.

References

1. Taflove, A., *Computational Electrodynamics: The Finite-Difference Time-Domain, Method*, Artech House, Norwood, MA, 1995.
2. Taflove, A. and Hagness, S., *Computational Electrodynamics: The Finite-Difference Time-Domain, Method*, 2nd edn., Artech House Inc., Norwood, MA, 2000.
3. Sullivan, D. M., *Electromagnetic Simulation Using FDTD Method*, IEEE Press, New York, 2000.
4. Yee, K., Numerical solution of initial boundary value problems involving Maxwell equations in isotropic media, *IEEE Trans. Ant. Prop.*, 14(3), 302–307, 1966.
5. Kunz, K. S. and Luebbers, R. J., *Finite Differences Time Domain Method for Electromagnetics*, CRC Press LLC, Boca Raton, FL, 1993.

20

Finite-Difference Time-Domain Method Simulation of Electromagnetic Pulse Interaction with a Switched Plasma Slab*

20.1 Introduction

The interaction of an electromagnetic wave with a plasma slab is experimentally more realizable than an unbounded plasma medium. When an incident wave enters a pre-existing plasma slab, the wave will experience space discontinuity. If the plasma frequency is lower than the incident wave frequency, then the incident wave will be partially reflected and transmitted. When the plasma frequency is higher than that of the incident wave, the wave is totally reflected because the dielectric constant in the plasma is less than zero. However, if the width of the plasma slab is sufficiently thin, the wave can then be transmitted, which is known as tunneling effect [1]. For this time-invariant plasma, the reflected and transmitted waves have the same frequency as the source wave frequency and we call these A waves. The wave inside the plasma has a different wave number but the same frequency due to the requirement of the boundary conditions.

When a source wave is propagating in free space and suddenly a plasma slab is created, the wave inside will experience a time discontinuity in the properties of the medium. Hence, the switching action generates new waves whose frequencies are upshifted and then the waves propagate out of the slab. We call these B waves. The phenomenon is illustrated in Figure 20.1. This figure shows the source wave of the frequency ω_0 propagating in the free space.

At $t = 0$, a plasma slab of the plasma frequency ω_p is created. The A waves in Figure 20.1 have the same frequency as that of the source wave. The B waves are newly created due to the creation of the plasma and upshifted to $\omega_1 = \sqrt{\omega_0^2 + \omega_p^2} = -\omega_2$. The negative value for the frequency of the second B wave shows that it is backward-propagating. These waves, however, have the same wave number as that of the source wave as long as they remain in the slab. As the B waves come out of the slab, the waves will encounter space discontinuity and therefore the wave number will change accordingly. The B waves are only created at the time of switching and, hence, exist for a finite time.

Analytical solution for a sudden switching of a plasma slab requires computationally expensive numerical inversion of Laplace transform. It is extensively studied as Kalluri has published a theoretical work on this problem using a Laplace transform method [2,3]. However, extension of such methods to the more realistic problems of plasma profiles of finite rise time and spatial inhomogeneities becomes very cumbersome. The finite-difference time-domain (FDTD) method discussed in this chapter handles such practical profiles with ease.

* Sections 20.1 through 20.4: Reprinted from *Int. J. Infrared Millim. Waves*, 24(3), 349–365, March 2003. With permission.

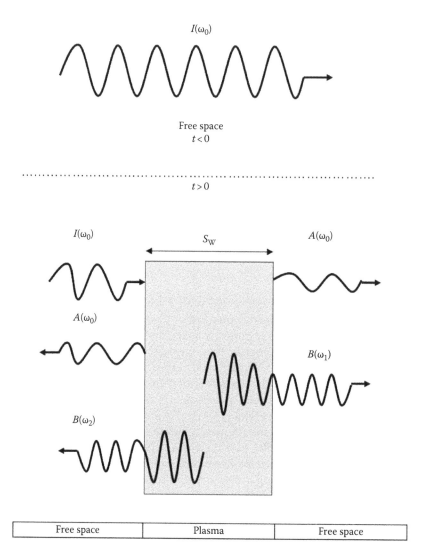

FIGURE 20.1
Effect of switching an isotropic plasma slab. *A* waves have the same frequency as the incident wave frequency (ω_0) and *B* waves have upshifted frequency $\omega_1 = \left(\omega_0^2 + \omega_p^2\right)^{1/2} = -\omega_2$. (Adapted from Kalluri, D.K., *IEEE Trans. Plasma Sci.*, 16(1), 11, 1988.)

20.2 Development of FDTD Equations

Consider a continuous source wave of frequency ω_0 propagating in free space. At $t = 0$, a plasma slab is created with a spatial distribution. For this problem, the condition of an infinite unbounded space cannot be assumed since after switching, the plasma medium is of finite extent with a defined spatial plasma density profile. Hence, space and time domain considerations must be taken into account in FDTD simulation.

20.2.1 Total-Field and Scattered-Field Formulation

The consideration of the space formulation in the FDTD method brings out the realization of a continuous source wave. Unless the source is located at a very distant place, the reflected wave by the plasma slab will eventually arrive at the source and corrupt the source wave. Consider a sinusoidal source wave at a grid location *s*:

$$E\big|_s^n = \exp\big[j\big(\omega_0 n\Delta t - k_0 s\Delta z\big)\big]. \tag{20.1}$$

This wave propagates to the region of interest and is eventually reflected back to the source location unless the source wave is located at a very distant position. The source wave behaves as a *hard* source [3] and prevents the movement of the reflected wave from passing and propagating toward infinity. Consequently, the hard source causes nonphysical reflection of the scattered wave to the system of interest. In order to simulate the source condition properly, the source must be hidden to the reflected wave, yet continuously feed the system. This is realized by introducing the concept of total-field/scattered-field formulation [4]. This approach is based on the linearity of Maxwell's equations and decomposition of the electric field and magnetic field as

$$E_{\text{tot}} = E_{\text{inc}} + E_{\text{scat}}, \tag{20.2}$$

$$H_{\text{tot}} = H_{\text{inc}} + H_{\text{scat}}, \tag{20.3}$$

where E_{inc} and H_{inc} are incident fields and assumed to be known at all space points of the FDTD grid at all time steps. E_{scat} and H_{scat} are scattered fields (either reflected wave or transmitted wave) and unknown initially. To implement this idea, we divide the computational domain into two regions, Regions 1 and 2, as shown in Figure 20.2.

Region 1 is the total-field region and the plasma slab is embedded within this region. The incident wave is initially assigned to the total field. Region 2 is the scattered-field region and is located in the free space.

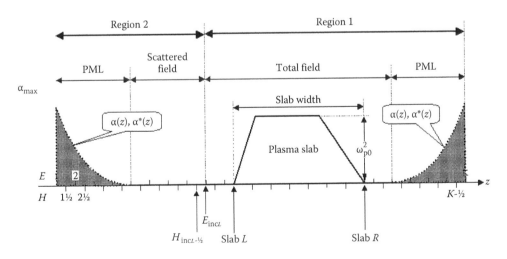

FIGURE 20.2
Configuration of geometry.

The FDTD formulation at the interface of Regions 1 and 2 requires various field components E and H from both regions to advance one time step. Let us assume that the interface belongs to Region 1 and the electric field component $E_\text{tot}|_L^{n+1}$ lies on the interface as indicated by $E_{\text{inc}L}$ in Figure 20.2. The FDTD equation for this component is given by

$$E_\text{tot}\Big|_L^{n+1} = E_\text{tot}\Big|_L^n - \frac{\Delta t}{\varepsilon_0 \Delta z}\left(H_\text{tot}\Big|_{L+(1/2)}^{n+(1/2)} - H_\text{tot}\Big|_{L-(1/2)}^{n+(1/2)}\right). \tag{20.4}$$

However, $H_\text{tot}|_{L-(1/2)}^{n+(1/2)}$ is located in Region 2 and is not defined. At this location $(L-(1)/(2))$, the scattered field $H_\text{scat}|_{L-(1/2)}^{n+(1/2)}$ is available. Hence, Equation 20.4 needs some modification in such a way that

$$E_\text{tot}\Big|_L^{n+1} = E_\text{tot}\Big|_L^n - \frac{\Delta t}{\varepsilon_0 \Delta z}\left(H_\text{tot}\Big|_{L+(1/2)}^{n+(1/2)} - H_\text{scat}\Big|_{L-(1/2)}^{n+(1/2)}\right) + \frac{\Delta t}{\varepsilon_0 \Delta z}H_\text{inc}\Big|_{L-(1/2)}^{n+(1/2)}. \tag{20.5}$$

Since

$$H_\text{tot}\Big|_{L-(1/2)}^{n+(1/2)} = H_\text{scat}\Big|_{L-(1/2)}^{n+(1/2)} + H_\text{inc}\Big|_{L-(1/2)}^{n+(1/2)}, \tag{20.6}$$

where $H_\text{inc}|_{L-(1/2)}^{n+(1/2)}$ is assumed to be known as mentioned earlier.

In a similar way, the formulation for the H field component at $L-(2)/(2)$ needs some modification. At this space point, H field component is the scattered field because it is in Region 2. The FDTD equation is written as

$$H_\text{scat}\Big|_{L-(1/2)}^{n+(1/2)} = H_\text{scat}\Big|_{L-(1/2)}^{n-(1/2)} - \frac{\Delta t}{\mu_0 \Delta z}\left(E_\text{scat}\Big|_L^n - E_\text{scat}\Big|_{L-1}^n\right). \tag{20.7}$$

Again, $E_\text{scat}|_L^n$ is located in Region 1 and is therefore not defined. The modification of Equation 20.7 is made using known quantities by

$$H_\text{scat}\Big|_{L-(1/2)}^{n+(1/2)} = H_\text{scat}\Big|_{L-(1/2)}^{n-(1/2)} - \frac{\Delta t}{\mu_0 \Delta z}\left(E_\text{tot}\Big|_L^n - E_\text{scat}\Big|_{L-1}^n\right) + \frac{\Delta t}{\mu_0 \Delta z}E_\text{inc}\Big|_L^n. \tag{20.8}$$

Since

$$E_\text{scat}\Big|_L^n = E_\text{tot}\Big|_L^n - E_\text{inc}\Big|_L^n. \tag{20.9}$$

In short, the concept of splitting the computational domain into Regions 1 (total-field region) and 2 (scattered-field region) is to provide the incident field (source wave) to Region 1 (where materials of interest are embedded) and to separate the scattered-field out from the total field by canceling it with the incident field at the interface. In this simulation, Region 1 extends to the end of the PML on the RHS.

20.2.2 Lattice Truncation: PML

The reflected and transmitted waves will propagate to the ends of the computational boundaries. To prevent numerical reflection of the waves from the boundaries of the computational domain, an absorption boundary condition such as PML is needed. The concept of the absorption boundary condition is that when an electromagnetic wave enters a lossy medium, reflection by the lossy medium will not occur if the impedance matching condition is satisfied. Consider Maxwell's equations in a lossy medium for an R wave propagation:

$$\frac{\partial H}{\partial t} + \sigma^* H = -\frac{1}{\mu_0} \frac{\partial E}{\partial z}, \tag{20.10}$$

$$\frac{\partial E}{\partial t} + \sigma E = -\frac{1}{\varepsilon_0} \frac{\partial H}{\partial z}, \tag{20.11}$$

where σ is the electric conductivity and σ^* the magnetic conductivity. The magnetic conductivity σ^* is not a physical quantity but is introduced for the requirement of impedance matching. Since the PMLs are positioned in the boundaries that are in the free space, the introduction of the magnetic conductivity will not change the physics of the region of interest. The impedance matching condition [5] is well known: $\sigma/\varepsilon_0 = \sigma^*/\mu_0$. To simplify computation, we assume that $\varepsilon_0 = \mu_0 = 1$. This assumption implies normalization of the variables. For example, the velocities are all normalized with respect to the velocity of light in the free space $c = 3 \times 10^8$ m/s. The impedance matching condition then simplifies to $\sigma = \sigma^*$.

As the wave propagates in the lossy medium, the wave will decay. However, if the PML is finite, then the wave will arrive at the boundary and reflect back to the system. This reflection is generally very weak and can be ignored. In numerical calculation, however, there will be numerical reflection at each space step since the unit space and time steps are finite and therefore the source wave will experience sudden changes in the conductivity, resulting in numerical reflection. To reduce this reflection, the conductivity must vary smoothly from zero to a maximum value and numerical simulation shows that the following quadratic form gives very low reflection [5]:

$$\sigma = \sigma^* = \sigma_{max} \left(\frac{z}{W} \right)^2, \tag{20.12}$$

where z is measured from the beginning of the PML from inside direction and W is the width of the PML. The maximum value of the conductivity is determined to obtain the minimum reflection from PML.

20.2.3 FDTD Formulation for an *R* Wave in a Switched Plasma Slab

Maxwell's equations containing a damping constant ν for an R wave [6] propagation in a magnetized plasma medium are written as

$$\frac{\partial H}{\partial t} + \sigma H = -\frac{1}{\mu_0} \frac{\partial E}{\partial z}, \tag{20.13}$$

$$\frac{\partial E}{\partial t} + \sigma E = -\frac{1}{\varepsilon_0} \frac{\partial H}{\partial z} - \frac{1}{\varepsilon_0} J, \tag{20.14}$$

$$\frac{dJ}{dt} + \nu J = \varepsilon_0 \omega_p^2 (z,t) E. \tag{20.15}$$

In the above, ω_p is the plasma frequency and ν the collision frequency [1].

The FDTD equations of Equations 20.13 through 20.15 can be written as

$$H\big|_{k+(1/2)}^{n+(1/2)} = e^{-\sigma\Delta t}\, H\big|_{k+(1/2)}^{n-(1/2)} - \frac{1}{\mu_0\sigma\Delta z}\left(1-e^{-\sigma\Delta t}\right)\left(E\big|_{k+1}^{n} - E\big|_{k}^{n}\right),\tag{20.16}$$

$$J\big|_{k}^{n+(1/2)} = e^{(-\nu)\Delta t}\, J\big|_{k}^{n-(1/2)} + \frac{\varepsilon_0}{\nu}\left(1-e^{(-\nu)\Delta t}\right)\omega_p^2\big|_{k}^{n}\, E\big|_{k}^{n},\tag{20.17}$$

$$E\big|_{k}^{n+1} = e^{-\sigma\Delta t}\, E\big|_{k}^{n} - \frac{1}{\varepsilon_0\sigma}\left(1-e^{-\sigma\Delta t}\right)\left[\frac{1}{\Delta z}\left(H\big|_{k+(1/2)}^{n+(1/2)} - H\big|_{k-(1/2)}^{n+(1/2)}\right) + J\big|_{k}^{n+(1/2)}\right].\tag{20.18}$$

In Equation 20.17, $\nu = 0$ when the plasma is considered lossless. Numerical results obtained in the next two sections are based on such a consideration.

The algorithm given in Equations 20.16 through 20.18 is implemented in various media as follows:

$$\text{Free space}: \sigma = 0, \quad J = 0, \quad \omega_p^2 = 0,$$
$$\text{Plasma slab}: \sigma = 0,$$
$$\text{PML}: J = 0, \quad \omega_p^2 = 0.$$

It is noted that the following approximation can be made for small σ:

$$\lim_{\sigma\to 0}\frac{\left(1-e^{-\sigma\Delta t}\right)}{\sigma} = \Delta t.\tag{20.19}$$

20.3 Interaction of a Continuous Wave with a Switched Plasma Slab

Figure 20.3 shows profiles of plasma creation in space and time as (a) plasma frequency increases with a rise time T_r to the maximum values ω_{p0}^2 and (b) spatial distribution of the square of the plasma frequency, constant over a width $(S_W - S_L - S_R)$, a linear rising profile of width (S_L), and a linear declining profile of width (S_R).

A result of the transient wave on the RHS of the slab of $\omega_0 = 0.8$, $\omega_p = 1.0$, the slab width $S_W = 4.0\pi/\omega_p$, $S_L = 0$, $S_R = 0$, and the rise time $T_r = 0$ is shown in Figure 20.4. The real part of the electric field component at "SlabR" in Figure 20.2 is shown and this wave will disappear quickly. Since the source wave frequency is less than the plasma frequency and the

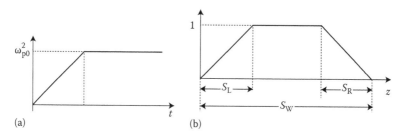

FIGURE 20.3
Plasma profiles in space and time domain. (a) Time profile and (b) space profile.

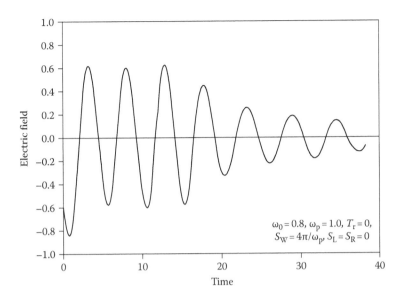

FIGURE 20.4
A transient wave due to the creation of a plasma slab.

slab is sufficiently large, only the transmitted wave exists and soon propagates away. The result is in good agreement with [1]. The angular frequency of this wave is $\sqrt{\omega_0^2 + \omega_p^2} = 1.28$, as expected.

20.4 Interaction of a Pulsed Wave with a Switched Plasma Slab

For the case of pulsed signals, the interaction of newly created plasma slab is easily shown. Consider a Gaussian pulse with the frequency ω_0 in the free space as

$$E(z,t) = \exp\left[-\left(\frac{\omega_0 t - k_0(z - z_0)}{k_0 z_w / 2}\right)^2 + j(\omega_0 t - k_0 z)\right], \qquad (20.20)$$

where z_0 is the center of the pulse at $t = 0$ and z_w the pulse width. When a pulsed wave is propagating in the free space along the z-axis, a plasma slab is created, which encloses the whole pulse. The wave will then interact with the plasma slab. The interaction will split the pulse into a forward-propagating wave and a backward-propagating wave. The magnetic field component of the pulse will have an additional mode, that is, the wiggler mode whose frequency is zero. Some of the results for the sudden creation of the plasma slab are shown in Figures 20.5 and 20.6. Parameters are given as $\omega_{p0} = 1.5\omega_0$, $T_r = 0$, $S_W = 16\lambda_0$, $S_L = 0.2S_W$, $S_R = 0.3S_W$, and the pulse width $Z_W = 2\lambda_0$. The spatial distribution of the plasma slab after its creation is illustrated in a dotted line. The electric field component is shown in Figure 20.5. The electric field in space domain at different times is presented. The electric field at $t = 0^+$ is the initial pulse shape. At $t = 5T_0$ (here, $T_0 = 2\pi/\omega_0$), the pulse is split into two waves, where one is a forward-propagating wave and the other is a backward-propagating wave.

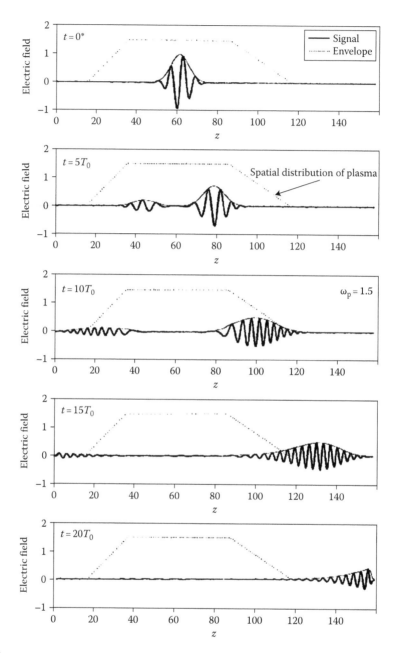

FIGURE 20.5
The time evolution of the electric field after sudden creation of plasma. $T_r = 0$, $\omega_p = 15\omega_0$, $z_w = \lambda_0$, $S_w = 16\lambda_0$, $S_L = 0.25S_W$, and $S_R = 0.35S_W$.

They appear to have same spatial frequency as the initial wave. When the waves reach the plasma slab boundaries, their frequencies get upshifted ($t = 10T_0$ and $t = 15T_0$). The reflection of the pulse at the plasma boundaries is not noticeable due to the gradual changes of plasma slab in space. It is noted that the pulses broaden as they come out of the slab. The upshifted frequency of the pulse is given by $\sqrt{\omega_0^2 + \omega_p^2} = 1.8$, which is the same as that of the

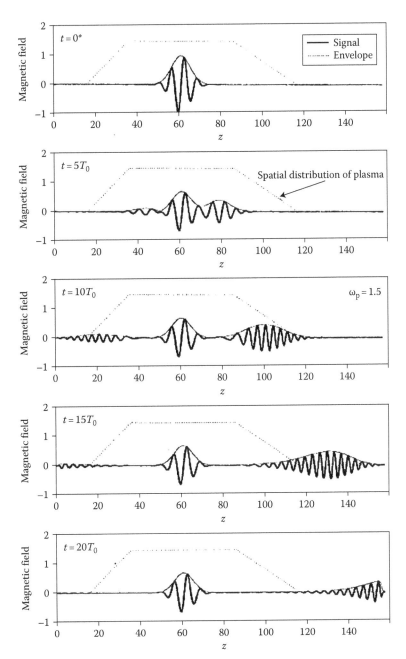

FIGURE 20.6
The time evolution of the magnetic field after sudden creation of plasma. $T_r = 0$, $\omega_p = 15\omega_0$, $z_w = \lambda_0$, $S_W = 16\lambda_0$, $S_L = 0.2S_W$, and $S_R = 0.3S_W$.

unbounded plasma case. The magnetic component of the pulse in Figure 20.6 has an extra wave sitting at the initial position. This wave is a wiggler mode whose frequency is zero.

Figure 20.7 shows the electric fields of the forward-propagating waves in time outside the slab on the RHS for various plasma frequencies, $\omega_{p0} = 0$, $\omega_{p0} = 0.5\omega_0$, $\omega_{p0} = 1\omega_0$, $\omega_{p0} = 2\omega_0$, and $\omega_{p0} = 4\omega_0$ from top to bottom, respectively, and pulse width $z_w = 2\lambda_0$. The slab parameters are

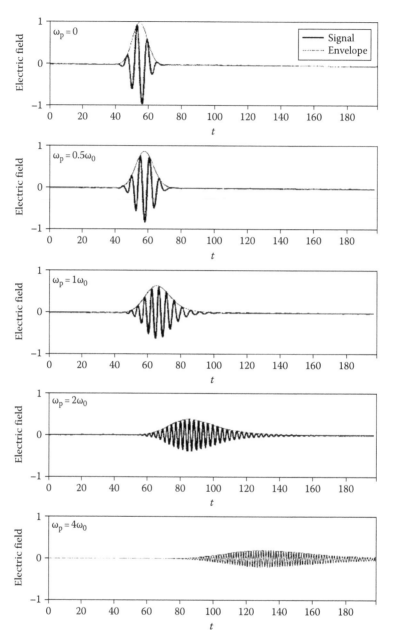

FIGURE 20.7
The electric fields on the RHS of the plasma slab for various plasma frequencies. $T_r = 2T_0$, $z_w = \lambda_0$, $S_W = 16\lambda_0$, $S_L = 0.2S_W$, and $S_R = 0.3S_W$.

$S_W = 16\lambda_0$, $S_L = 0.2S_W$, and $S_R = 0.3S_W$, where the gradual changes in the boundaries of the slab reduce the reflection of the pulse. The plasma slab is created with rise time $T_r = 2T_0$ and this will suppress the generation of backward propagation of the pulse. With these parameters, the pulse will keep its shape but become a higher-frequency pulse as the plasma frequency increases. The widths of the pulse decrease as ω_{p0} increases and it can be understood as pulse broadening. The pulse broadening shows the clear nature of the dispersive characteristics of

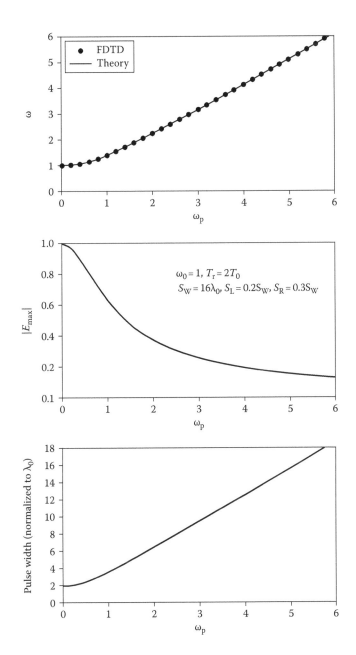

FIGURE 20.8
Characteristics of pulses of the electric fields shown in Figure 20.7.

plasma medium. Since the group velocity depends on the plasma frequency as $v_g = c\omega_0 / \sqrt{\omega_0^2 + \omega_p^2}$ and as the plasma frequency becomes higher the pulse is more delayed accordingly. The plasma creation does not provide any additional energy to the pulse and therefore the amplitude of the wave decreases as the pulse broadens. The frequency, the peak amplitude, and the pulse width of the electric field component of the pulse outside at the right edge of the plasma slab are shown in Figure 20.8. The pulse duration is measured by locating points on the envelope function which are e^{-1} ($= 0.368$) of the maximum value. The frequency of the pulse is the same as that of the unbounded plasma creation, that is, $\omega = \sqrt{\omega_0^2 + \omega_{p0}^2}$.

20.5 Interaction of a Pulsed Wave with a Switched Magnetoplasma Slab

Assume that a static magnetic field exists in the z-direction before the plasma slab is switched on in a manner similar to the previous section. For R wave propagation, in the absence of collisions (from 11.17 of Volume 1) Equation 20.15 gets modified as given in (20.21):

$$\frac{dJ}{dt} = \varepsilon_0 \omega_p^2 (z,t) E + j\omega_b J. \tag{20.21}$$

Consequently (20.17) gets modified as [6]

$$J\Big|_k^{n+\frac{1}{2}} = e^{j\omega_b \Delta t} J\Big|_k^{n-\frac{1}{2}} + \Delta t \varepsilon_0 \left(e^{j\omega_b \Delta t/2}\right) \operatorname{sin} c\left(\frac{\omega_b \Delta t}{2}\right) \omega_p^2\Big|_k^n E\Big|_k^n. \tag{20.22}$$

The plasma slab is instantaneously created ($T_r = 0$) at time $t = 0$ when the pulse is at the center of the slab location. The movement of the various wave modes is analyzed as a function of the background magnetic field.

Figure 20.9 shows the electric field and magnetic field as a function of position at various time snapshots for $\omega_b = 0.5$. Comparing Figures 20.5 and 20.6 with Figure 20.9, it can

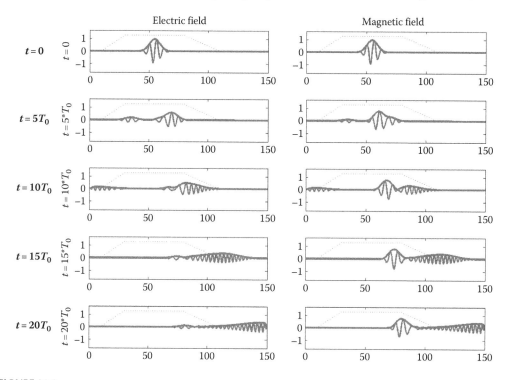

FIGURE 20.9
Electromagnetic pulse interaction with a suddenly created magnetoplasma slab ($\omega_b = 0.5$, R-wave), E and H fields versus space and time.

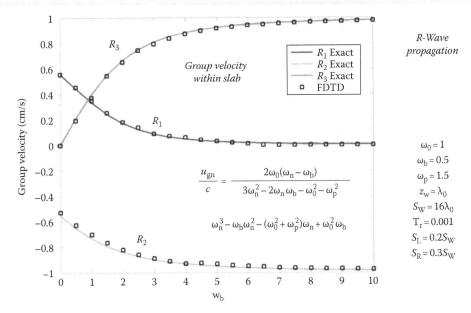

Electromagnetic pulse interaction with a suddenly created
magnetoplasma slab (group velocity vs. ω_b)

Analysis demonstrates that FDTD analysis of plasma slab creation agrees very well with the theory of sudden plasma creation in an unbounded medium

FIGURE 20.10
Comparison of pulse group velocity as determined by FDTD against theoretically calculated values.

be seen that the wiggler magnetic field has transitioned to a propagating whistler wave when the switched medium is a magnetoplasma [7].

The group velocity of each of the R waves was computed using the FDTD analysis and compared to theoretical values [1] in a magnetoplasma as a function of ω_b.

The results are shown to agree very well with theory as shown in Figure 20.10 [7]. In computing the group velocity of each of the modes, ω_n is obtained as the roots of the polynomial in ω_n given in the figure [8, Appendix B].

References

1. Kalluri, D. K., *Electromagnetics of Complex Media*, CRC Press, Boca Raton, FL, 1998.
2. Kalluri, D. K., On reflection from a suddenly created plasma half-space: Transient solution, *IEEE Trans. Plasma Sci.*, 16(1), 11–16, 1988.
3. Kalluri, D. K. and Goteti, V. R., Frequency shifting of electromagnetic radiation by sudden-creation of a plasma slab, *J. Appl. Phys.*, 72(10), 4575–4580, 1992.
4. Taflove, A., *Computational Electrodynamics: The Finite Difference Time Domain Method*, Artech House, Boston, MA, 1995.
5. Berenger, J. P., A perfectly matched layer for the absorption of electromagnetic waves, *J. Comput. Phys.*, 114, 185–200, 1994.

6. Lee, J. H., Kalluri, D. K., and Nigg, G. C., FDTD simulation of electromagnetic transformation in a dynamic magnetized plasma, *Int. J. Infrared Millimeter Waves*, 21(8), 1223–1253, 2000.
7. Lade, R. K., Electromagnetic wave interactions in a time varying medium: Appropriate and different models for a building-up and collapsing magnetoplasma medium, PhD thesis, University of Massachusetts Lowell, Lowell, MA, 2012.
8. Kalluri, D. K., *Electromagnetics of Time Varying Complex Media*, 2nd edition, CRC Press, Taylor & Francis Group, Boca Raton, FL, 2010.

21

Radiation and Scattering: Exterior Problems

21.1 Introduction

In Sections 7.5 through 7.7 of Volume 1, we discussed radiation, scattering, and diffraction from a theoretical viewpoint. These classes of problems are called exterior problems in the sense that the field region of interest includes the sphere of infinite radius. However, we assume that the sources and obstructing objects/scatterers can all be included in a sphere of finite radius (in three-dimensional problems). Theoretical and implied boundary conditions at the radius r equal to infinity are called Sommerfeld radiation boundary conditions; in simple language, they state that at infinity you can have only outgoing waves. Two-dimensional versions of these boundary conditions (BC) are given in Example 17.3 of Section 17.4.2.

Computer solutions of exterior problems by FEM or FDTD require the determination of a computational domain whose edge is the outer limit of the points included in the field region of the electromagnetic problem under investigation. If we impose the Dirichlet BC at the edge of the computational domain, it is equivalent to having a perfect electric conductor (PEC) at the edge of the computational domain giving rise to reflections that are not warranted by physics of the problem; these reflections (incoming waves) occur due to the termination of the computational domain and are part of the numerical errors of computer simulation on computers with finite memory. Instead of Dirichlet BC at the edge, if we are able to come up with an additional medium of some thickness whose inner wall is the edge of the computational domain and if there is impedance match at this edge, the wave transmits into the additional medium without reflection. As it propagates into the additional medium whose conductivity gradually increases with propagating distance, the wave will be absorbed. A Dirichlet BC at the outer edge of the additional medium will then cause very little reflection. Even if there is residual amplitude, the reflected wave traveling back into the computational domain is further absorbed by the additional medium, thus minimizing the corruption of the numerical data. The additional medium facilitates the absorption of the wave and implements numerically the Sommerfeld radiation boundary condition. In other words, the additional medium implements the absorbing boundary condition (ABC). This additional medium is called perfectly matched layer (PML).

Design of the PML for one-dimensional problems is much easier than for two-dimensional and three-dimensional problems. Illustration of one-dimensional PML is given in Section 20.2.2. In Section 21.2, we will develop the current theory of PML for designing two-dimensional and three-dimensional PML.

A remark on the advantage of using moment method (MM) over FEM or FDTD method for exterior problems is in order. Radiation BC is automatically taken care of in MM method as illustrated in Section 16.8. However, the disadvantage of MM is that the matrix to be inverted to get numerical results will be a full matrix (have a large number of nonzero elements). For example, the solution of Problem P16.15 (b) requires the inversion of a full matrix of the size (72×72).

21.2 Perfectly Matched Layer

Let us review briefly the concept of perfectly matched layer (PML) using the one-dimensional PML used in Section 20.2.2. The PML has conductivity that is zero at $z = 0$. Since the PML layer is in free space and the medium to the left of PML is free space, there is a perfect match of impedances at $z = 0$. The conductivity increases with z as the wave travels into the layer as given in (20.12). Even if we terminate the outer edge of PML with a Dirichlet BC at $z = W$, it does not matter since the wave is essentially absorbed by the time it reaches the outer PML boundary. No doubt, the Dirichlet BC causes perfect reflection of the highly attenuated wave, which, however, will be further attenuated on its travel toward the inner edge of the PML layer ($z = 0$). One can adjust W and the power to which z/W in (20.12) is raised to accommodate the required attenuation of the wave that reaches the outer boundary of PML.

The problem in Section 20.2 dealt with normal incidence. The PML for oblique incidence, for example of a p wave, is more involved than that of the normal incidence. As discussed in Section 2.12 of Volume 1, total transmission at the inner edge of PML can occur only at one angle of incidence (Brewster angle). So, one can have matched impedance at the inner edge of the PML of only one ray that is incident on the interface at Brewster angle. PML of two-dimensional and three-dimensional require impedance match that is independent of wave polarization, frequency, and the angle of incidence. Berenger [1] in the seminal paper in 1994 came up with such a solution proposing for the two-dimensional PML, an artificial anisotropic absorbing medium. He used the concept of split fields for the Cartesian components. Many advancements have taken place since then, leading to the current versions of the popular implementation of PML. A limited list of references leading to the PML explained in this book is given in [2–8]. Sections 21.2.1 and 21.2.2 deal with such explanations.

21.2.1 Uniaxial Perfectly Matched Layer

The purpose of the PML layer is to provide a material layer with which the computational domain can be terminated. The layer can be an artificial layer with proposed material properties that provide the absorption of the EM wave without causing reflections. In this section we explain the properties of uniaxial perfectly matched layer (UPML) [4,5,8] and show how such a layer can achieve the desired properties within a prescribed numerical error.

Wave propagation in a uniaxial crystal, in the context of photonic crystals, is discussed at length in Chapter 12 of Volume 1. Let us assume that the *artificial* PML is a biaxial

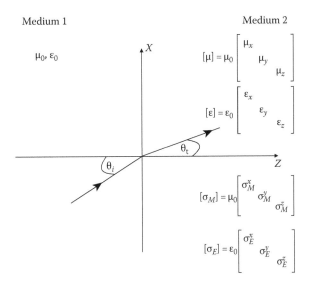

FIGURE 21.1
Medium 2 is a biaxial anisotropic medium, which also has a magnetic conductivity and acts an artificial perfectly matched layer.

anisotropic medium, which also has magnetic conductivity. The parameters of such a medium are marked as those of medium 2 in Figure 21.1.

Maxwell's equations in the frequency domain of such a medium are given in (21.1) through (21.4):

$$\nabla \cdot [\varepsilon] \tilde{\mathbf{E}} = 0 \tag{21.1}$$

$$\nabla \cdot [\mu] \tilde{\mathbf{H}} = 0 \tag{21.2}$$

$$\nabla \times \tilde{\mathbf{E}} = -j\omega [\mu] \tilde{\mathbf{H}} - [\sigma_M] \tilde{\mathbf{H}} \tag{21.3}$$

$$\nabla \times \tilde{\mathbf{H}} = j\omega [\varepsilon] \tilde{\mathbf{E}} + [\sigma_E] \tilde{\mathbf{E}} \tag{21.4}$$

In the above, $[p]$ stands for biaxial electromagnetic parameter tensor, a diagonal matrix of p, and the symbol p stands for permeability μ in (21.2) or permittivity ε in (21.1) or magnetic conductivity σ_M in (21.3) or electric conductivity σ_E in (21.4). See Figure 21.1.

Let

$$[\tilde{\mu}] = \mu_0 \begin{bmatrix} \mu_x + \dfrac{\sigma_M^x}{j\omega} & 0 & 0 \\ 0 & \mu_y + \dfrac{\sigma_M^y}{j\omega} & 0 \\ 0 & 0 & \mu_z + \dfrac{\sigma_M^z}{j\omega} \end{bmatrix} \tag{21.5}$$

and

$$[\breve{\varepsilon}] = \varepsilon_0 \begin{bmatrix} \varepsilon_x + \dfrac{\sigma_E^x}{j\omega} & 0 & 0 \\ 0 & \varepsilon_y + \dfrac{\sigma_E^y}{j\omega} & 0 \\ 0 & 0 & \varepsilon_z + \dfrac{\sigma_E^z}{j\omega} \end{bmatrix}. \tag{21.6}$$

The parameter in [.] in (21.5) is the equivalent complex relative permeability and the parameter [.] in (21.6) is the complex relative permittivity of the proposed biaxial medium for the PML. Equations 21.3 and 21.4 can be rewritten more compactly using (21.5) and (21.6):

$$\nabla \times \tilde{\mathbf{E}} = -j\omega [\breve{\mu}] \tilde{\mathbf{H}} \tag{21.7}$$

$$\nabla \times \tilde{\mathbf{H}} = j\omega [\breve{\varepsilon}] \tilde{\mathbf{E}}. \tag{21.8}$$

The necessary condition of impedance matching for zero reflection from the PML will be met by assuming

$$\frac{[\breve{\varepsilon}]}{\varepsilon_0} = \frac{[\breve{\mu}]}{\mu_0} = [s] = \begin{bmatrix} a & 0 & 0 \\ 0 & b & 0 \\ 0 & 0 & c \end{bmatrix}. \tag{21.9}$$

Propagation of plane waves in such a medium can be studied by assuming

$$\tilde{\mathbf{F}} = Fe^{-jk \cdot r}, \tag{21.10}$$

where $\tilde{\mathbf{F}}$ stands for the electric field in (21.7) and the magnetic field in (21.8). One can now obtain the dispersion relation [4]

$$\frac{k_x^2}{s_x^2} + \frac{k_y^2}{s_y^2} + \frac{k_z^2}{s_z^2} = \omega^2 \mu \varepsilon, \tag{21.11}$$

where k_0 is the free space wave number and

$$k_x = k_0 \sqrt{bc} \sin\theta \cos\phi \tag{21.12}$$

$$k_y = k_0 \sqrt{ac} \sin\theta \sin\phi \tag{21.13}$$

$$k_z = k_0 \sqrt{ab} \cos\theta. \tag{21.14}$$

Let us now illustrate the use of the analysis given above to compute the reflection coefficient for oblique incidence. Figure 21.1 gives the geometry of the problem. Medium 1 is free space and medium 2 is the biaxial PML under discussion. The incident wave propagating in the x–z plane is incident on the interface $z = 0$ at an incident angle of θ_i. Snell's law requires

$$k_0 \sin \theta_i = k_{x1} = k_{x2} = k_0 \sqrt{bc} \sin \theta_t, \tag{21.15}$$

where θ_t is the angle of reflection. From (21.15), for a choice of $b = 1/c$, the refraction angle will become equal to the incident angle:

$$\theta_i = \theta_t = \theta, \quad \text{if } b = \frac{1}{c}. \tag{21.16}$$

Sections 2.12 and 2.13 of Volume 1 deal with the reflection coefficients for the p wave and the s wave, respectively, when the medium 2 is an isotropic medium. The reflection coefficients when the medium 2 is biaxial can be obtained on similar lines and discussed in [4]. The results are

$$\Gamma_p = \frac{\sqrt{b/a} \cos \theta_t - \cos \theta_i}{\sqrt{b/a} \cos \theta_t + \cos \theta_i} \tag{21.17}$$

$$\Gamma_s = \frac{\sqrt{b/a}/\cos \theta_t - 1/\cos \theta_i}{\sqrt{b/a}/\cos \theta_t + 1/\cos \theta_i}. \tag{21.18}$$

Note that the reflection coefficients become zero if $a = b$, which is true for a uniaxial medium. We thus show that if [s] in (21.9) to be denoted by [s_z] is given by

$$[s_z] = \begin{bmatrix} s_z & 0 & 0 \\ 0 & s_z & 0 \\ 0 & 0 & 1/s_z \end{bmatrix}, \tag{21.19}$$

a ray incident on the interface $z = 0$ will be totally transmitted into medium 2 irrespective of its angle of incidence or polarization.

The second requirement of the PML is that the totally transmitted wave be absorbed as the wave travels to the outer edge of the PML. The wave attenuation is tied to the complex nature of s_z in (21.19) given by

$$s_z = \alpha - j\beta. \tag{21.20}$$

The field variable $\tilde{\mathbf{F}}$ in (21.10) can now be written as

$$\tilde{\mathbf{F}} = Fe^{-k_0 \beta z \cos \theta} e^{-jk_0 (x \sin \theta + \alpha z \cos \theta)}. \tag{21.21}$$

Thus, β controls the damping of the wave and α determines the wavelength of the wave in the PML. If we now assume that the outer edge of the PML is terminated by a PEC at $z = W$, the wave is thus damped by a factor $R(\theta)$ when it reenters the computational domain through the inner edge of the PML at $z = 0$:

$$R(\theta) = e^{-2k_0 \beta W \cos \theta}. \tag{21.22}$$

The tensorial material properties $[\varepsilon]$, $[\mu]$, $[\sigma_E]$, $[\sigma_M]$ in (21.3) and (21.4) of the uniaxial PML can be written [4] by separating the real and imaginary parts of $[s_z]$:

$$[\varepsilon] = \varepsilon_0 [P] \tag{21.23}$$

$$[\mu] = \mu_0 [P] \tag{21.24}$$

$$[\sigma_E] = \omega \varepsilon_0 [Q] \tag{21.25}$$

$$[\sigma_M] = \omega \mu_0 [Q], \tag{21.26}$$

where

$$[P] = \begin{bmatrix} \alpha & 0 & 0 \\ 0 & \alpha & 0 \\ 0 & 0 & \dfrac{\alpha}{\alpha^2 + \beta^2} \end{bmatrix} \tag{21.27}$$

and

$$[Q] = \begin{bmatrix} \beta & 0 & 0 \\ 0 & \beta & 0 \\ 0 & 0 & -\dfrac{\beta}{\alpha^2 + \beta^2} \end{bmatrix}. \tag{21.28}$$

The negative sign in the element Q_{33} shows that the z component of the conductivity tensors is negative requiring the imagination of dependent sources in the PML. Since UPML is a layer of an artificial medium imagined for the purpose of terminating the computational domain, physical realizability is not a problem.

For total transmission across the interface at $x = 0$, we need an uniaxial PML whose $[s]$ factor, to be denoted by $[s_x]$, is given by

$$[s_x] = \begin{bmatrix} s_x & 0 & 0 \\ 0 & s_x & 0 \\ 0 & 0 & 1/s_x \end{bmatrix} \tag{21.29}$$

and for total transmission across the interface $y = 0$, we need an uniaxial PML whose $[s]$ factor, to be denoted by $[s_y]$, is given by

$$[s_y] = \begin{bmatrix} s_y & 0 & 0 \\ 0 & s_y & 0 \\ 0 & 0 & 1/s_y \end{bmatrix}. \tag{21.30}$$

21.2.2 Stretched Coordinate Perfectly Matched Layer

UPML is a layer of artificial medium with assigned tensorial material properties attached to a computational domain to act as a reflectionless absorbing layer to simulate Sommerfeld radiation conditions. The absorption occurs since the k vector in (21.10) is complex. For the specific case of a PML interface at $z = 0$, k_z is complex and k_x is real, giving rise to the damping term $k_0 \beta z \cos \theta$ in (21.21). An alternative way of obtaining the damping term is to keep k real in (21.10) but consider the position vector to be complex, that is, replace $\mathbf{r}$ by $\check{\mathbf{r}}$ in (21.10). This removes the need of construing an artificial uniaxial medium but requires a change into complex space of the spatial coordinates in the PML layer. We will show next that the change of variables given in (21.31) will satisfy [3] the requirement of absorption without reflection in the PM layer:

$$\check{x} = \int_0^x s_x(x')dx' \tag{21.31}$$

$$\check{y} = \int_0^y s_y(y')dy' \tag{21.32}$$

$$\check{z} = \int_0^z s_z(z')dz'. \tag{21.33}$$

From (21.31), the variable $\check{x}$, which is a definite integral, is the stretched variable of x and can be complex if the stretch parameter s_x is complex. Similar remarks apply to (21.32) and (21.33).

The gradient operator $\check{\nabla}$ in complex coordinate space can be written as

$$\check{\nabla} = \hat{x}\frac{1}{s_x}\frac{\partial}{\partial x} + \hat{y}\frac{1}{s_y}\frac{\partial}{\partial y} + \hat{z}\frac{1}{s_z}\frac{\partial}{\partial z}. \tag{21.34}$$

Equations 21.1 through 21.4 in the complex coordinate system in an ordinary material medium with permittivity ε and permeability μ can be written as

$$\check{\nabla} \cdot \varepsilon \tilde{\mathbf{E}} = 0 \tag{21.35}$$

$$\check{\nabla} \cdot \mu \tilde{\mathbf{H}} = 0 \tag{21.36}$$

$$\check{\nabla} \times \tilde{\mathbf{E}} = -j\omega\mu\tilde{\mathbf{H}} \tag{21.37}$$

$$\check{\nabla} \times \tilde{\mathbf{H}} = j\omega\varepsilon\tilde{\mathbf{E}}. \tag{21.38}$$

It can be shown that Equations 21.35 and 21.36 can be obtained from (21.37) and (21.38) provided the stretch factor s_i is a function of the i coordinate only, where i stands for x or y or z.

Plane wave solution of (21.35) through (21.38) of the type given by (21.10) can be obtained and it can be shown that [3]

$$\frac{k_x^2}{s_x^2} + \frac{k_y^2}{s_y^2} + \frac{k_z^2}{s_z^2} = \omega^2 \mu \varepsilon \tag{21.39}$$

and

$$\frac{E}{H} = \sqrt{\frac{\mu}{\varepsilon}}. \tag{21.40}$$

21.2.3 Complex Frequency-Shifted Perfectly Matched Layer

The UPML described in Section 21.2.1 can be viewed as an anisotropic medium and in the frequency domain, the constitutive relation can be written as

$$\tilde{\mathbf{D}}(\omega) = \left[\breve{\varepsilon}(\omega) \right] \tilde{\mathbf{E}}(\omega) \tag{21.41}$$

$$\tilde{\mathbf{B}}(\omega) = \left[\breve{\mu}(\omega) \right] \tilde{\mathbf{H}}(\omega), \tag{21.42}$$

where the expressions for $\left[\breve{\varepsilon}(\omega) \right]$ and $\left[\breve{\mu}(\omega) \right]$ can be obtained from (21.9). For the UPML with $z = 0$ interface the s factor s_z is given by (21.20); thus

$$\alpha = \varepsilon_z \tag{21.43}$$

$$\beta = \frac{\sigma_E^z}{\omega}. \tag{21.44}$$

For $\omega = 0$, β blows up; thus we expect problems when the signal to be absorbed has very low frequency components. Moreover, the real and imaginary parts of the constitutive relations of the medium need to satisfy the Kramer–Kronig relations given by (8.56) and (8.57) of Section 8.5 of Volume 1 to ensure the satisfaction of the causality [9].

A modification of the s_z factor given below,

$$s_z = \varepsilon_z + \frac{\sigma_E^z}{d_z + j\omega}, \tag{21.45}$$

solves several problems associated with the original UPML [9–11] including the absorption of evanescent waves as well as the low-frequency waves. Since the introduction of the new parameter d_z is equivalent to shifting to a complex frequency, the method is called complex frequency shifted-perfectly matched layer (CFS-PML).

21.3 Near-Field to Far-Field Transformation

The PML discussed in Section 21.2 allows us to cut down the computational domain considerably. However, the calculated fields inside the computational domain can still be in the near-field region of the radiating/scattering object. For example, we can be

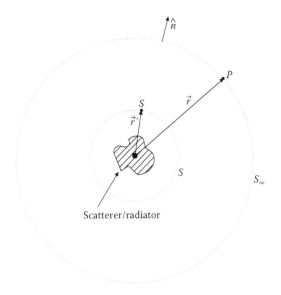

FIGURE 21.2
Geometry for near-field to far-field (NFTFF) transformation. From the known field on *s*, which falls in the near-field zone of the scatterer/radiator, the transformation facilitates calculation of fields in the far-field zone. In the figure *s* is shown to be a sphere; however, in practice one can choose any closed surface that is suitable based on the available data of the fields on the surface.

interested in the radar cross section of the scatterer as defined in (7.140) or (7.141) of Volume 1 of this book.

The topic of this section is to come up with the recipe for determining the far field, from the computed values of the electric and magnetic fields in the near field of the scatterer/radiator. References 12–15 can be consulted for more detailed discussions.

Figure 21.2 shows the geometry for discussing the transformation.

The electromagnetic theory needed is given earlier in Section 7.7 of Volume 1 of this book. Some of the equations given in that section are repeated here for convenience. If the electric field $\tilde{E}(r')$ and the magnetic field $\tilde{H}(r')$ at S on the surface s are given, then $\tilde{A}(r)$ and $\tilde{F}(r)$ can be obtained at the field point r:

$$\tilde{\mathbf{A}}(r) = \frac{\mu}{4\pi} \iint\limits_{s} \tilde{K}(r') \frac{e^{-jk|r-r'|}}{|r-r'|} ds' \tag{7.153}$$

$$\tilde{\mathbf{F}}(r) = \frac{\varepsilon}{4\pi} \iint\limits_{s} \tilde{M}(r') \frac{e^{-jk|r-r'|}}{|r-r'|} ds', \tag{7.157}$$

where

$$\tilde{K}(r') = \hat{n} \times \tilde{H}(r')$$

$$\tilde{M}(r') = -\hat{n} \times \tilde{E}(r').$$

The electric and magnetic fields at P are then given by

$$\tilde{\mathbf{E}} = -j\omega\tilde{\mathbf{A}} - \frac{j\omega}{k^2}\nabla\left(\nabla\cdot\tilde{\mathbf{A}}\right) - \frac{1}{\varepsilon}\nabla\times\tilde{\mathbf{F}} \tag{7.158}$$

$$\tilde{\mathbf{H}} = -j\omega\tilde{\mathbf{F}} - \frac{j\omega}{k^2}\nabla\left(\nabla\cdot\tilde{\mathbf{F}}\right) + \frac{1}{\mu}\nabla\times\tilde{\mathbf{A}} \tag{7.159}$$

In the far zone, we make the following approximations in (7.153) and (7.154) on $|r-r'|$, the distance SP between the source point and the field point:

$$|r-r'| \approx r - \hat{r}\cdot r' \quad \text{(in the exponent)}$$

$$|r-r'| \approx r \quad \text{(in the denominator)}$$

$$\hat{r}\cdot r' = r'\cos\psi \tag{7.171}$$

$$\cos\psi = \cos\theta\cos\theta' + \sin\theta\sin\theta'\cos\left(\phi-\phi'\right). \tag{7.172}$$

By substituting these approximations, one can obtain the following expression for the electric field [12] in the far zone:

$$\tilde{\mathbf{E}}_{\text{far}}\left(r\right) = -jk\frac{e^{-jkr}}{4\pi r}\oiint_s\left[\eta\tilde{\mathbf{K}}\left(r'\right) + \left(\tilde{\mathbf{M}}\left(r'\right)\times\hat{r}\right)\right]e^{jk\left(\hat{r}\cdot r'\right)}ds' \tag{21.46}$$

Numerical evaluation of (21.46) can be done by dividing s into several triangular elements N. A direct method of calculation is given in [12] based on the following:

$$\int_{-1}^{1}f\left(x\right)dx \approx \sum_{i=1}^{p}w_i f\left(x_i\right), \tag{21.47}$$

that is, the integral is approximated by a weighted sum. The examples of such a weighted sum formulas are *a.* Simpson's rule and *b.* Gaussian 5 point formula. Equation 21.47 can be generalized to two dimensions:

$$\iint_{(e)}f\left(x,y\right)ds = \sum_{i=1}^{p}\sum_{j=1}^{3}w_i\varsigma_j^{(e)}f_j^{(e)}, \tag{21.48}$$

where $\varsigma_j^{(e)}$ is the shape function discussed in Section 16.3.

Equation 21.46 can now be computed by approximating it, based on (21.47) and (21.48):

$$\tilde{\mathbf{E}}_{\text{far}}\left(r\right) = -jk\frac{e^{-jkr}}{4\pi r}\sum_{(e)=1}^{N}\sum_{i=1}^{p}\sum_{j=1}^{3}w_i\varsigma_j^{(e)}\left[\eta\tilde{\mathbf{K}}_j + \tilde{\mathbf{M}}_j\times\hat{r}\right]e^{jk\left(\hat{r}\cdot r'\right)}. \tag{21.49}$$

In the above, $\tilde{K}_j$ and $\tilde{M}_j$ can be considered as the values of these currents at vertex j. The computation of (21.49) consists of choosing a direction $\hat{r}$ (θ and ϕ in spherical coordinates) and evaluating the sum on the right side of (21.49). To get a complete picture of the far-field pattern, one needs to repeat the above for various values of θ and ϕ.

Reference 12 describes "fast far-field transform" based on partitioning the scattering/radiating object into several groups.

Reference 13 uses a similar transformation to calculate the far field due to the source of an aperture field. The implementation is done by FDTD method by converting the frequency domain expression into the time domain by noting

$$f\left(x',y',z'\right)e^{-j\frac{\omega}{c}R} \rightarrow f\left(x',y',z',t-\frac{R}{c}\right)$$

$$j\frac{\omega}{c}f\left(x',y',z'\right) \rightarrow \frac{1}{c}\frac{d}{dt}f\left(x',y',z'\right).$$

Reference 14 uses Green's function approach in formulating the near-field to far-field (NFTFF) transformation.

Reference 15 discusses the theory and gives three examples of using FDTD and NFTFF transformation to calculate the far fields of the scattered waves. The examples are as follows: (1) double-slit aperture radiation, (2) scattering by a circular cylinder, and (3) scattering by a dielectric sphere.

21.4 Finite Element Boundary Integral Method

Moment method is a boundary integral method. It is well suited for open boundaries problems. The implied boundary condition at infinity is automatically incorporated since the fields scattered by an object are determined by solving an integral equation, whose kernel is a Green's function. The integral equation also contains the incident known field, which of course automatically satisfies the essential boundary condition at infinity. To bring concreteness to the above statement, let us review some of the examples discussed earlier in this book.

In Section 16.7, Equation 16.193 is an integral equation with the unknown ρ_s in the integral and the kernel is the Laplacian Green's function for free space. The incident potential is the constant voltage V_0 on the plate, zero at infinity, and the integral is performed over the area of the plate which is the scatterer of the incident potential. In Section 16.8, we discussed the two-dimensional wave-scattering problem. The scatterer is a PEC cylinder. Equation 16.238 is the two-dimensional TM^z electric field integral equation (EFIE). Its kernel is the free space two-dimensional Green's function that satisfies the required essential condition at infinity. Since, in this example, the cylindrical object being a PEC, the boundary condition that the total electric field component E_z is zero on C gave us the EFIE (16.238). Reading through Section 16.8 including the example, one realizes that the solution involves the inversion of a full matrix (most of the elements of the matrix are nonzero). The size of the matrix to be inverted equals the number of segments N into which the curve C is broken into, in approximating it. For example, the Problem P16.15(b) requires N to be 72.

The moment method thus becomes computationally intensive for large-scale problems, and this is the disadvantage of the moment method. Moreover, it is well suited for the cases where the scatterer is not a PEC. We have seen that FEM is well suited for lossy inhomogeneous dielectrics, but it does not automatically incorporate the essential conditions at infinity. The previous sections suggested various ways to terminate the computational domain to overcome the difficulty of the FEM method. One can formulate the finite element boundary integral method (FEM-BIM) to take advantage of the attractive features of boundary integral method (also called boundary element method, BEM) to terminate the computational domain, while using the FEM method inside the computational domain to take advantage of its attractive feature of handling with ease the inhomogeneous, dispersive, lossy materials.

21.4.1 Two-Dimensional FEM-BIM Method

Figure 21.3 shows the geometry that explains this hybrid method.

The area S is bound by the closed curves Γ_1 and Γ_2. The area is divided into, say, N triangular elements, a few of those are shown in the figure. We use the FEM method to obtain the fields at the vertices of the triangle using the techniques discussed in Chapter 16. Suppose Γ_1 is a PEC boundary. Then we apply Dirichlet boundary conditions on Γ_1. The closed curve Γ_2 is a fictitious boundary and our goal will be to find a relationship between the field and its normal derivative for any point on Γ_2. The relationship comes as an integral involving Green's function and its derivative. Such an equation is called boundary integral. We will then use the continuity condition at the vertices of the finite elements that fall on Γ_2 boundary to complete the process and obtain the necessary algebraic equations to solve the problem.

To illustrate the technique of obtaining the boundary integral, we will consider a two-dimensional scattering problem of a conductor coated with an electromagnetic medium of permittivity ε_2 and permeability μ_2. Let the incident wave have E_z polarization and denote it by Φ as the potential for the problem. In the free space region exterior to the boundary Γ_2, the potential satisfies the equation

$$\nabla^2 \Phi + k_0^2 \Phi = f(\rho) \tag{21.50}$$

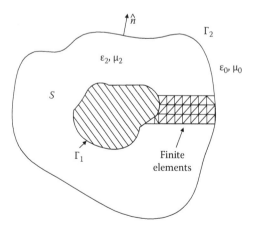

FIGURE 21.3
Geometry to explain two-dimensional FEM-BIM method.

The free space Green's function $G_0(\boldsymbol{\rho}, \boldsymbol{\rho}')$, relevant to the problem, satisfies the equation

$$\nabla^2 G_0(\boldsymbol{\rho}, \boldsymbol{\rho}') + k_0^2 G_0(\boldsymbol{\rho}, \boldsymbol{\rho}') = -\delta(\boldsymbol{\rho}, \boldsymbol{\rho}') \tag{21.51}$$

The solution of (21.51) that satisfies Sommerfeld's radiation condition is obtained before and is given by

$$G_0(\boldsymbol{\rho}, \boldsymbol{\rho}') = \frac{1}{4j} H_0^{(2)}(k_0 |\boldsymbol{\rho} - \boldsymbol{\rho}'|) \tag{21.52}$$

From (21.50), (21.51), and Green's second identity (7A.44 of Volume 1 of this book), it can be shown that [16]

$$\Phi(\boldsymbol{\rho}) = \Phi^{inc}(\boldsymbol{\rho}) + \oint_{\Gamma_2} \left[\Phi(\boldsymbol{\rho}') \frac{\partial G_0(\boldsymbol{\rho}, \boldsymbol{\rho}')}{\partial n'} - G_0(\boldsymbol{\rho}, \boldsymbol{\rho}') \frac{\partial \Phi(\boldsymbol{\rho}')}{\partial n'} \right] d\Gamma', \tag{21.53}$$

where

$$\Phi^{inc}(\boldsymbol{\rho}) = -\iint_S f(\boldsymbol{\rho}') G_0(\boldsymbol{\rho}, \boldsymbol{\rho}') dS'. \tag{21.54}$$

Equation 21.53 when enforced on Γ_2 relates the field with its normal derivative and when coupled to the FEM solution on Γ_2 leads to the required algebraic equations suitable for a numerical solution. The above is the principle and the concept of the FEM-BIM. Note that the computational domain is terminated by Γ_2 and we do not need PML.

References 16–19 give the examples, the required extension of the theory, and the details to implement FEM-BIM for two- and three-dimensional electromagnetic problems.

References

1. Berenger, J. P., A perfectly matched layer for the absorption of electromagnetic waves, *J. Comput. Phys.*, 114, 185–200, 1994.
2. Chew, W. C. and Weedon, W. H., A 3D perfectly matched medium from modified Maxwell's equations with stretched coordinates, *Microwave Opt. Technol. Lett.*, 7(13), 599–604, 1994.
3. Chew, W. C., Jin, J. M., and Michielssen, E., Complex coordinate stretching as a generalized absorbing boundary condition, *Microwave Opt. Technol. Lett.*, 15(6), 363–369, 1997.
4. Sacks, Z. S., Kingsland, D. M., Lee, R., and Lee, J.-F., A perfectly matched anisotropic absorber for use as an absorbing boundary condition, *IEEE Trans. Antennas Propag.*, 43(12), 1460–1463, 1995.
5. Gedney, S. D., An anisotropic perfectly matched layer-absorbing medium for the truncation of FDTD lattices, *IEEE Trans. Antennas Propag.*, 44(12), 1630–1639, 1996.
6. Elsherberi, A. and Demir, V., *The Finite Difference Time Domain Method for Electromagnetics: With MATLAB Simulations*, SciTech Publishing, Raleigh, NC, 2009.
7. Schneider, J. B., Understanding the finite-difference—time-domain method, e-book, available from Internet, 2016. http://www.eecs.wsu.edu/~schneidj/ufdtd/ufdtd.pdf. Accessed on August 13, 2017.

8. Rumpf, R., Computational electromagnetics, e-course, available from Internet, 2017. http://emlab.utep.edu/ee5390cem/. Accessed on August 13, 2017.
9. Kuzuoglu, M. and Mitra, R., Frequency dependence of the constitutive parameters of causal perfectly matched anisotropic absorbers, *IEEE Microw. Guided Wave Lett.*, 6(12), 447–449, 1996.
10. Berenger, J.-P., Application of the CFS-PML to the absorption of evanescent waves in waveguides, *IEEE Microw. Wireless Compon. Lett.*, 12(6), 218–220, 2002.
11. Gedney, S. D., *Introduction to Finite-Difference Time-Domain (FDTD) Method to Electromagnetics*, Morgan and Claypool Publishers, San Rafael, CA, 2011.
12. Stephanson, M. B., A fast Near-To-Far-Field transform algorithm, Distinction thesis, The Ohio State University, Columbus, OH, 2007, available from Internet. https://kb.osu.edu/dspace/bitstream/handle/1811/24537/stephanm_thesis.pdf?sequence=1. Accessed on August 13, 2017.
13. Sullivan, D. M., *Electromagnetic Simulation Using the FDTD Method*, IEEE Press, New York, 2000.
14. Taflove, A., *Computational Electrodynamics*, Artech House, Norwood, MA, 1995.
15. Schneider, J., Near-To-Far-Field transformation, Chapter 14, available from Internet URL, www.eecs.wsu.edu/~schneidj/ufdtd.chap14.pdf.
16. Jin, J., *The Finite Element Method in Electromagnetics*, 2nd edition, Wiley, New York, 2002.
17. Volakis, J. L., Chatterjee, A., and Kempel, L. E., *Finite Element Method for Electromagnetics*, IEEE Press, New York, 1998.
18. Silvester, P. P. and Pelosi, G. (Eds.), *Finite Elements for Wave Electromagnetics*, IEEE Press, New York, 1994.
19. Jin, J.-M., Volakis, J. L., and Collins, J. D., A finite element-boundary—Integral method for scattering and radiation by two- and three-dimensional structures, *IEEE Antennas Propag. Magazine*, 33(3), June 1991.

22

Approximate Analytical Methods Based on Perturbation and Variational Techniques*

Numerical methods such as finite element method (FEM), FDTD method, moment method, and the electromagnetic software based on these methods are extensively used in the industry. However, whenever exact analytical solutions are not possible, approximate analytical solutions serve an important role in giving a "feel for the problem." In addition, in many cases, they reduce the "computational burden" by giving the approximate acceptable solution with less amount of computation. Any how, they have a rich history dating back to the times when one did not entirely depend on computers.

A sample of such approximate analytical techniques is compiled in this chapter. References 1–4 give many more aspects of such solutions.

22.1 Perturbation of a Cavity

A tuned microwave source consists of a microwave generator attached to a cavity resonator. The tuned frequency is the resonant frequency of the cavity. The tuned frequency can be altered to some extent by altering the cavity walls, that is, by perturbing the cavity walls.

Figure 22.1 shows two positions of the metal screw. Position 1 at

$$P_1\left(\frac{a}{2}, b, \frac{d}{2}\right)$$

and position 2 at

$$P_2\left(a, \frac{b}{2}, \frac{d}{2}\right).$$

By screwing-in the metal screw at position 1, we will show that the resonant frequency of the TE_{101} mode of the rectangular cavity will be lowered. By doing the same at position 2, the resonant frequency will be increased. We can show these results from perturbation technique.

22.1.1 Theory for Cavity Wall Perturbations

Suppose we have a cavity with a PEC wall S, bounding a volume τ with a resonant-frequency ω_0, and fields $\tilde{E}_0$ and $\tilde{H}_0$ (Figure 22.2a).

---

* For chapter appendices, see 22A in Appendices section.

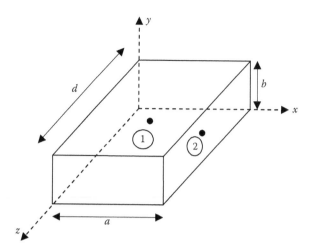

FIGURE 22.1
Rectangular cavity with a metal screw.

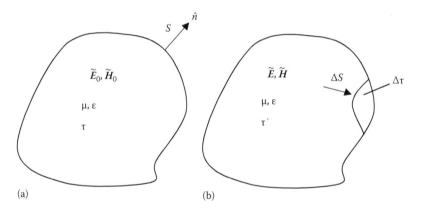

FIGURE 22.2
Cavity wall perturbation. (a) Original cavity and (b) perturbed cavity.

Figure 22.2b represents deformation of the cavity so that the PEC now covers $S' = S - \Delta S$ and encloses, where ΔS is a closed surface and encloses a volume $\Delta \tau$. Let $\tilde{E}$ and $\tilde{H}$ be the fields in the perturbed cavity. We wish to determine the resonant frequency ω of the perturbed cavity.

In both cases, the field equations have to be satisfied:

$$\nabla \nabla - \times \tilde{E}_0 = j\omega_0 \mu \tilde{H}_0, \tag{22.1a}$$

$$\nabla \times \tilde{H}_0 = j\omega_0 \varepsilon \tilde{E}_0, \tag{22.1b}$$

$$\nabla \nabla \times \tilde{E} = j\omega \mu \tilde{H}, \tag{22.2a}$$

$$\nabla \times \tilde{H} = j\omega_0 \varepsilon \tilde{E}. \tag{22.2b}$$

From Equations 22.2b and 22.1a,

$$\tilde{E}_0^* \cdot \nabla \times \tilde{H} = j\omega\varepsilon\tilde{E}_0^* \cdot \tilde{E}, \tag{22.3a}$$

$$-\tilde{H} \cdot \nabla \times \tilde{E}_0^* = -j\omega_0\mu\tilde{H} \cdot \tilde{H}_0^*. \tag{22.3b}$$

The vector identity

$$\nabla \cdot (A \times B) = B \cdot \nabla \times A - A \cdot \nabla \times B \tag{22.4}$$

may now be applied to the expression resulting from addition of Equation 22.3b to 22.3a:

$$\tilde{E}_0^* \cdot \nabla \times \tilde{H} - \tilde{H} \cdot \nabla \times \tilde{E}_0^* = j\omega\varepsilon\tilde{E} \cdot \tilde{E}_0^* - j\omega_0\mu\tilde{H}_0^* \cdot \tilde{H},$$
$$\nabla \cdot \left(\tilde{H} \times \tilde{E}_0^*\right) = j\omega\varepsilon\tilde{E} \cdot \tilde{E}_0^* - j\omega_0\mu\tilde{H} \cdot \tilde{H}_0^*. \tag{22.5}$$

By similar operations on Equations 22.2a and 22.1b, we get

$$\nabla \cdot \left(\tilde{H}_0^* \times \tilde{E}\right) = j\omega\mu\tilde{H} \cdot \tilde{H}_0^* - j\omega_0\varepsilon\tilde{E}_0^* \cdot \tilde{E}. \tag{22.6}$$

By adding Equations 22.5 and 22.6, integrating over τ, and using the divergence theorem

$$\iiint_v (\nabla \cdot A)\,\mathrm{d}v = \oiint_s A \cdot \mathrm{d}s, \tag{22.7}$$

we obtain the expression

$$\oiint_{S'} \left(\tilde{H} \times \tilde{E}_0^*\right)\overline{\mathrm{d}s} + \underbrace{\oiint_{S'} \left(\tilde{H}_0^* \times \tilde{E}\right)\overline{\mathrm{d}s}}_{\text{term 1}}$$
$$= \iiint_{\tau'} j\varepsilon(\omega - \omega_0)\tilde{E} \cdot \tilde{E}^*\,\mathrm{d}\tau + \iiint_{\tau'} j\mu(\omega - \omega_0)\tilde{H} \cdot \tilde{H}_0^*\,\mathrm{d}\tau. \tag{22.8}$$

Term 1 in Equation 22.8 is zero since

$$\tilde{H}_0^* \times \tilde{E} \cdot \mathrm{d}s = \mathrm{d}s\left(\tilde{H}_0^* \times \tilde{E} \cdot \hat{n}\right)$$

(exchange ($\cdot$) and ($\times$) in triple scalar product)

$$= \mathrm{d}s\left(\tilde{H}_0^* \cdot \tilde{E} \times \hat{n}\right)$$

and $\left(\tilde{E} \times \hat{n}\right)$ is the tangential component of $\tilde{E}$, which is zero on PEC.

Hence Equation 22.8 becomes

$$\oiint_{S'}\left(\tilde{H}\times\tilde{E}_0^*\right)\cdot ds = j\left(\omega-\omega_0\right)\iiint_{\tau'}\left[\varepsilon\tilde{E}\cdot\tilde{E}_0^* + \mu\tilde{H}\cdot\tilde{H}_0^*\right]d\tau. \tag{22.9}$$

Note that

$$\iint_{S}\tilde{H}\times\tilde{E}_0^*\,ds = \iint_{S}\tilde{H}\cdot\left(\tilde{E}_0^*\times\hat{n}\right)ds = 0. \tag{22.10}$$

Note: $\left(\tilde{E}_0\times\hat{n}\right)$ is the tangential component of $\tilde{E}_0$ on PEC which is zero, hence Equation 22.10 is equal to zero.

From Equations 22.9 and 22.10,

$$\oiint_{S'-S}\left(\tilde{H}_0\times\tilde{E}_0^*\right)ds = j\left(\omega-\omega_0\right)\iiint_{\tau'}\left[\varepsilon\tilde{E}\cdot\tilde{E}_0^* + \mu\tilde{H}\cdot\tilde{H}_0^*\right]d\tau. \tag{22.11}$$

Noting $S' = S - \Delta S$, Equation 22.11 becomes

$$\left(\omega-\omega_0\right) = \frac{\left(-(1)/(j)\right)\iint_{\Delta S}\left(\tilde{H}\times\tilde{E}_0^*\right)ds}{\iiint_{\tau'}\left[\left(\varepsilon\tilde{E}\cdot\tilde{E}_0\right)+\left(\mu\tilde{H}\cdot\tilde{H}_0^*\right)\right]d\tau}. \tag{22.12}$$

Equation 22.12 is an exact formula.

The crudest approximation in Equation 22.12 is to replace $\tilde{E}$ and $\tilde{H}$ in the numerator and denominator by $\tilde{E}_0$ and $\tilde{H}_0$, respectively. For small perturbations, this is reasonable in the denominator. The replacement in the numerator is not a bad approximation if the deformation is shallow and smooth:

$$\oiint_{\Delta S}\left(\tilde{H}\times\tilde{E}_0^*\,ds\right) \approx \oiint_{\Delta S}\tilde{H}_0\times\tilde{E}_0^*\,ds. \tag{22.13}$$

From complex Poynting theorem, given in Appendix 22A,

$$\oiint_{\Delta S}\left(\tilde{H}\times\tilde{E}_0^*\right)\bar{d}s = j\omega_0\iiint_{\Delta\tau'}\left[\varepsilon\left|\tilde{E}_0\right|^2 - \mu\left|\tilde{H}_0\right|^2\right]d\tau. \tag{22.14}$$

Thus, we get

$$\frac{\left(\omega-\omega_0\right)}{\omega_0} \approx \frac{\iiint_{\Delta\tau}\left[\mu\left|\tilde{H}_0\right|^2 - \varepsilon\left|\tilde{E}_0\right|^2\right]d\tau}{\iiint_{\tau}\left[\varepsilon\left|\tilde{E}_0\right|^2 + \mu\left|\tilde{H}_0\right|^2\right]d\tau}. \tag{22.15}$$

Equation 22.15 can be written as

$$\frac{\Delta\omega}{\omega_0} = \frac{\langle\Delta W_m\rangle - \langle\Delta W_e\rangle}{\langle W\rangle}. \tag{22.16}$$

In Equation 22.16, $\Delta\omega$ is the change in the resonance frequency, $\langle\Delta W_m\rangle$, $\langle\Delta W_e\rangle$ are proportional to the time-averaged electric and magnetic energies originally contained in $\Delta\tau$, and the denominator is proportional to the total energy stored in the original cavity. If $\Delta\tau$ is of small extent, one can write Equation 22.16 as

$$\frac{(\omega - \omega_0)}{\omega_0} \approx C\frac{\Delta\tau}{\tau}, \tag{22.17a}$$

$$C = \frac{\langle w_m\rangle - \langle w_e\rangle}{\langle w_\tau\rangle}, \tag{22.17b}$$

where $\langle w_\tau\rangle$ is $\langle w\rangle/\tau$, which is the space average. In the above, the lower case w is used to denote energy densities.

Note that we derived the above, assuming $\Delta\tau$ positive for an inward perturbation. The constant C is positive if $\langle w_m\rangle > \langle w_e\rangle$, that is, the perturbation is made at a point of large H and hence C is positive and $\omega > \omega_0$; the inward perturbation at such a point will raise the resonant frequency. If it is made at a point of large E, then the inward perturbation lowers it. The opposite behavior results from outward perturbation.

Let us use the above information to determine the frequency change that will take place by introducing an inward perturbation at the position (1) in Figure 22.1. Let the rectangular cavity be operating in TE_{101} mode. The fields are

$$\tilde{E}_y = D\sin\frac{\pi x}{a}\sin\frac{\pi z}{d}, \tag{22.18a}$$

$$\tilde{H}_z = A\cos\frac{\pi x}{a}\sin\frac{\pi z}{d}, \tag{22.18b}$$

$$\tilde{H}_x = B\sin\frac{\pi x}{a}\cos\frac{\pi z}{d}. \tag{22.18c}$$

For position 1,

$$
\left.
\begin{aligned}
x &= \frac{a}{2}, \quad z = \frac{d}{2}, \\
\tilde{E}_y\left(\frac{a}{2}, b, \frac{d}{2}\right) &= D \\
\tilde{H}_z\left(\frac{a}{2}, b, \frac{d}{2}\right) &= 0 \\
\tilde{H}_x\left(\frac{a}{2}, b, \frac{d}{2}\right) &= 0
\end{aligned}
\right\} \quad \text{at } P_1.
\tag{22.19}
$$

The electric field is maximum and the magnetic field is zero. So the inward perturbation reduces the resonant frequency. On the other hand, at

$$P_2\left(a,\frac{b}{2},\frac{d}{2}\right),\text{ we have}$$

$$\left.\begin{array}{l}\tilde{E}_y\left(a,\dfrac{b}{2},\dfrac{d}{2}\right)=0,\\[2mm]\tilde{H}_z=\left(a,\dfrac{b}{2},\dfrac{d}{2}\right)=-A\\[2mm]\tilde{H}_x=\left(a,\dfrac{b}{2},\dfrac{d}{2}\right)=0.\end{array}\right\}\tag{22.20}$$

Since the time-averaged magnetic energy density at this point is large and the electric energy density is zero at this point, the inward perturbation at this point increases the resonant frequency.

It can be shown that $C = -2$, from Equation 22.17b, for the point P_1.

22.1.2 Cavity Material Perturbation [1]

Figure 22.3 defines the problem. We start with the equations

$$\nabla\times\tilde{E}_0=-j\omega_0\mu\tilde{H}_0,\tag{22.21a}$$

$$\nabla\times\tilde{H}_0=j\omega_0\varepsilon\tilde{E}_0,\tag{22.21b}$$

$$\nabla\times\tilde{E}=-j\omega\left(\mu+\Delta\mu\right)\tilde{H},\tag{22.22a}$$

$$\nabla\times\tilde{H}=j\omega\left(\varepsilon+\Delta\varepsilon\right)\tilde{E}.\tag{22.22b}$$

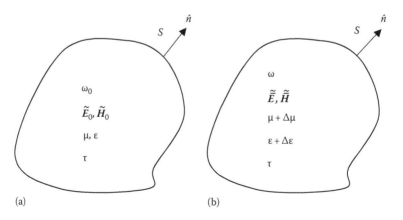

FIGURE 22.3
Cavity material perturbation. (a) Original cavity and (b) perturbation of matter in a cavity.

We can show, for PEC walls, that

$$\frac{(\omega - \omega_0)}{\omega} = \frac{-\iiint_\tau \left(\Delta\varepsilon \tilde{\boldsymbol{E}} \cdot \tilde{\boldsymbol{E}}_0^* + \Delta\mu \tilde{\boldsymbol{H}} \cdot \tilde{\boldsymbol{H}}_0^* \right) d\tau}{\iiint_\tau \left[\left(\varepsilon \tilde{\boldsymbol{E}} \cdot \tilde{\boldsymbol{E}}_0 + \mu \tilde{\boldsymbol{H}} \cdot \tilde{\boldsymbol{H}}_0^* \right) \right] d\tau}. \tag{22.23}$$

Once again for small changes of $\Delta\varepsilon$ and $\Delta\mu$, we can replace $\tilde{\boldsymbol{E}}$ by $\tilde{\boldsymbol{E}}_0$, $\tilde{\boldsymbol{H}}$ by $\tilde{\boldsymbol{H}}_0$ in Equation 22.23:

$$\frac{(\omega - \omega_0)}{\omega_0} \approx \frac{-\iiint_\tau \left[\Delta\varepsilon \left| \tilde{\boldsymbol{E}}_0 \right|^2 + \Delta\mu \left| \tilde{\boldsymbol{H}}_0 \right|^2 \right] d\tau}{\iiint_\tau \left[\varepsilon \left| \tilde{\boldsymbol{E}}_0 \right|^2 + \mu \left| \tilde{\boldsymbol{H}}_0 \right|^2 \right] d\tau}. \tag{22.24}$$

Any small increase in $\Delta\varepsilon$ and $\Delta\mu$ can only decrease resonant frequency.

Suppose we wish to consider a big change in ε and μ in a small region $\Delta\tau$ provided the region is small compared to the wavelength, one can use quasistatic approximation since the Helmholtz equation can be approximated by the Laplace equation.

Example

Let us consider a TE_{101} mode in a rectangular cavity. A thin dielectric slab of thickness δ is placed at the bottom of the cavity, that is,

$$\varepsilon = \varepsilon_0 \varepsilon_r, \quad 0 < y < \delta,$$
$$\varepsilon = \varepsilon_0, \quad \delta < y < b.$$

In Equation 22.23, $\Delta\varepsilon$ is zero outside $0 < y < \delta$, and $\Delta\mu$ is zero everywhere Figure 22.4. Therefore, the region of the numerator will be integrated over the dielectric plate. $\tilde{\boldsymbol{E}}$ the electric field to be used in the numerator is the perturbed electric field in the dielectric.

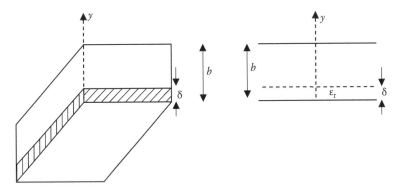

FIGURE 22.4
A thin dielectric slab in the cavity.

In the electrostatic approximation, we can get this electric field by using the boundary condition at $y = \delta$:

$$D_{n1} = D_{n2},$$

$$\varepsilon_r \varepsilon_0 E_d = \varepsilon_0 E_0; \quad E_d = \frac{E_0}{\varepsilon_r}. \tag{22.25}$$

Replace E by E_d in the numerator and replace E and H in the denominator by E_0 and H_0, respectively, since the perturbed part is a small part of the cavity.

The denominator is (Denom) then $2\varepsilon_0 \iiint_\tau \left| \tilde{E}_0 \right|^2 d\tau$:

$$\text{Denom} = 2\varepsilon_0 \iiint_\tau \left| \tilde{E}_0 \right|^2 d\tau, \tag{22.25a}$$

whereas the numerator (*Num*) is

$$\text{Num} = -\iiint_{\Delta\tau} \Delta\varepsilon \left(E_d \cdot E_0^* \right) d\tau. \tag{22.25b}$$

In $\Delta\tau$,

$$\Delta\varepsilon = \varepsilon_0 \varepsilon_r - \varepsilon_0 = \varepsilon_0 \left(\varepsilon_r - 1 \right). \tag{22.25c}$$

Thus,

$$\text{Num} = -\left(\text{constant} \right) \frac{\varepsilon_r - 1}{\varepsilon_r} \delta.$$

Denom = 2 (same constant as the one in num) b:

$$\frac{\omega - \omega_0}{\omega} \approx \frac{\omega - \omega_0}{\omega_0} = -\frac{1}{2} \frac{\varepsilon_r - 1}{\varepsilon_r} \frac{\delta}{b}. \tag{22.26}$$

In the above, the constant is $\varepsilon_0 E_0^2$ multiplied by the base area:

$$\text{constant} = \left(ad \right) \varepsilon_0 E_0^2. \tag{22.27}$$

22.2 Variational Techniques and Stationary Formulas

22.2.1 Rayleigh Quotient

Let X be an eigenvector of a matrix A, that is,

$$AX = \lambda X. \tag{22.28}$$

Here λ is the eigenvalue.

The Rayleigh quotient is given by

$$\lambda_R = \frac{X^T A X}{X^T X}. \tag{22.29}$$

where X is a column vector, X^T a row vector, and A a square matrix.

If we substitute Equation 22.28 into Equation 22.29, then we have

$$\lambda_R = \frac{X^T \lambda X}{X^T X} = \lambda \frac{X^T X}{X^T X} = \lambda. \tag{22.30}$$

The Rayleigh quotient is the same as the eigenvalue when X is an exact eigenvector. Suppose we have an approximation to X, called V, which in some sense has an error of ε. In this case, we show that the Rayleigh quotient approximates λ with an error of ε^2. That is, if V has an error of 0.1 (10%), then

$$\lambda_R = \frac{V^T \lambda V}{V^T V}, \tag{22.31}$$

has an error of $(0.1)^2 = 0.01(1\%)$.

The Rayleigh quotient is a stationary formula for the eigenvalue of a matrix. We can prove this in the following way.

Let,

$$V = a_1 V_1 + a_2 V_2 + \cdots + a_n V_n, \tag{22.32}$$

where V is a approximate eigenvector and $V_1, V_2, \ldots, V_n$ are the exact eigenvectors of this n dimensional space

$$AV = a_1 A V_1 + a_2 A V_2 + \cdots + a_n A V_n$$
$$= a_1 \lambda_1 V_1 + a_2 \lambda_2 V_2 + \cdots + a_n \lambda_n V_n, \tag{22.33}$$

$$V^T = a_1 V_1^T + a_2 V_2^T + \cdots + a_n V_n^T \tag{22.34}$$

$$\lambda_R = \frac{\left\{ \Sigma_{i=1}^n \left(a_i V_i^T \right) \right\} \{A\} \left\{ \Sigma_{j=1}^n \left(a_j V_j \right) \right\}}{\left(\Sigma_{i=1}^n \left(a_i V_i^T \right) \right) \left(\Sigma_{j=1}^n \left(a_j V_j \right) \right)}. \tag{22.35}$$

The orthogonality property of the eigenvector is to be noted

$$V_i^T V_j = 0, \quad i \neq j,$$
$$V_i^T V_j = 1, \quad i = j \quad \text{if } V\text{'s are normalized.} \tag{22.36}$$

In evaluating the numerator of Equation 22.8, we can use the following simplifications. A typical term in the sum is

$$a_i V_i^T A a_j V_j = a_i a_j \lambda_j V_i^T V_j$$
$$= 0, \quad i \neq j,$$
$$= a_i^2 \lambda_i, \quad i = j.$$

Hence Equation 22.35 becomes

$$\lambda_R = \frac{\sum_{i=1}^{n}\lambda_i a_i^2}{\sum_{i=1}^{n}a_i^2} \tag{22.37}$$

of V is close to V_1 with an error of the order of ε, it means

$$\frac{a_j}{a_1}, \quad j = 2,\ldots,n \text{ are of order } \varepsilon, \quad \text{and} \quad \left(\frac{a_j}{a_1}\right)^2 \text{ is of order } \varepsilon^2$$

and

$$\lambda_R = \lambda_1 \frac{a_1^2\left(1+O\left(\varepsilon^2\right)\right)}{a_1^2\left(1+O\left(\varepsilon^2\right)\right)},$$

$$\lambda_R = \lambda_1\left[1+O\left(\varepsilon^2\right)\right].$$

22.2.2 Variational Formulation: Scalar Helmholtz Equation

Let Φ be a scalar field, then the scalar Helmholtz equation

$$\nabla^2\Phi + \lambda\Phi = 0 \tag{22.38}$$

is an eigenvalue problem. This can be converted into a variational form by multiplying Equation 22.38 by Φ and integrating over the volume τ:

$$\iiint_\tau \Phi\nabla^2\Phi d\tau + \lambda\iiint_\tau \Phi^2 d\tau = 0, \tag{22.39}$$

$$\lambda = -\frac{\iiint_\tau \Phi\nabla^2\Phi\, d\tau}{\iiint_\tau \Phi^2\, d\tau}. \tag{22.40}$$

We can modify this equation using

$$\nabla\cdot\left[\Phi - \Phi\right] = \Phi\nabla^2\Phi + -\Phi\cdot -\Phi. \tag{22.41}$$

Then we get

$$\lambda = -\frac{\iiint_\tau \left\{\nabla\cdot\left[\Phi - \Phi\right] - \left|-\Phi\right|^2\right\}d\tau}{\iiint_\tau \Phi^2\, d\tau}.$$

Using the divergence theorem

$$\lambda = -\frac{\iiint_{\tau}|\Phi|^2\,d\tau - \oiint_{S}\Phi - \Phi\cdot\bar{d}s}{\iiint_{v}\Phi^2\,d\tau}, \tag{22.42}$$

where the closed surface S bound τ. For the case of homogeneous BC, say

$$\Phi(r)\big|_{S} = 0 \quad \text{or} \quad \frac{\partial\Phi}{\partial n}\bigg|_{S} = 0, \tag{22.43}$$

the surface integral in Equation 22.42 vanishes.
    Then

$$\lambda = \frac{\iiint_{\tau}|-\Phi|^2\,d\tau}{\iiint_{\tau}\Phi^2\,d\tau}. \tag{22.44}$$

It can be shown that Equation 22.44 is a variational principle, that is, if Φ is an eigenfunction, then λ is the corresponding eigenvalue. Also if Φ_{trial} is an approximate eigenvector with an error of ε, then the λ_{trial} computed from Equation 22.44, will be the corresponding eigenvalue with an error of ε^2. Let us illustrate the application of Equation 22.44 through a straightforward example for which we have an exact solution.

Example

In this example, we compute the eigenvalue of the dominant TM mode of the rectangular waveguide shown in Figure 22.5.
    In Section 3.2, we investigated TM modes by writing

$$\tilde{E}_z(x,y,z) = F(x,y)e^{-\gamma z},$$

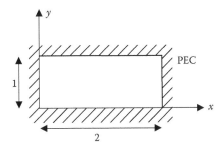

FIGURE 22.5
Cross section of rectangular waveguide.

where $F(x, y)$ satisfies the equation

$$\frac{\partial^2 F}{\partial x^2} + \frac{\partial^2 F}{\partial y^2} + k_c^2 F = 0, \tag{22.45}$$

and

$$k_c^2 = k^2 + \gamma^2. \tag{22.46}$$

Comparing Equation 22.45 with Equation 22.39, we note that k_c^2 is the eigenvalue λ, and we have a two-dimensional problem in rectangular coordinates, that is, we replace ∇^2 by $\nabla_t^2 = (\partial^2)/(\partial x^2) + (\partial^2)/(\partial y^2)$. Thus, Equation 22.44 becomes

$$k_c^2 = \lambda = \frac{\iint_S |\nabla F|^2 \, ds}{\iint_S F^2 \, ds}. \tag{22.47}$$

For TM modes, the boundary condition $E_z = 0$ on the walls $x = 0$, or $y = 0$ or 1 translates to $F = 0$; $x = 0$, or 2; $F = 0$; $y = 0$ or 1.

A trial function F_{trial} that satisfies these boundary conditions can be written as

$$\begin{aligned} F_{\text{trial}} &= cx(2-x)y(1-y) \\ &= c(2x - x^2)(y - y^2), \end{aligned} \tag{22.48}$$

where c is a constant.

$$\begin{aligned} \frac{F_{\text{trial}}}{\partial x} &= c(2 - 2x)(y - y^2) \\ \frac{F_{\text{trial}}}{\partial y} &= c(2x - 2x^2)(1 - 2y), \\ |-F_{\text{trial}}|^2 &= \left(\frac{F_{\text{trial}}}{\partial x}\right)^2 + \left(\frac{F_{\text{trial}}}{\partial y}\right)^2 \\ &= c^2\left[4y^2(1-x)^2(1-y)^2 + (1-2y)^2 x^2 (2-x)^2\right], \\ F_{\text{trial}}^2 &= c^2 x^2 (2-x)^2 y^2 (1-y)^2. \end{aligned} \tag{22.49}$$

We obtain

$$\lambda_{\text{trial}} \approx 12.5. \tag{22.50}$$

Exact value for λ value can be computed from

$$k_c^2 = \lambda = \left(\frac{m\pi}{a}\right)^2 + \left(\frac{n\pi}{b}\right)^2, \quad m = 1, n = 1, a = 2, b = 1,$$

$$\lambda = 12.337. \tag{22.51}$$

Because Equation 22.47 is a variational principle, we have obtained λ_{trial} within 1.3% of the exact value.

22.2.3 Variational Formulation: Vector Helmholtz Equation

Let us start with the vector Helmholtz equation where $\varepsilon(r)$ could be an inhomogeneous dielectric.

The electric field equation is

$$\nabla \tilde{E} - \omega^2 \mu \varepsilon \tilde{E} = 0. \tag{22.52}$$

Take the dot product of Equation 22.52 with $\tilde{E}$ and integrate. We can now write Equation 22.52 in the variational form

$$\omega_r^2 = \frac{\iiint_\tau \tilde{E} \cdot \left(\nabla \tilde{E} \right) d\tau}{\mu \iiint_\tau \varepsilon \tilde{E}^2 \, d\tau}. \tag{22.53}$$

We can show that Equation 22.53 is a variational principle by the following technique.
Let

$$\tilde{E}_{\text{trial}} = \tilde{E} + \Delta \tilde{E} = \tilde{E} + p \tilde{e}, \tag{22.54}$$

where $\tilde{E}$ is the true field and the trial field is expressed in terms of an arbitrary parameter p. Substituting Equation 22.54 into Equation 22.53, where we use $\tilde{E}_{\text{trial}}$,

$$\omega^2(p) = \frac{\iiint_\tau \left(\tilde{E} + p \tilde{e} \right) \cdot \left(\nabla \left(\tilde{E} + p \tilde{e} \right) \right) d\tau}{\mu \iiint_\tau \varepsilon \left(\tilde{E} + p \tilde{e} \right) \cdot \left(\tilde{E} + p \tilde{e} \right) d\tau}, \tag{22.55}$$

where we show ω^2 as a function of p for fixed $\tilde{e}$. The Maclaurin expansion of ω^2 is

$$\omega^2(p) = \omega_r^2 + p \frac{\partial}{\partial p} \left(\omega^2 \right) \bigg|_{p=0} + \frac{p^2}{2!} \frac{\partial^2}{\partial p^2} \left(\omega^2 \right) \bigg|_{p=0} + \cdots. \tag{22.56}$$

The first term is the true resonant frequency because $\omega^2(0) = \omega_r^2$. In the variational notation, Equation 22.56 is written as

$$\omega^2(p) = \omega_r^2 + p \delta \left(\omega^2 \right) + \frac{p^2}{2!} \delta^2 \left(\omega^2 \right) + \cdots. \tag{22.57}$$

The term $\delta(\omega^2)$ is called the first variation of ω^2, the term $\delta^2(\omega^2)$ is the second variation of ω^2, and so on.

A formula for ω^2 is said to be stationary if the first variation of ω^2 vanishes, that is,

$$\frac{\partial}{\partial p} \left(\omega^2 \right) \bigg|_{p=0} = 0. \tag{22.58}$$

Considering Equation 22.55 as N/D, Equation 22.58 may be evaluated as

$$\left.\frac{\partial}{\partial p}\left(\omega^2\right)\right|_{p=0} = \frac{D(0)N'(0) - N(0)D'(0)}{D^2(0)}$$ (22.59)

$$N'(0) = \iiint_\tau \left(\tilde{E}\cdot -\times -\times \tilde{e} + \tilde{e}\cdot -\times -\times \tilde{E}\right)d\tau.$$

From the vector identity,

$$\iiint_\tau (\tilde{E}\cdot -\times -\times \tilde{e}\, d\tau = \iiint_\tau \left(-\times \tilde{e}\right)\cdot\left(-\times \tilde{E}\right)d\tau$$

$$+ \oiint_S \left[\left(-\times \tilde{e}\right)\times \tilde{E}\right]\overline{d}s.$$

The last term on the RHS is zero since

$$\oiint_S \left(-\times \tilde{e}\right)\times \tilde{E}\cdot \hat{n}\, ds = \oiint_S \left(-\times \tilde{e}\right)\cdot\left(\tilde{E}\times \hat{n}\right)ds,$$

and the PEC boundary condition on s is $\hat{n}\times \tilde{E} = 0$.

The first term on the RHS is

$$\iiint_\tau \left(-\times \tilde{e}\right)\cdot\left(-\times \tilde{E}\right)d\tau = \iiint_\tau \left(e\cdot -\times -\times \tilde{E}\right)d\tau$$

$$- \oiint_S \left(-\times \tilde{E}\right)\times \tilde{e}\cdot \hat{n}\, ds.$$

Note from Equation 22.52, $-\times -\times \tilde{E} = \omega_r^2 \mu \varepsilon$.

Putting all this information together, we obtain

$$N'(0) = 2\omega_r^2 \mu \iiint_\tau \varepsilon \tilde{e}\cdot \tilde{E}\, d\tau - \iiint_\tau \left(-\times \tilde{E}\right)\cdot\left(\tilde{e}\times \hat{n}\right)ds.$$ (22.60)

We can also show that

$$D'(0) = 2\mu \iiint \varepsilon \tilde{e}\cdot \tilde{E}\, d\tau.$$

From Equation 22.59 [1],

$$\left.\frac{\partial}{\partial p}\left(\omega^2\right)\right|_{p=0} = \frac{D(0)N'(0) - N(0)D'(0)}{D^2(0)}$$

$$= -\frac{\oiint_s \left(-\times \tilde{E}\right)\cdot\left(\tilde{e}\times \hat{n}\right)ds}{\mu \iiint_\tau \varepsilon E^2\, d\tau}.$$ (22.61)

The above equation vanishes when $\hat{n} \times \tilde{e} = 0$ on S which in turn means $\hat{n} \times \tilde{E}_{\text{trial}} = 0$ on S.

Hence Equation 22.53 is a stationary formula for the resonant frequency if the tangential components of the trial $\tilde{E}$ vanish on the cavity walls.

Note that Equation 22.53 can be put in a more symmetric form:

$$\omega_r^2 = \frac{\iiint_\tau \left| -\times \tilde{E} \right|^2 d\tau}{\mu \iiint_\tau \varepsilon E^2 \, d\tau}. \tag{22.62}$$

Example [1]

A circular cylindrical cavity of height d and radius a is shown. For $d < 2a$, the dominant mode is TM_{010} mode with (Figure 22.6)

$$E_z = c_1 J_0\left(\frac{2.405\rho}{a}\right),$$

$$H_\phi = c_2 \frac{2.405}{a} J_1\left(\frac{2.405\rho}{a}\right),$$

$$\omega_r = \frac{2.405}{a\sqrt{\mu\varepsilon}}.$$

Use variational principle to obtain an estimate for ω_r.

Choose

$$\tilde{E}_{\text{trial}} = \hat{z}\left(1 - \frac{\rho}{a}\right).$$

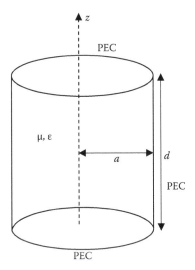

FIGURE 22.6
Cylindrical cavity.

Note that the trial function satisfies the boundary condition that $\tilde{E}_{tan} = 0$ on the walls of the cavity.

$$-\times \tilde{E}_{trial} = \hat{\phi}\,\frac{1}{a}.$$

From Equation 22.61,

$$\omega_r^2 = \frac{\int_{\rho=0}^{a}\int_{\phi=0}^{2\pi}\int_{z=0}^{d}\frac{1}{a^2}\rho\,d\rho\,d\phi\,dz}{\mu\varepsilon\int_{\rho=0}^{a}\int_{\phi=0}^{2\pi}\int_{z=0}^{d}\left(1-\rho/a\right)^2\rho\,d\rho\,d\phi\,dz}$$

$$= \frac{\int_{0}^{a}\frac{\rho}{a^2}\,d\rho}{\int_{0}^{a}\left(1-\rho/a\right)^2\rho\,d\rho} = \frac{6}{\mu\varepsilon a^2},$$

$$\omega_r \approx \frac{\sqrt{6}}{a\sqrt{\mu\varepsilon}} = \frac{2.449}{a\sqrt{\mu\varepsilon}}; \quad Error = \frac{2.449-2.405}{2.405}.$$

The error using this trial function is 1.8%.

References

1. Harrington, R. F., *Field Computation by Moment Methods*, Macmillan, New York, 1968.
2. Van Bladel, J., *Electromagnetic Fields*, McGraw-Hill, New York, 1964.
3. Khalifeh, A. F., Perturbation technique as applied to space-varying and time-varying electromagnetic systems, Doctoral thesis, University of Massachusetts Lowell, Lowell, MA, 2005.
4. Ji, C., Approximate analytical techniques in the study of quadruple-ridged waveguide (QRW) and its modifications, Doctoral thesis, University of Massachusetts Lowell, Lowell, MA, 2007.

23

Miscellaneous Topics on Electromagnetic Computation

In this concluding chapter, we will consider some aspects of electromagnetic computation not yet discussed in this book, including those computational techniques that are in development or the extensions to the well-known computational techniques. Sections 23.4 through 23.9 discuss integral transforms and their use in electromagnetic computation. Those with some background, normally acquired in the undergraduate signals and systems course, will find Section 23.4 as a review of the integral transform methods.

23.1 Intuitive and Qualitative Solution

Before embarking on a computer solution through commercially available electromagnetic software or through your own developed software, it is important to simplify the problem to the bare minimum, yet retaining the underlying physics, so that one can obtain an intuitive and qualitative solution to the problem to act as a check on the proper computer simulation of the problem.

The time scales and the frequencies of interest including or excluding the possibilities of cutoff or resonant frequencies, the length scales and the wavelengths of interest, approximation of the boundary to the closest regular geometric shapes for which analytical solutions are easy to obtain, the reduction of the dimensionality of the problem in obtaining the qualitative solution, the qualitative solutions based on the limiting values of the electromagnetic parameters, the dominant effect due to the kind of complexity of the electromagnetic medium (Volume 1 emphasizes the electromagnetic theory from this view point), and the zero- or first-order models leading to a circuit approximation should be useful to get an idea of the expected solution. Two examples of a qualitative solution that can be attempted before a computer simulation are briefly mentioned next.

23.1.1 Field Sketching, Curvilinear Squares, and Solution of Laplace Equation

Before the advent of computers, many static potential problems with irregular boundaries were solved approximately by a hand-drawn field map, of orthogonally intersecting field lines and equipotential contours (in two dimensions), called curvilinear squares. The URLs of free downloads of some literature from the internet are given below:

http://onlinelibrary.wiley.com/doi/10.1029/2007WR006221/pdf

http://web.mit.edu/6.013_book/www/chapter7/7.6.html

https://theses.lib.vt.edu/theses/available/etd-07092004-123926/unrestricted/Thesis.pdf

23.1.2 Scatterers in Time-Harmonic Fields and Circuit Approximations

The frequency selective surface (FSS) is a good example of getting a good qualitative picture of it acting as a band-pass filter or a band-reject filter. An equivalent transmission line circuit with the placement of lumped elements L and C across the transmission line is discussed in the URL given below, which is a free download from the internet. For band-reject, L and C are in series and for band-pass they are in parallel.

https://en.wikipedia.org/wiki/Frequency_selective_surface

Section 7.4, Problems P7.12 and P7.13 of Volume 1 can be reviewed to understand the one-dimensional Floquet modes involved in a periodic loading of a transmission line with either L or C. Problems P7.14 and P7.15 explore the band-reject and band-pass filters with L and C series and parallel combinations, respectively.

An article "Everything You Ever Wanted to Know About Frequency-Selective Surface Filters but Were Afraid to Ask" by Benjamin Hooberman has an illuminating write-up on FSS, starting from the radiation by an electron accelerated by the electric field of a plane wave. A free download of it from the internet can be obtained at the following URL: http://cosmology.phys.columbia.edu/group_web/about_us/memos/hooberman_filters_memo.pdf

A spectral domain numerical method for FSS is discussed in Section 23.9.

23.2 Transverse Resonance Method (TRM)

Analytical solutions for wave propagation in the z-direction in a rectangular waveguide with perfect electric conductor (PEC) boundary conditions on the walls were discussed earlier using E_z as the potential for TM^z modes and H_z as the potential for TE^z modes. The solution involved trigonometric functions for describing the field distributions in the transverse plane. See for example (3.20) or (3.24) of Volume 1. Trigonometric functions signify standing waves (Table 2.3). A transmission line with short-circuit or open-circuit or reactive load has standing wave solutions and the impedance transformation as you move on such a transmission line can be determined from (2.67), where $Z_L = 0$, for short circuit, $Z_L = \infty$ for open circuit, or $Z_L = jX_L$, X_L real for a reactive element. Equations 2.72 and 2.73 give expressions for the input impedance for the short circuit and the open circuit, respectively. If one uses Smith chart (Appendix 2C) for calculating the input impedance, the outer circle contains all the points on the transmission line when the termination is open circuit/short circuit. This shows that the fields in the transverse plane can be represented by a transmission line.

It is useful now to recall the transmission line representation of a one-dimensional cavity. A resonant transmission line shown in Figure 23.1a and a one-dimensional cavity shown in Figure 23.1b are equivalent in resonant frequencies.

$$\tilde{V}(x) = V_0 \sin \beta_r x, \quad 0 < x < a \tag{23.1a}$$

$$\tilde{E}_y(x) = E_0 \sin \beta_r x, \quad 0 < x < a \tag{23.2a}$$

$$\tilde{V}(x) = V_0 \sin \frac{n\pi x}{a}, \quad n = 1, 2, 3, \dots \tag{23.1b}$$

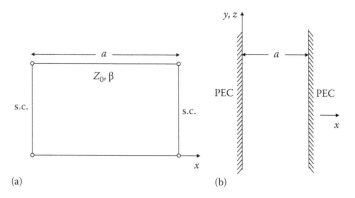

FIGURE 23.1
Equivalence of (a) a short-circuited (S.C.) transmission line, (b) one-dimensional cavity with PEC walls.

$$\tilde{E}_y(x) = E_0 \sin \beta_r x, \quad n = 1, 2, 3, \ldots \tag{23.2b}$$

$$\beta = \beta_r = \frac{n\pi}{a} \tag{23.1c}$$

$$\beta = \beta_r = \frac{n\pi}{a} \tag{23.2c}$$

It can be noted that the given solution exists only at a discrete set but infinite number of frequencies.

$$\frac{n\pi}{l} = \beta = k = \frac{\omega_r}{v} = \omega_r \sqrt{L'C'} \tag{23.3}$$

$$f_r = \frac{n\pi}{2na\sqrt{L'C'}} = \frac{n}{2a\sqrt{L'C'}} \tag{23.4}$$

The idea of resonance can be expressed in a different way by considering the input impedance of the short-circuited transmission line shown in Figure 23.2.

$$Z_{in}^{+x} = jZ_0 \tan \beta d \tag{23.5}$$

$$Z_{in}^{-x} = jZ_0 \tan \beta (a - d) \tag{23.6}$$

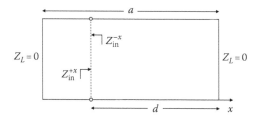

FIGURE 23.2
Resonance condition expressed in terms of input impedances to the right and the left at a point $x = -d$ on the transmission line.

For resonance to take place

$$Z_{in}^{+x} = -Z_{in}^{-x},$$ (23.7)

which gives the resonance condition.
   Thus we get the equation

$$\tan \beta d + \tan \beta (a - d) = 0,$$ (23.8)

which simplifies to

$$\sin \beta a = 0.$$ (23.9)

Equation 23.7 expresses the condition for resonance, which in the example above is the same transmission line to the left as well as to the right.
   Let us now take a slightly different example (Figure 23.3), where the transmission line to the left has different characteristic impedance Z_{01} than the one to the right (Z_{02}).

$$Z_{in}^{+x} = jZ_{02} \tan \beta_2 d$$ (23.10)

$$Z_{in}^{-x} = jZ_{01} \tan \beta_1 (a - d).$$ (23.11)

The resonant condition is given by the transcendental equation

$$Z_{02} \tan \beta_2 d + Z_{01} \tan \beta_1 (a - d) = 0.$$ (23.12)

Suppose the transmission line to the left has μ_1, ε_1 as the medium parameters and the transmission line to the right has μ_2, ε_2 as the medium parameters, then

$$Z_{02} = \eta_0 \sqrt{\frac{\mu_{r2}}{\varepsilon_{r2}}}$$ (23.13)

$$Z_{01} = \eta_0 \sqrt{\frac{\mu_{r1}}{\varepsilon_{r1}}}$$ (23.14)

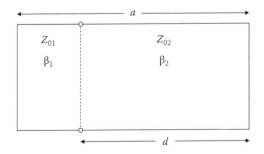

FIGURE 23.3
The transmission lines to the right and the left of $x = -d$ are different.

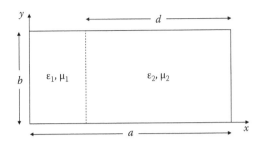

FIGURE 23.4
Cross section of a rectangular waveguide loaded with different media.

$$\beta_1 = \omega\sqrt{\mu_1\varepsilon_1} \tag{23.15}$$

$$\beta_2 = \omega\sqrt{\mu_2\varepsilon_2} \tag{23.16}$$

The only unknown in (23.12) is ω, the angular frequency, which is a resonant frequency. It has infinite number of discrete values of frequencies as solutions corresponding to the resonant frequencies of various modes.

Let us next consider a rectangular waveguide (instead of a transmission line) [1], whose cross section is as shown in Figure 23.4. If β_z is now the propagation constant, then the transverse wave numbers β_{t1} and β_{t2} are related to β_z by

$$\beta_{t1}^2 + \beta_z^2 = \omega^2\mu_1\varepsilon_1 \tag{23.17}$$

$$\beta_{t2}^2 + \beta_z^2 = \omega^2\mu_2\varepsilon_2 \tag{23.18}$$

$$Z_{in}^{+x} = jZ_{02}\tan\beta_{t2}d \tag{23.19}$$

$$Z_{in}^{-x} = jZ_{01}\tan\beta_{t1}(a-d) \tag{23.20}$$

and the transverse resonance equation is given by

$$Z_{02}\tan\beta_{t2}d + Z_{01}\tan\beta_{t1}(a-d) = 0 \tag{23.21}$$

Let us be more specific by considering TE^x modes of a rectangular waveguide loaded as shown in Figure 23.4.

$$Z_{02} = \frac{\omega\mu_2}{\beta_{x2}}, \quad Z_{01} = \frac{\omega\mu_1}{\beta_{x1}}, \quad \beta_{t2} = \beta_{x2}, \quad \beta_{t1} = \beta_{x1}$$

The Equation 23.21 becomes

$$\frac{\beta_{x2}}{\mu_2}\cot\beta_{x2}d + \frac{\beta_{x1}}{\mu_1}\cot\beta_{x1}(a-d) = 0 \tag{23.22}$$

Equation 23.22 can be used to determine the cutoff frequencies, but it does not easily give information on the field distribution in the cross section. On the other hand, a problem similar to the problem on hand but even more general is treated by using the finite element method (FEM) under Section 17.3 and another example in Section 18.2. The FEM not only gives the cutoff frequencies but also the field distributions. However, the TRM is computationally much lighter than the FEM since the computational load is only the solution of a transcendental equation like (23.22).

Ji [2] used TRM method to obtain the cutoff frequencies of the quadruple-ridged waveguide and verified the results by the FEM method.

23.3 Mode-Matching and Field-Matching Techniques

In many electromagnetic problems, two or more field regions are connected by apertures [3–11]. The geometry of the problem can be such that the region to the left of the aperture admits solutions in terms of the known normal modes of its structure. So the electromagnetic fields can be expressed as an expansion of its normal modes. So will be the fields in the region to the right of the aperture. The coefficients of expansion can be determined by enforcing the continuity conditions of the fields at the aperture. This technique is called the mode-matching technique [5,6]. Of course, the above remarks equally apply if the aperture is common to the top and bottom regions of the aperture. The geometry for an example of application of such a technique is given in Figure 23.5.

A related technique called field-matching consists of solving the field equations in each subregion by the method of separation of variables and then imposing the continuity conditions at the boundary between adjacent regions. The geometry for a typical application of the field-matching technique is the shielded microstrip (Figure 23.6) discussed in [7].

In the field-matching technique, for some of the subregions, the normal modes are not readily available, but on the other hand, the fields can be expressed as functions of the coordinates of the geometry by separation of variables technique, and the expansion coefficients associated with each subregion are related by enforcing the boundary conditions

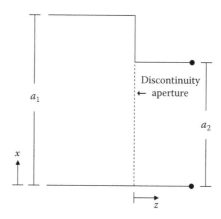

FIGURE 23.5
A rectangular waveguide with abrupt change of the width of the guide.

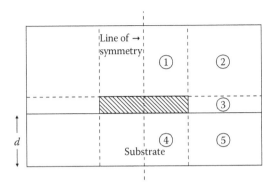

FIGURE 23.6
Five subregions of the cross section of a shielded microstrip line.

at various boundaries between the subregions. An example [5] of such a technique is given in Figure 23.6.

In the general literature, the distinction between the two is not maintained and for some problems a combination of these two techniques is used and still referred to as the mode-matching technique [3].

Since the expansions involve infinite series, a numerical solution needs truncation by a finite number of terms, say N. For some problems with slow convergence of the series, N could be large [6]. A solution technique to overcome the convergence problem is given in [7] by solving in the spectral domain and using "modified residue calculus" technique to avoid ill-conditioned matrices in the spectral domain. Comprehensive summary and seminal papers on mode-matching and field-matching techniques are given in [3]. We will discuss two examples given below.

Example 23.1: Inductive Iris in a Waveguide [8,9]

The cross section of a waveguide ($z = 0$ plane) and the iris in that plane is shown in Figure 23.7.

Let the phasor electric field of the incident wave of TE_{10} mode, is given by

$$E_y^i(x,z) = \sin\frac{\pi x}{a} e^{-\gamma_1 z}, \quad z < 0 \tag{23.23a}$$

where

$$\gamma_1 = \sqrt{\frac{\pi^2}{a^2} - k_0^2} \tag{23.23b}$$

$$k_0^2 = \omega^2 \mu_0 \varepsilon_0. \tag{23.23c}$$

The associated magnetic field is given by

$$H_x^i(x,z) = -\frac{\gamma_1}{j\omega\mu_0} \sin\frac{\pi x}{a} e^{-\gamma_1 z}, \quad z < 0 \tag{23.24}$$

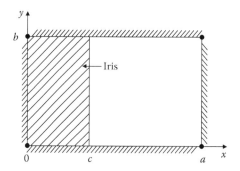

FIGURE 23.7
Inductive iris in a waveguide. The hatched part is a thin rectangular PEC patch acting as an iris. The cross section in the $z = 0$ plane is shown.

The reflected (scattered fields for $z < 0$) are

$$E_y^r(x,z) = \sum_{n=1}^{\infty} E_n \sin\frac{n\pi x}{a} e^{\gamma_n z}, \quad z < 0 \tag{23.25}$$

$$H_x^r(x,z) = \sum_{n=1}^{\infty} \frac{\gamma_n}{j\omega\mu_0} E_n \sin\frac{n\pi x}{a} e^{\gamma_n z}, \quad z < 0 \tag{23.26}$$

$$\gamma_n = \sqrt{\frac{\pi^2}{n^2 a^2} - k_0^2}. \tag{23.27}$$

Since the PEC iris boundary condition at $z = 0$, and $0 < x < c$, is that E_y is zero, we get the equation

$$E_y^i(x,0) + E_y^r(x, 0) = 0, \quad 0 < x < c \tag{23.28}$$

$$\sum_{n=1}^{\infty} E_n \sin\frac{n\pi x}{a} = -\sin\frac{\pi x}{a}, \quad 0 < x < c \tag{23.29}$$

The continuity condition on H_x at $z = 0$ for the domain $c < x < a$ gives another equation:

$$\sum_{n=1}^{\infty} \frac{\gamma_n}{j\omega\mu_0} E_n \sin\frac{n\pi x}{a}, \quad 0 < x < c \tag{23.30}$$

Numerical methods to solve (23.29) and (23.30) simultaneously can be done by first terminating the summation to a finite number, say N, in (23.29) and (23.30) and generating the N required number of algebraic equations using the weighted residual method (WRM) discussed in Section 15.3.2. For example, one can choose N_1 weight pulse functions in the subinterval $(0, c)$ and $(N - N_1)$ in the subinterval (c, a).

Example 23.2: Field Distribution in Double-Ridged Waveguide

While TRM method discussed in Section 23.2 is useful in getting the cutoff k_c information of a double-ridged waveguide and quadruple-ridged waveguide for various modes [11], it does not give the field distribution in the guide. Having determined k_c, one can use the matching technique of this section to get an accurate computation of the field distribution. Ji [2], using this technique, computed the field distribution in double-ridged as well as quadruple-ridged wave guide for TE_{10}, TE_{20}, and TE_{11} modes and verified the solution by a commercial software that uses the FEM. As an example of such a solution, we give below the field distribution computation of TE_{10} mode of a double-ridged waveguide, which is also discussed in [10,11].

It is worthwhile to refresh your knowledge on the choice of admissible functions from a template for a rectangular waveguide in Chapter 3 and the restricted Fourier series expansion in Appendix 6A of Volume 1. We give here a detailed derivation of the fields in a double-ridged waveguide as an example of solving such problems. Because of the symmetry involved, we need to consider only half of the cross section as shown in Figure 23.8. In this Example 23.2, we will use a bar on the top to indicate a vector.

Let us start with region II. We make the approximation that in this region the electric field of a single TE_{10} mode will adequately represent the electric field. For this mode, the electric field has only y component and it can be written directly by inspection by the requirement that the electric field is maximum at the center of the ridge $x = a/2$; it is symmetric with respect to $x = a/2$. We assume a unit amplitude for the electric field. Note that the PEC boundary condition (BC) on the faces of the ridges is automatically satisfied.

$$\tilde{\bar{E}}_{II} = \hat{y}\cos k_c\left(\frac{a}{2} - x\right)e^{-j\beta z} \tag{23.31}$$

$$\beta^2 = k_0^2 - k_c^2 \tag{23.32}$$

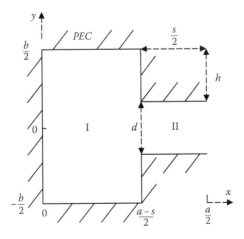

FIGURE 23.8
Left half of the cross section of a double-ridged waveguide.

The corresponding magnetic field is obtained from Maxwell's equation

$$\bar{\nabla} \times \tilde{\bar{E}} = -j\omega\mu_0 \tilde{\bar{H}} \tag{23.33}$$

$$\begin{vmatrix} \hat{x} & \hat{y} & \hat{z} \\ \dfrac{\partial}{\partial x} & 0 & -j\beta \\ 0 & E_y & 0 \end{vmatrix} = -j\omega\mu_0 \tilde{\bar{H}} \tag{23.34}$$

$$\hat{x}\left(0 + j\beta E_y\right) - \hat{y}\left(0-0\right) + \hat{z}\frac{\partial E_y}{\partial x} = -j\omega\mu_0\left(\tilde{H}_x + \tilde{H}_y + \tilde{H}_z\right) \tag{23.35}$$

$$\tilde{H}_x = \frac{j\beta}{-j\omega\mu_0}\tilde{E}_y = -\frac{\beta}{\omega\mu_0}\cos k_c\left(\frac{a}{2}-x\right)e^{-j\beta z} \tag{23.36}$$

$$\tilde{H}_z = -\frac{1}{j\omega\mu_0}\left[-\sin k_c\left(\frac{a}{2}-x\right)\right]\left(-k_c\right)e^{-j\beta z}. \tag{23.37}$$

Thus,

$$\tilde{\bar{H}}_{\text{II}} = -\frac{\beta}{\omega\mu_0}\left\{\hat{x}\cos k_c\left(\frac{a}{2}-x\right) + \hat{z}\frac{k_c}{j\beta}\sin k_c\left(\frac{a}{2}-x\right)\right\}e^{-j\beta z}. \tag{23.38}$$

Now let us look at region I (Figure 23.9) and consider the form of E_y component.

$$\nabla^2\tilde{E}_y + k_0^2\tilde{E}_y = 0 \tag{23.39}$$

$$\tilde{E}_y = G_y\left(x,y\right)e^{-j\beta z} \tag{23.40}$$

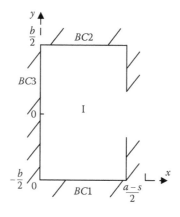

FIGURE 23.9
Region I of the double-ridged waveguide.

From (23.39) and (23.40),

$$\nabla_t^2 G_y + \left[k_0^2 + \left(-j\beta \right)^2 \right] G_y = \nabla_t^2 G_y + k_c^2 G_y = 0$$

Assuming a product solution for G_y,

$$G_y = f_1(x) f_2(y).$$

From BC1 and BC2 (Figure 23.9) at the PEC boundary,

$$\frac{\partial G_y}{\partial y} = 0, \quad y = -\frac{b}{2}, \frac{b}{2} \tag{23.41a}$$

we note

$$f_2(y) \sim \cos \frac{2n\pi y}{b}, \quad n = 0, 1, 2, \cdots \tag{23.41b}$$

From BC3 at the PEC boundary,

$$E_y = E_{\text{tan}} = 0, x = 0; \quad G_y = 0, x = 0 \tag{23.42}$$

$$f_1(x) \sim \sin \alpha_n x, \tag{23.43}$$

where

$$\alpha_n^2 = k_c^2 - \left(\frac{2n\pi}{b} \right)^2$$

So $\tilde{E}_y$ is of the form

$$\tilde{E}_y = \sum_{n=1,2,\ldots}^{\infty} E_{yn} \sin \alpha_n x \cos \frac{2n\pi y}{b} e^{-j\beta z} \tag{23.44}$$

Now we will determine the form of $\tilde{E}_x$.
   Let

$$\tilde{E}_x = G_x(x,y) e^{-j\beta z} = f_1(x) f_2(y) e^{-j\beta z}$$

$$\nabla_t^2 G_x + k_c^2 G_x = 0$$

Let us look at the boundary conditions to determine the form of $f_1(x)$ and $f_2(y)$.
   The BC1 and BC2 for E_x are

$$E_x = E_{\text{tan}} = 0, \quad y = -\frac{b}{2} \text{ or } \frac{b}{2}. \tag{23.45}$$

Hence,

$$f_2(y) \sim \sin\frac{2n\pi y}{b}, \quad n = 1, 2, \ldots.$$

From BC3,

$$\frac{\partial E_x}{\partial x} = \frac{\partial E_n}{\partial n} = 0, \quad x = 0. \tag{23.46}$$

Hence,

$$f_1(x) \sim \cos\alpha_n x.$$

Therefore,

$$\tilde{E}_x = \sum_{n=0,1,2,\ldots}^{\infty} E_{xn} \cos\alpha_n x \sin\frac{2n\pi y}{b} e^{-j\beta z}. \tag{23.47}$$

To determine the unknown amplitudes of the electric field, we use the equation

$$\bar{\nabla} \cdot \tilde{E} = 0 \tag{23.48}$$

$$\frac{\partial \tilde{E}_x}{\partial x} + \frac{\partial \tilde{E}_y}{\partial y} = 0 \tag{23.49}$$

$$(-\alpha_n) E_{xn} + \left(-\frac{2n\pi}{b}\right) E_{yn} = 0$$

$$E_{yn} = -\alpha_n A_n \tag{23.50}$$

$$E_{xn} = \left(\frac{2n\pi}{b}\right) A_n. \tag{23.51}$$

Thus, the electric field is described with common unknown amplitude (A_n):

$$\tilde{E}_y = \sum_{n=1,2,\ldots}^{\infty} A_n(-\alpha_n) \sin\alpha_n x \cos\frac{2n\pi y}{b} e^{-j\beta z} \tag{23.52}$$

$$\tilde{E}_x = \sum_{n=0,1,2,\ldots}^{\infty} \left(\frac{2n\pi}{b}\right) A_n \cos\alpha_n x \sin\frac{n\pi y}{b} e^{-j\beta z}. \tag{23.53}$$

Finally,

$$\tilde{E}_l = \sum_{n=0}^{\infty} A_n \left[\hat{x}\frac{2n\pi}{b}\cos\alpha_n x \sin\frac{2n\pi y}{b} - \hat{y}\alpha_n \sin\alpha_n x \cos\frac{2n\pi y}{b}\right] e^{-j\beta z}. \tag{23.54}$$

The unknown parameter (A_n) is derived from the boundary condition of continuity of $\tilde{E}_y$ at $x=\ell$. By equating (matching) $\tilde{E}_y$ of both regions at $x=\ell$,

$$\psi(y) = \sum_{n=0}^{\infty} A_n(-\alpha_n)\sin\alpha_n\ell\cos\frac{2n\pi y}{b} = \cos k_c\left(\frac{a}{2}-\ell\right)$$

$$= 0, \quad -\frac{d}{2} < y < \frac{d}{2}, \quad -\frac{b}{2} < y < -\frac{d}{2}, \quad \frac{d}{2} < y < \frac{b}{2}.$$

Assume

$$B_n = A_n(-\alpha_n)\sin\alpha_n\ell \tag{23.55}$$

$$A_n = \frac{B_n}{(-\alpha_n)\sin\alpha_n\ell} \tag{23.56}$$

$$\psi(y) = \sum_{n=0,1,2,\ldots}^{\infty} B_n\cos\frac{2n\pi y}{b} = \cos k_c\left(\frac{a}{2}-\ell\right) = \cos\frac{k_c s}{2}.$$

The Fourier coefficient B_n can be now determined from the orthogonality property:

$$B_0 = \frac{1}{b}\int_{-\frac{d}{2}}^{\frac{d}{2}}\cos\frac{k_c s}{2}dy = \frac{d}{b}\cos\frac{k_c s}{2}$$

$$B_n = \frac{2}{b}\int_{-\frac{d}{2}}^{\frac{d}{2}}\cos\frac{k_c s}{2}\cos\frac{2n\pi y}{b}dy = \frac{2\cos\frac{k_c s}{2}}{b}\left(\frac{b}{2n\pi}\right)\sin\frac{2n\pi y}{b}\Big|_{-\frac{d}{2}}^{\frac{d}{2}}$$

$$= \frac{2\cos\frac{k_c s}{2}}{b}\left(\frac{b}{2n\pi}\right)\left[\sin\frac{2n\pi}{b}\left(\frac{d}{2}\right) - \sin\frac{2n\pi}{b}\left(-\frac{d}{2}\right)\right] = \frac{2\cos\frac{k_c s}{2}}{n\pi}\sin\frac{n\pi d}{b}$$

Therefore,

$$A_n = \frac{B_n}{(-\alpha_n)\sin\alpha_n\ell} = \frac{\Gamma_n\cos\frac{k_c s}{2}}{n\pi(-\alpha_n)\sin\alpha_n\ell}\sin\frac{n\pi d}{b} \qquad \begin{array}{l}\Gamma_n = 1, \ n = 0 \\ \Gamma_n = 2, \ n = 1,2,\ldots\end{array}, \tag{23.57}$$

and the electric field in the region I can now be computed from (23.54).
The corresponding magnetic field is obtained from Maxwell's equations

$$\bar{\nabla}\times\tilde{E} = -j\omega\mu_0\tilde{H} \tag{23.58}$$

$$\begin{vmatrix} \hat{x} & \hat{y} & \hat{z} \\ \dfrac{\partial}{\partial x} & \dfrac{\partial}{\partial y} & -j\beta \\ E_x & E_y & 0 \end{vmatrix} = -j\omega\mu_0 \tilde{\tilde{H}}$$

(23.59)

$$\hat{x}\left(0 + j\beta E_y\right) - \hat{y}\left(0 + j\beta E_x\right) + \hat{z}\left(\frac{\partial E_y}{\partial x} - \frac{\partial E_x}{\partial y}\right) = -j\omega\mu_0\left(\tilde{H}_x + \tilde{H}_y + \tilde{H}_z\right)$$

(23.60)

$$\tilde{H}_x = \frac{j\beta}{-j\omega\mu_0}\tilde{E}_y = \frac{\beta}{\omega\mu_0}\sum_{n=0}^{\infty} A_n \alpha_n \sin\alpha_n x \cos\frac{2n\pi y}{b}$$

(23.61)

$$\tilde{H}_y = \frac{-j\beta}{-j\omega\mu_0}\tilde{E}_x = \frac{\beta}{\omega\mu_0}\sum_{n=0}^{\infty} A_n \frac{2n\pi}{b}\cos\alpha_n x \sin\frac{2n\pi y}{b}$$

(23.62)

$$\tilde{H}_z = -\frac{1}{j\omega\mu_0}\left(\frac{\partial E_y}{\partial x} - \frac{\partial E_x}{\partial y}\right) = -\frac{1}{j\omega\mu_0}\left[\sum_{n=0}^{\infty} A_n\alpha_n^2 \cos\alpha_n x \cos\frac{2n\pi y}{b} - \sum_{n=0}^{\infty} A_n\left(\frac{2n\pi}{b}\right)^2 \cos\alpha_n x \cos\frac{2n\pi y}{b}\right]$$

(23.63)

$$\hat{\tilde{H}} = \frac{\beta}{\omega\mu_0}\left\{\begin{aligned}&\hat{x}\sum_{n=0}^{\infty} A_n\alpha_n \sin\alpha_n x\cos\frac{2n\pi y}{b} + \hat{y}\sum_{n=0}^{\infty} A_n \frac{2n\pi}{b}\cos\alpha_n x\sin\frac{2n\pi y}{b}\\ &-\hat{z}\frac{1}{j\beta}\left[\sum_{n=0}^{\infty} A_n\alpha_n^2 \cos\alpha_n x\cos\frac{2n\pi y}{b} - \sum_{n=0}^{\infty} A_n\left(\frac{2n\pi}{b}\right)^2\cos\alpha_n x\cos\frac{2n\pi y}{b}\right]\end{aligned}\right\}e^{-j\beta z}$$

(23.64)

with A_n as given in (23.57).

The electric field distribution of TE_{10} mode is depicted in Figure 23.10. Part (a) is based on the Equations 23.31, 23.54, and 23.57. Part (b) is based on the simulation using a commercial simulation package.

Similar derivations for the TE_{20} and TE_{11} modes are a little more involved and given as problems at the end of the chapter.

A perturbation technique, as discussed in Chapter 22, is added to the analysis to get the results for modified DRW and QRW and illustrated by chamfering the ridges of the wave-guide in [2].

23.4 Orthogonal Functions and Integral Transforms

Integral transform methods are extensively used in engineering and physics to solve a variety of problems [12–17]. They include Laplace transform, Fourier transform, z transform, weighted Laguerre polynomial (WLP) transform, Hankel transform, Mellin transform, Legendre transform, etc. Their main advantage is conversion of time/space domain differential, integral, difference equations into algebraic equations in the so-called spectral (transformed) domain thus reducing to solving algebraic operations instead of solving

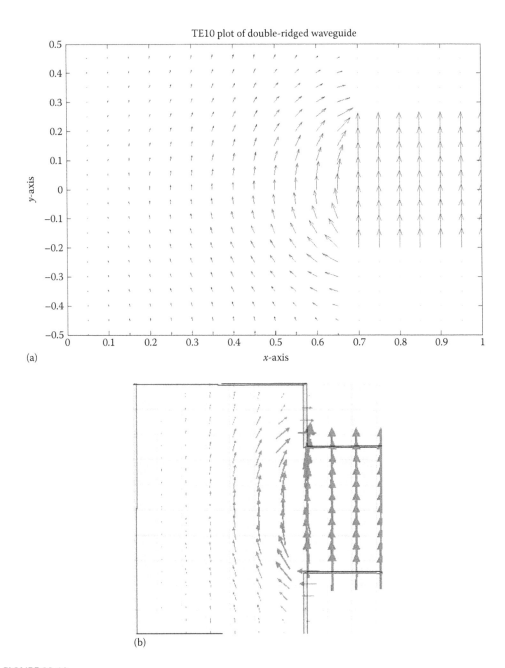

FIGURE 23.10
TE$_{10}$ plot of the electric field of the double-ridged waveguide ($S = 0.6$, $D = 0.5$) (a) based on Equations 23.31, 23.54, and 23.57, (b) HFSS simulation.

differential-integral-convolutional-difference type of operations in one or more variables as an intermediate step. The inverse transform step gives the solution in the time–space domain. If the inversion can be done by manipulating the spectral solution (for example, by partial fraction expansion) into forms of the inversion tables of standard type of functions in the spectral domain, extensively available [13,17], considerable effort is saved.

In the context of electromagnetics and Maxwell's equations, we have partial differential equations with a maximum of three spatial coordinates and time, a total of four independent variables. One can convert one or more of these variables into spectral domains suitable for each of those variables.

We will review these transforms quickly mainly to introduce the notation. Their applications in electromagnetic computation are given in Sections 23.6 and 23.7.

23.4.1 Laplace Transform

The Laplace transform of $f(t)$ is defined by the integral

$$L\left[f(t)\right]=F(s)=\int_0^\infty f(t)e^{-st}\,dt \tag{23.65}$$

and its inverse by the integral

$$L^{-1}\left[F(s)=f(t)=\frac{1}{2\pi j}\left[\int_{\sigma_1-j\infty}^{\sigma_1+j\infty}F(s)e^{st}\,ds\right]\right. \tag{23.66}$$

The Laplace transform of the derivative of $f(t)$ is given by

$$L\left[\frac{df}{dt}\right]=sF(s)-f(0) \tag{23.67}$$

From (23.67) we note that the differentiation in the time domain is equivalent to the multiplication in the spectral s-domain. Moreover, the initial condition $f(0)$ is automatically introduced into the algebraic equation in the s-domain, when converting differential equations in time domain to the s-domain.

Another important property is the conversion of the convolution operation in the time domain into the algebraic operation of multiplication in the s-domain as given in (23.68)

$$L\left[\int_0^1 f_1(t)f_2(t-\tau)d\tau\right]=L\left[f_1(t)*f_2(t)\right]+F_1(s)F_2(s) \tag{23.68}$$

For an example of using the Laplace transform technique in electromagnetics, see Section 13.3 for obtaining the transient solution for the voltage on a lossy transmission line.

23.4.2 Fourier Transform

In signal processing applications, one uses Fourier transform defined by

$$F(\omega)=\left[\int_{-\infty}^{+\infty}f(t)e^{-j\omega t}\,dt\right]. \tag{23.69}$$

Its inverse transform is given by

$$f(t) = \frac{1}{2\pi}\left[\int_{-\infty}^{+\infty} f(\omega)e^{j\omega t}\,d\omega\right]. \tag{23.70}$$

The operation of differentiation in time domain transforms into multiplication with $j\omega$ in the frequency domain.

$$FT\left(\frac{df(t)}{dt}\right) = j\omega F \tag{23.71}$$

In the two applications of transient analysis and signal processing extensively taught in the undergraduate courses, the independent variable is t. In system theory, the black box relates the output $f_o(t)$ to the input $f_i(t)$ by the system function or transfer function. The relationships between these entities in the s-domain or ω domain are given below:

$$\left(\frac{L\{f_o(t)\}}{L\{f_i(t)\}}\right) = H(s); \quad \left(\frac{FT\{f_o(t)\}}{FT\{f_i(t)\}}\right) = H(j\omega) \tag{23.72}$$

If the input is an impulse, the output is $h(t)$, the Laplace inverse of $H(s)$, and is called the impulse response. The impulse response $h(t)$ essentially characterizes the system in the black box so that the response to an arbitrary input function $f_i(t)$ can be obtained as

$$f_o(t) = \left[\int_0^t f_i(\tau)h(t-\tau)d\tau\right] \tag{23.73}$$

$$f_o(t) = \left[f_i(t) * h(t)\right] \tag{23.74}$$

where the symbol (*) denotes convolution. In the spectral domain,

$$F_o(s) = H(s)F_i(s) \tag{23.75}$$

Another way of looking at (23.73) and (23.75) is that the former requires an integration operation in time domain, whereas (23.75) requires an algebraic operation of multiplication. The output function f_o in time domain is then obtained by inverting the Laplace transform.

In electromagnetics, the Green's function discussed in Section 7.2 plays the same role as the impulse response. In Section 7.2 we obtained $G(r, r')$ in spatial domain. The output function, using Green's function is

$$f_o(r) = \left[\int_{\text{source}} G(r,r')f_i(r')\,dS'\right] \tag{23.76}$$

The evaluation of this integral may be difficult since the Green's function may have a difficult type of singularity not easily integrable. If one can define a suitable transform so that in the transformed spectral domain one can obtain an algebraic equation like (23.75),

$$F_o(k) = G(k, k') F_i(k') \tag{23.77}$$

where k is the transforming variable that converts the convolution in the spatial domain into multiplication in the spectral domain. Of course, since we still would like to have the solution in the spatial domain, converting $F_o(k)$ to $f_o(r)$ could involve the evaluation of an integral. If you are lucky, you may be able to invert the transform by looking up or making use of tables of integral transforms, like in [13]. Any case, if one has to do the evaluation of the integral, hopefully $G(k, k')$ might involve a more easily integral singularity than the integral involving $G(r, r')$. Obviously the choice of the transform type appropriate to the problem on hand is important. The Green's functions in the spatial domain and its corresponding spectral domain form for some of the important cases are discussed in Section 23.5.

23.4.3 Z Transform

Z transform deals with $f(n)$ instead of $f(t)$, where $f(n)$ are the discrete values of a function at integer values of the independent variable n. Its z transform is defined in (23.78), and its inverse in (23.79).

$$ZT[f(n)] = F(z) = \sum_{n=0}^{\infty} f(n) z^{-n} \tag{23.78}$$

$$ZT^{-1}[F(z)] = f(n) = \frac{1}{2\pi j}\left[\oint F(z) z^{n-1} dz\right] \tag{23.79}$$

The operation similar to differentiation in discrete domain is shifting $f(n+1)$ and its z transform is given by

$$ZT[f(n+1)] = zF(z) - zf(0). \tag{23.80}$$

The convolution operation in discrete domain (n domain) is equivalent to the multiplication in the z domain.

$$ZT\left[\sum_{k=0}^{n} f_1(k) f_2(n+k)\right] = F_1(z) F_2(z) \tag{23.81}$$

Equation 23.80 is useful in converting a difference equation into an algebraic equation. Thus, it is useful in studying the FDTD solution. Section 23.6 shows such an application.

23.4.4 Mellin Transform

It is defined by

$$F_M(s) = MT(f(r)) = \int_0^{\infty} f(r) r^{s-1} dr \tag{23.82}$$

The Mellin transform of $(r\,df/dr)$ is given by

$$MT\left(r\frac{df}{dr}\right) = -sF_M(s)$$ (23.83)

Obviously the differential equation of the type

$$A_n r^n \frac{d^n f}{dr^n} + A_{n-1} r^{n-1} \frac{d^{n-1} f}{dr^{n-1}} + \cdots + A_0 f(r) = g(r)$$ (23.84)

can be converted to an algebraic equation by using Mellin transform.

23.4.5 Hankel Transform

In solving boundary value problems in cylindrical coordinates, one comes across a differential operator D defined by (23.85)

$$\frac{1}{\rho}\frac{d}{d\rho}\left(\rho\frac{df(\rho)}{d\rho}\right) - \frac{n^2}{\rho^2} f(\rho) = Df(\rho).$$ (23.85)

By defining a Hankel transform,

$$F_H(s) = HK(f(\rho)) = \int_0^\infty f\rho J_v(s\rho)\,d\rho$$ (23.86)

One can Hankel transform (23.85) into (23.87):

$$HK\left[Df(\rho)\right] = -s^2 F_H(s)$$ (23.87)

23.4.6 Legendre Transform

Legendre transform is defined by

$$LGR\left[f(z)\right] = F(n) = \int_{-1}^{1} f(z) P_n(z)\,dz,$$ (23.88)

where P_n is a Legendre polynomial. Equation 23.89 brings out the advantage of using (23.88) in converting a certain type of derivative operation that occurs in spherical coordinates.

$$LGR\left[\frac{d}{dz}\left\{(1-z^2)\frac{df(z)}{dz}\right\}\right] = -n(n+1)F(n).$$ (23.89)

It can be noted that the factor $C(\theta)$ given by

$$C(\theta) = \frac{1}{\sin\theta}\frac{d}{d\theta}\left(\sin\theta\frac{df}{d\theta}\right)$$ (23.90)

is the same as the factor in the square bracket [] on the left side of (23.89), if $z = \cos\theta$. In fact Legendre polynomials can be shown to be the solutions of the differential equation

$$\left[\frac{d}{dz}\left\{\left(1-z^2\right)\frac{df(z)}{dz}\right\}\right] + n(n+1)f(z) = 0 \tag{23.91}$$

Legendre transformation is defined to convert differential equations in spherical coordinates in spatial domain into the algebraic equations in spectral domain.

23.4.7 Finite Fourier Transform

Equations 23.92 and 23.93 define Fourier finite sine transform and Fourier finite cosine transform, respectively.

$$FS[f(x)] = F_s(m) = \int_0^L f(x)\sin\frac{m\pi x}{L}dx \tag{23.92}$$

$$FC[f(x)] = F_c(m) = \int_0^L f(x)\cos\frac{m\pi x}{L}dx \tag{23.93}$$

The transformations of (df/dx) for these two transforms are given by

$$FS\left[\frac{df}{dx}\right] = -\frac{m\pi}{L}F_c \tag{23.94}$$

$$FC\left[\frac{df}{dx}\right] = \frac{m\pi}{L}F_s + (-1)^m f(L) - f(0) \tag{23.95}$$

It is easy to derive the transformations of the higher-order derivatives. Some of the examples given in Chapter 6 can be solved by using these transforms.

23.4.8 WLP Transform

Laguerre polynomials in their simplest form are defined as [15]

$$L_p(t) = \frac{e^t}{p!}\frac{d^p}{dt^p}\left(t^p e^{-t}\right) \quad p \geq 0, \ t \geq 0 \tag{23.96}$$

In the above, p is the order of this polynomial. Low-order L_p's are listed below:

$$L_0 = 1 \tag{23.97}$$

$$L_1 = 1 - t \tag{23.98}$$

They satisfy the differential equation

$$p\frac{d^2 L_p}{dt^2} - (1-t)\frac{dL_p}{dt} + pL_p = 0 \tag{23.99}$$

and a recursive relation

$$pL_p(t) = (2p-1-t)L_{p-1}(t) - (p-1)L_{p-2}(t). \tag{23.100}$$

The Laguerre polynomials are orthogonal with respect to the weight function e^{-t}, that is,

$$\int_0^\infty e^{-t}L_p(t)L_q(t)\,\mathrm{d}t = \delta_{pq} \tag{23.101}$$

An alternative definition of Laguerre polynomials given in [16]

$$L_m(\varsigma) = \frac{e^{\varsigma/2}}{m!}\frac{\mathrm{d}^m}{\mathrm{d}\varsigma^m}\left(e^{-\varsigma}\varsigma^m\right) \tag{23.102}$$

absorbs the weight function given in (23.99) in the definition of the polynomial itself. This definition given in (23.102) thus becomes an orthonormal set and is used in [16] to discuss Laguerre optics of ultrashort transients in dispersive media. Laguerre optics has become important since it has now become possible to generate single-cycle waveforms.

Laguerre polynomials given in [15] have assumed great importance in giving an implicit FDTD algorithm, discussed in Section 23.6. In this context, the weighted Laguerre polynomials (WLP) are defined as *follows* [15]:

$$\Phi_p(t,s) = \frac{e^{-s.t/2}}{m!}L_p(s.t), \tag{23.103}$$

where $s > 0$ is a scale factor. Defining the scaled time

$$\breve{t} = s.t \tag{23.104}$$

it is easy to see

$$\int_0^\infty \Phi_p(\breve{t})\Phi_q(\breve{t})\,\mathrm{d}\breve{t} = \delta_{pq}. \tag{23.105}$$

The complete WLP set can be used to expand a causal function $U(t)$ in WLP series:

$$U(t) = \sum_{p=0}^\infty U_p\Phi_p(\breve{t}) \tag{23.106}$$

The derivative of U can be shown to be [17]

$$\frac{\mathrm{d}U(t)}{\mathrm{d}t} = s\left(\sum_{p=0}^\infty 0.5U_p - \sum_{k=0,p>0}^{p-1} U_k\right)\Phi_p(\breve{t}) \tag{23.107}$$

Its application in developing WLP–FDTD method, overcoming the Courant–Friedrichs–Lewy (CFL) limit in the choice of the time step Δt, is discussed in Section 23.6.

23.5 Examples of Green's Functions in Spatial and Spectral Domains

Section 7.2 and Appendix 7A of Volume 1 discuss the Green's function in the spatial domain extensively. In this section, we discuss two examples of Green's functions in the spatial domain and their corresponding spectral domain functions. The two-dimensional (x–y plane) free space spatial domain Green's function $\tilde{G}(x,y)$ is given by

$$\tilde{G}(x,y) = \frac{e^{-jk\sqrt{x^2+y^2}}}{4\pi\sqrt{x^2+y^2}} \tag{23.108}$$

Its Fourier transform $G(\alpha,\beta)$ is defined by

$$G(\alpha,\beta) = \int\limits_{-\infty}^{\infty}\int\limits_{-\infty}^{\infty}\tilde{G}(x,y)e^{-j(\alpha x+\beta y)}\,\mathrm{d}x\mathrm{d}y \tag{23.109}$$

Substituting (23.108) in (23.109), it can be shown [18] that

$$G(\alpha,\beta) = \frac{-j}{2\sqrt{k^2-\alpha^2-\beta^2}}\quad \left(k^2 > \alpha^2+\beta^2\right) \tag{23.110a}$$

$$G(\alpha,\beta) = \frac{1}{2\sqrt{\alpha^2+\beta^2-k^2}}\quad \left(k^2 < \alpha^2+\beta^2\right) \tag{23.110b}$$

In obtaining (23.110) from (23.109), the following identity is used [19]:

$$\frac{1}{2\pi}\int\limits_{0}^{2\pi}e^{-jk_\rho\rho\cos(\phi-\phi_k)}\,\mathrm{d}\phi = J_0\left(k_\rho\rho\right), \tag{23.111}$$

where

$$\alpha = k_\rho\cos\phi_k \tag{23.112a}$$

$$\beta = k_\rho\sin\phi_k \tag{23.112b}$$

$$x = \rho\cos\phi \tag{23.113a}$$

$$y = \rho\sin\phi. \tag{23.113b}$$

Another example of spatial domain and corresponding spectral domain Green's functions are those useful in solving thin wire(radius $a \ll$ wavelength λ) dipole antenna radiation/scattering problems. The spatial domain Green's function $\tilde{G}(a,z)$ is given by

$$\tilde{G}(a,z) = \frac{e^{-jk\sqrt{a^2+z^2}}}{4\pi\sqrt{a^2+z^2}}, \tag{23.114}$$

where

$$\sqrt{a^2 + z^2} = r \tag{23.115}$$

is the distance from the origin of a point on the surface of the cylindrical wire.

The Fourier-spectral-domain function $G(a, k_z)$, from definition, is given by

$$G(a, k_z) = \int_{-\infty}^{\infty} \tilde{G}(a, z) e^{-jk_z z} \, dz \tag{23.116}$$

Substituting (23.114) in (23.116) and evaluating, it can be shown [18] that

$$G(a, k_z) = \frac{1}{4j} H_0^{(2)} \left(a\sqrt{k^2 - k_z^2} \right) \tag{23.117}$$

In arriving at (23.117), the following identity [19] is used:

$$\frac{e^{-jkr}}{4\pi r} = \frac{1}{8\pi j} \int_{-\infty}^{\infty} H_0^{(2)} \left(a\sqrt{k^2 - w^2} \right) e^{jwz} \, dw \tag{23.118}$$

23.6 WLP-FDTD Method

The conventional FDTD introduced in Chapter 19 is an *explicit* method leading to step-by-step leap-frog algorithm. The only restriction on it is the CFL $(c \Delta t / \Delta x)$ limit on the time step Δt in relation to the space step for ensuring the stability of the algorithm. For problems with big changes in the electromagnetic properties in space, the space step has to be small, and consequently, Δt has to be small. It could take many iterations N_L to march the solution to the required value of t, say t_F. The implicit methods allow independent choice of Δt and Δx with no concern for the stability of the method. Alternating direction implicit (ADI) method [20] is one of them. However, the ADI scheme suffers from the increase in numerical dispersion error with the increase in the size of the time step. A better method with unconditionally stable scheme called weighted Laguerre polynomial (WLP)-FDTD method is well explained in a seminal paper [15] using a two-dimensional formulation. This WLP-FDTD technique subsequently expanded and applied to many problems can be found in [21–25].

In Section 23.4, we discussed the orthonormal WLP. From that discussion, we surmise that they are completely convergent to zero as t tends to infinity. This property allows the time-varying part of the fields to be expanded in WLP's and separate the space variation of the fields. Therein lies the property of unconditional stability of the WLP-FDTD.

To quantify these ideas and understand the concept of WLP-FDTD, we will simplify the example of TE^z of parallel plate waveguide in [15] to a one-dimensional FDTD considered in Section 19.4. The final step of LU factorization of a tridiagonal matrix is well known and well suited to explain the concept than the seven diagonal matrix dealt with in [15].

Let

$$\mathbf{E} = \hat{z}E \tag{23.119}$$

$$\mathbf{H} = \hat{y}H \tag{23.120}$$

Then, the first order differential equations for E and H are

$$\dot{E} = \frac{1}{\varepsilon}\frac{\partial H}{\partial x} \tag{23.121}$$

$$\dot{H} = \frac{1}{\mu}\frac{\partial E}{\partial x}. \tag{23.122}$$

Expanding these fields in terms of WLP set

$$E(x,t) = \sum_{p=0}^{\infty} E_p(x)\Phi_p(\breve{t}) \tag{23.123}$$

$$H(x,t) = \sum_{p=0}^{\infty} H_p(x)\Phi_p(\breve{t}). \tag{23.124}$$

From the general transformation equation of the derivative given in Equation 23.107

$$\dot{U}(x,t) = s\left\{\sum_{p=0}^{\infty} 0.5U_p(x) + \sum_{k=0}^{p-1} U_k(x)\right\}\Phi_p(\breve{t}) \tag{23.125}$$

we can write (23.123) and (23.124)

$$s\left\{\sum_{p=0}^{\infty} 0.5E^p(x) + \sum_{k=0,p>0}^{p-1} E^k(x)\right\}\Phi_p(\breve{t}) = \frac{1}{\varepsilon}\left[\sum_{p=0}^{\infty} \frac{\partial H^p(x)}{\partial x}\right]\Phi_p(\breve{t}) \tag{23.126}$$

$$s\left\{\sum_{p=0}^{\infty} 0.5H^p(x) + \sum_{k=0,p>0}^{p-1} H^k(x)\right\}\Phi_p(\breve{t}) = \frac{1}{\varepsilon}\left[\sum_{p=0}^{\infty} \frac{\partial E^p(x)}{\partial x}\right]\Phi_p(\breve{t}) \tag{23.127}$$

Orthonormal property of WLP is stated in (23.105) and will be used to process the equations further.

Multiply (23.126) and (23.127) by $\Phi_q(\breve{t})$, integrate with respect to $\breve{t}$ from 0 to ∞, and use (23.105) to obtain (23.128) and (23.129):

$$s\left\{0.5E^q(x) + \sum_{k=0}^{q-1} E^k(x)\right\} = \frac{1}{\varepsilon(x)}\frac{\partial H^q(x)}{\partial x} \tag{23.128}$$

$$s\left\{0.5H^q(x) + \sum_{k=0}^{q-1} H^k(x)\right\} = \frac{1}{\mu(x)}\frac{\partial E^q(x)}{\partial x} \tag{23.129}$$

Rewrite (23.128) and (23.129) using central difference approximation for the partial derivative with respect to x and obtain (23.130), (23.132), and (23.134):

$$E_i^q = C_i^E \left[H_{i+\frac{1}{2}}^q - H_{i-\frac{1}{2}}^q \right] - 2\sum_{k=0}^{q-1} E_i^k \qquad (23.130)$$

$$C_i^E = \frac{2}{s\varepsilon_i \Delta x} \qquad (23.131)$$

$$H_{i+\frac{1}{2}}^q = C_{i+\frac{1}{2}}^H \left[E_{i+1}^q - E_i^q \right] - 2\sum_{k=0}^{q-1} H_{i+\frac{1}{2}}^k \qquad (23.132)$$

$$C_{i+\frac{1}{2}}^H = \frac{2}{s\mu_{i+\frac{1}{2}} \Delta x} \qquad (23.133)$$

From (23.132),

$$H_{i-\frac{1}{2}}^q = C_{i-\frac{1}{2}}^H \left[E_i^q - E_{i-1}^q \right] - 2\sum_{k=0}^{q-1} H_{i-\frac{1}{2}}^k \qquad (23.134)$$

Substituting (23.132) and (23.134) in (23.130), we obtain

$$E_i^q = C_i^E \left[C_{i+\frac{1}{2}}^H \left\{ E_{i+1}^q - E_i^q \right\} - 2\sum_{k=0}^{q-1} H_{i+\frac{1}{2}}^k - C_{i-\frac{1}{2}}^H \left\{ E_i^q - E_{i-1}^q \right\} - 2\sum_{k=0}^{q-1} H_{i-\frac{1}{2}}^k \right] - 2\sum_{k=0}^{q-1} E_i^k \qquad (23.135)$$

In the above, one can treat all those with $(q - 1)$ superscript as known from the previous iteration and be moved to the right side of the equation as the source term β^{q-1}. Equation 23.136 thus relates the three unknowns of E^q with the known β^{q-1}.

$$C_i^E C_{i-\frac{1}{2}}^H E_{i-1}^q + \left(1 + C_i^E C_{i+\frac{1}{2}}^H + C_i^E C_{i-\frac{1}{2}}^H \right) E_i^q - C_i^E C_{i+\frac{1}{2}}^H E_{i+1}^q = \beta^{q-1} \qquad (23.136)$$

$$\beta^{q-1} = C_i^E \left[-2\sum_{k=0}^{q-1} H_{i+\frac{1}{2}}^k - 2\sum_{k=0}^{q-1} H_{i-\frac{1}{2}}^k \right] - 2\sum_{k=0}^{q-1} E_i^k \qquad (23.137)$$

Equation 23.136 shows that each row will have three nonzero terms. Generalizing (23.136) for all i we get the matrix equation

$$[A]\{E^q\} = \{\beta^{q-1}\}. \qquad (23.138)$$

Had there been a source current in our formulation, that is,

$$J = \hat{z}J, \qquad (23.139)$$

then (23.121) gets modified as

$$\dot{E} = \frac{1}{\varepsilon}\left[\frac{\partial H}{\partial x} - J\right].$$ (23.140)

The corresponding modification in (23.138) is given by

$$[A]\{E^q\} = \{J^q\} + \{\beta^{q-1}\}$$ (23.141)

where

$$J^q(x) = \int_0^{T_f} J(x,\breve{t})\Phi_q(\breve{t})\,\mathrm{d}\breve{t}.$$ (23.142)

In (23.142), the upper limit infinity is replaced by a finite time interval, beyond which the source current may be considered as negligible.

It can be noted that the system matrix $[A]$ in (23.141) is a sparse tridiagonal matrix independent of the order q. The right side of (23.141) can be considered as excitation since $\{J_q\}$ are given and $\{\beta^{q-1}\}$, which is the summation term of the lower-order terms, are also known during the computation of the previous iteration. The important point to be noted is that the elements of the system matrix $[A]$ are independent of the order q of the basis function $\Phi(\breve{t})$. Thus, the stability is no longer affected by the time step size. In fact, the time step size only appears in (23.142), in calculating $J^q(x)$. To improve accuracy, one can compute these by choosing a small value for Δt to evaluate (23.142) at the start of the computation. Thus, $\{J_q\}$ are known ahead of the main computation.

For zero order (23.141) can be written as

$$[A]\{E^0\} = \{J^0\}$$ (23.143)

because $\{\beta^{-1}\} = 0$.

Since $[A]$ is a sparse matrix, independent of q, only once at the beginning of computation one can use LU factorization of $[A]$. Then one can solve (23.141) by using the back-substitution routine repeatedly. Of course the magnetic field for any order q is obtained from (23.132) and (23.134), knowing E^q and H^{q-1}.

At this point, we summarize the mechanics of the computation including the factors, as given in [15], that play a role in determining the number of times (23.141) is to be solved recursively to obtain an accurate solution. We already know T_f, the duration of interest in the time domain. So the real time signal $P(t)$ can be expressed in Fourier series given by

$$P(t) = \sum C_u e^{ju\omega_0 t}$$ (23.144)

where $\omega_0 = 2\pi/T_f$. If $P(t)$ is band limited to B Hertz, then the value of u in (23.144) can be expressed by the bounds

$$-B \le \frac{u}{T_f} \le B.$$ (23.145)

Hence, the minimum number N_L of the temporal functions needed is given by

$$N_L = 2BT_f + 1 \qquad (23.146)$$

Thus, N_L terms of Laguerre series are needed to completely characterize the temporal waveform of duration T_f and bandwidth 2B.

Conventional FDTD needs Δt small enough to satisfy *CFL*. In the two-dimensional example of a parallel plate waveguide with a thin slot in a conductor given in [15], conventional FDTD needed $\Delta t = 3.35\,$fs and 3,570,699 iterations. The *CPU* time is *214* s. The same problem solved by WLP-FDTD required $\Delta t = 10$ ps, $N_L = 151$, and *CPU* time of 2.75 s.

In [15] the authors have discussed the implementation of PEC boundary conditions and a dispersive boundary condition (DBC) as an absorbing boundary condition (ABC) to terminate the domain in the context of WLP-FDTD.

In [21] WLP-FDTD is applied to wave propagation in isotropic plasma. The improvements to WLP method explained in [21] are as follows:

1. Factorization splitting (FS) WLP scheme to solve the algebraic equations. Compared to solving large-size sparse matrices as in [15], the factorization splitting scheme improves the efficiency.

2. Implementation of perfectly matched layer (PML) for absorbing boundary conditions, in the context of WLP-FDTD for isotropic plasma.

The same authors gave the extension of [21] to wave propagation in a magnetized plasma [22] to solve numerically the Equations 12.4, 12.5, and 12.70 of Volume 1 by a different method than given in [23]. The WLP-FDTD method is extended to general dispersive media in [24] and to gyrotropic materials in [25].

23.7 Z-transform and FDTD Method

In Section 23.4, Equation 23.78 defines the z-transform [12,26], which converts difference equations into algebraic equations in the z-domain. In the FDTD method, we use finite difference approximations to derivatives to generate difference equations. The impulse response $h(t)$ will now be a discrete function $h(n)$ in the discrete domain n, and the convolution in n-domain will be multiplication in z-domain. Thus, the z-transform and the FDTD solutions are connected. Sullivan [26] had developed this connection and illustrated through several examples [27–29].

Let $f(t)$ be a real causal function and $f(n)$ define a sequence of real numbers obtained by sampling $f(t)$:

$$f(n) = f(t)\delta(t - nT), \quad 0 \le n \le \infty \qquad (23.147)$$

where T is the time period between numbers. The z-transform as defined in [26] is given by (23.148):

$$Z\big[f(n)\big] = F(z) = T\sum_{n=0}^{\infty} f(n)z^{-n} \qquad (23.148)$$

Note that the symbol for z-transform is Z in this section (instead of ZT used in the Section 23.4). Comparing (23.78) of Section 23.4, we have an extra T in front of the summation on the right side of (23.148). The T term introduced in the notation of [26] may be considered as a weight for the sequence $f(n)$.

An important property that follows from (23.148) is the shifting property

$$Z\left[f(n-1)T\right] = z^{-1}F(z). \tag{23.149}$$

Thus, z^{-1} represents a delay of one time period. This leads to an approximation to the time derivative in Maxwell's equations:

$$\frac{df(t)}{dt} \cong \frac{f(n \cdot \Delta t) - f(n-1)\Delta t}{\Delta t}. \tag{23.150}$$

From (23.149), the z-transform of (23.150) can be obtained as

$$Z\left[\frac{df(t)}{dt}\right] = \frac{1 - z^{-1}}{\Delta t}F(z). \tag{23.151}$$

The z-transform of an integral in time domain is given by [28]

$$Z\left[\int_0^t f(\tau)\,d\tau\right] = \frac{\Delta t}{1 - z^{-1}}F(z). \tag{23.152}$$

Equation 23.150 is a backward rectangular approximation for the derivative of a function in the time domain. If the constitutive relations are given in the frequency domain, we can connect the frequency domain description with the time domain by noting

$$\frac{d}{dt} \Rightarrow s = j\omega. \tag{23.153}$$

Thus, we can replace $j\omega$ in frequency description as per the rule

$$j\omega \Rightarrow \frac{1 - z^{-1}}{\Delta t}. \tag{23.154}$$

Equation 23.154 is equivalent to approximating the time-domain convolution by a rectangular approximation of the convolution integration. A better approximation is the trapezoidal approximation, called bilinear transform [26], and is given by

$$j\omega \Rightarrow \frac{2}{\Delta t}\frac{1 - z^{-1}}{1 + z^{-1}}. \tag{23.155}$$

Intuitively one can see that (23.155) will lead to better accuracy since the integral is evaluated by using trapezoidal approximation. However, it doubles the order of z in the z-domain.

Another aspect of importance is to consider the transformation of the convolution integral

$$y(t) = \int_0^\infty h(t-\tau)x(\tau)\,d\tau. \tag{23.156}$$

For a causal function, the upper limit in (23.156) can be replaced by t. In the discrete domain

$$y(n) \cong \sum_{i=0}^{n} h(n-i)x(i)\Delta t. \tag{23.157}$$

Taking the z-transform both sides [26]

$$\Delta t \sum_{n=0}^{\infty} y(n)z^{-n} = \Delta t \sum_{n=0}^{\infty} \sum_{i=0}^{n} h(n-i)x(i)z^{-n}\Delta t \tag{23.158}$$

$$Y(z) = \Delta t X(z)H(z). \tag{23.159}$$

Convolution in n-domain given by (23.157) translates into multiplication in z-domain given in (23.159).

However, the z-transform of a product of two sequences in the n-domain translates into complex convolution denoted by $\otimes$ in the Equation 23.160:

$$Z\big[f_1(n)f_2(n)\big] = F_1(z) \otimes F_2(z). \tag{23.160}$$

The rule given in [28] explains the symbol $\otimes$ used to denote complex convolution, different from the symbol * normally used for convolution. The symbol $\otimes$ indicates a term-by-term multiplication. In (23.160), it is not the multiplication of the z-transforms F_1 and F_2. Equation 23.161 gives the meaning of term-by-term multiplication:

$$Z\big[f_1(n)f_2(n)\big] = F_1(z) \otimes F_2(z) = f_1(0)f_2(0) + f_1(1)f_2(1)z^{-1} + f_1(2)f_2(2)z^{-2} + \cdots. \tag{23.161}$$

If

$$F_1(z) \otimes F_2(z) = G(z) \tag{23.162}$$

then

$$F_2(z) = \frac{G(z)}{F_1(z)} \tag{23.163}$$

with the meaning

$$F_2(z) = \frac{g(0)}{f_1(0)} + \frac{g(1)}{f_1(1)}z^{-1} + \frac{g(2)}{f_1(2)}z^{-2} + \cdots. \tag{23.164}$$

Though no special symbol is used in (23.163) for the inverse operation to $\otimes$, the operation denoted by the division symbol (numerator over denominator), from (23.164), it is clear that the meaning is term-by-term division.

$$f_2(0) = \frac{g(0)}{f_1(0)}, \quad f_2(1) = \frac{g(1)}{f_1(1)}, \quad f_2(2) = \frac{g(2)}{f_1(2)} \cdots. \tag{23.165}$$

If in (23.161) $f_1 = f_2 = f$, then

$$Z\big[f(n).f(n)\big] = f(0)f(0) + f(1)f(1)z^{-1} + f(2)f(2)z^{-2} + \cdots = F^2(z). \tag{23.166}$$

The right-most term in (23.166) is different from $[F(z)]^2$, which is the square of the z-transform of $f(n)$ [28].

The transforms like (23.166) will be useful to solve nonlinear problems [29].

23.7.1 Solution of Maxwell's Equations Using z-Transform and FDTD

The application of the z-transform technique using bilinear transform to wave propagation in anisotropic dispersive media, with magnetized plasma as an example, is illustrated in [30]. The major steps in formulating and obtaining the algorithm are given below.

Assuming that the background static magnetic field is in z-direction, the dielectric tensor of the magnetoplasma is derived and is given by (11.78) through (11.81) of Volume 1. Using D-H formulation ($B = \mu_0 H$), the relevant FDTD equations are [30]

$$D^n = D^{n-1} + \Delta t (\nabla \times H)^{n-1/2} \tag{23.167}$$

$$B^{n+1/2} = B^{n-1/2} - \Delta t (\nabla \times E)^n. \tag{23.168}$$

From the dielectric tensor,

$$D_x = \left[\frac{1 + \dfrac{\omega_p^2}{(j\omega)^2}\left(1 + \dfrac{\nu}{j\omega}\right)}{\left(1 + \dfrac{\nu}{j\omega}\right)^2 + \dfrac{\omega_b^2}{(j\omega)^2}} \right] E_x + \left[\frac{\dfrac{\omega_p^2}{(j\omega)^2}\dfrac{(-\omega_b)}{(j\omega)}}{\left(1 + \dfrac{\nu}{j\omega}\right)^2 + \dfrac{\omega_b^2}{(j\omega)^2}} \right] E_y \tag{23.169}$$

Equation for D_y can be obtained in a similar manner:

$$D_y = f_1\big(\omega, \omega_p^2, \omega_b, \nu\big) E_x + f_2\big(\omega, \omega_p^2, \omega_b, \nu\big) E_y \tag{23.170}$$

Replacing $j\omega$ by using the bilinear transform given by (23.155), one can obtain from (23.169) and (23.170), after converting z^{-1} as a time-shifting operator, two equations relating D_x, D_y with E_x, E_y. Solving those simultaneously, E_x^n, E_y^n can be expressed in terms of

$$D_x^{n-p}, \quad p = 0, 1, 2, 3$$

$$D_y^{n-p}, \quad p = 0, 1, 2, 3$$

$$E_x^{n-p}, \quad p = 1, 2, 3$$

$$E_y^{n-p}, \quad p = 1, 2, 3$$

Thus E_x and E_y at the current time step n can be expressed in terms of D_x, D_y at the current and previous three time steps, and E_x and E_y at the three previous time steps, as given in Equations 15 and 16 of Reference [30].

The z components of D and E are related by (23.171):

$$D_z = \varepsilon_0 \varepsilon_{r\|} E_z, \tag{23.171}$$

where $\varepsilon_{r\|}$ is given by (11.81) of Volume 1. One can thus obtain the relation between E_z at the current time step in terms of D_z at the current and the previous time steps and E_z at the previous time steps by using the bilateral z-transform as given in Equation 17 of Reference [30].

The algorithm thus consists of updating D at nth time step using (23.167), updating E at the current time step n from D's at the current and previous time steps, as well as E's at the previous time steps as mentioned above. Now we have all the required information of B's and E's on the right side of (23.168) to get updated values of B at the time step $(n + 1/2)$.

For open boundaries, one can employ PML implementation using stretched coordinates (SC) complex frequency shifted (CFS) factor discussed in Chapter 21, since for PML, Equations 23.167 and 23.168 can be applied without any additional considerations.

23.8 Laplace Transform and FDTD Method for Magnetized Plasma

An alternative formulation of the FDTD method for wave propagation in a magnetized plasma is discussed in [23], which is the same as Appendix 23A. The constitutive relation (11.70) of the Volume 1 is implemented by locating the components of the current variable J at the center of Yee's cube and expressing the first order time derivatives of the components of J in a state variable form. The solution of those three equations are obtained by the Laplace transform technique. The inverse transform gives the algorithm for J's at $(n + 1/2)$ time step and the values of E's at the nth time step. This algorithm leads to explicit computation for any direction and magnitude of the static magnetic field as well as the value of the collision frequency.

23.9 Frequency Selective Surfaces, Periodic Boundary Condition, and Fourier-Spectral Domain Method

As an example of the application of the spectral domain method (SDM) and its advantages over a spatial domain solution [18,31,32], we will discuss the so-called frequency selective surfaces (FSS), in particular a screen in the x–y plane with the periodic rectangular metal patches shown in Figures 23.11 and 23.12. A qualitative discussion of the expected results of FSS filters is mentioned in Section 23.1 and the URL of a free download from the internet given therein of an article by Hooberman will help to broaden the understanding of the FSS.

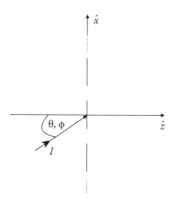

FIGURE 23.11
Scattering by a periodic planar metal screen in the x–y plane.

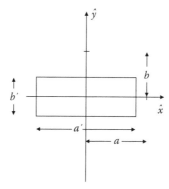

FIGURE 23.12
Periodic array of rectangular metal patches in the x–y plane. One metal patch in the unit cell is shown.

Let us first formulate the scattering problem. The electric field $\tilde{E}^I$ of an incident wave induces a surface current $\tilde{K}$ on the patches, which in turn give rise to a scattered field, whose electric field will be denoted by $\tilde{E}^S$. The boundary condition on the patches is given by

$$\left(\tilde{E}^S + \tilde{E}^I\right)\bigg|_{z=0} = 0. \tag{23.172}$$

The equation relating $\tilde{K}$ to $\tilde{E}^S$ can be obtained through the magnetic vector potential $\tilde{A}$ by using (1.50), (1.46), and (1.49) of Volume 1. Equation 1.50 of Volume 1 can be written as

$$\tilde{A}(r) = \mu_0 \iiint G(r,r') \tilde{J}(r') \mathrm{d}V' \tag{23.173}$$

$$\tilde{A}(r) = \mu_0 G(r) * \tilde{J}(r) \tag{23.174}$$

$$G(r) = \frac{e^{-jkr}}{4\pi r}. \tag{23.175}$$

For the problem on hand, the scattered fields are due to the surface currents on the patches. Thus in (23.173), we replace $\tilde{J}dV'$ by $\tilde{K}ds'$ and in (23.174) we replace the position vector by

$$r = \hat{x}x + \hat{y}y \tag{23.176}$$

$$r = \sqrt{x^2 + y^2}. \tag{23.177}$$

The vector potential in the $z=0$ plane is given by

$$\tilde{A}(x,y) = \mu_0 G(x,y) * \tilde{K}(x,y). \tag{23.178}$$

One can solve such a problem in the spatial domain but one has to carefully evaluate the integrals like (23.173) which contain a singularity at $r=r'$. In fact, the example in Section 16.8.1, where the integral equation containing the singularity, was solved by the moment method (MM). The direct relation between $\tilde{E}$ and $\tilde{A}$ can be obtained by eliminating $\tilde{\Phi}$ from (1.46) of Volume 1 using (1.49) of Volume 1:

$$\tilde{E} = \frac{1}{j\omega\mu_0\varepsilon_0} \left[k_0^2\tilde{A} + \nabla\left(\nabla \cdot \tilde{A}\right) \right]. \tag{23.179}$$

In the $z=0$ plane,

$$\begin{Bmatrix} \tilde{E}_x^s(x,y) \\ \tilde{E}_y^s(x,y) \end{Bmatrix} = \frac{1}{j\omega\mu_0\varepsilon_0} \begin{bmatrix} \dfrac{\partial}{\partial x^2} + k_0^2 & \dfrac{\partial}{\partial x}\dfrac{\partial}{\partial y} \\ \dfrac{\partial}{\partial x}\dfrac{\partial}{\partial y} & \dfrac{\partial}{\partial y^2} + k_0^2 \end{bmatrix} \begin{bmatrix} \tilde{A}_x(x,y) \\ \tilde{A}_y(x,y) \end{bmatrix}. \tag{23.180}$$

Fourier-transforming (23.180) with

$$\frac{\partial}{\partial x} \Rightarrow -j\alpha, \quad \frac{\partial}{\partial y} \Rightarrow -j\beta, \tag{23.181}$$

we obtain

$$\begin{Bmatrix} E_x^s(\alpha,\beta) \\ E_y^s(\alpha,\beta) \end{Bmatrix} = \frac{1}{j\omega\mu_0\varepsilon_0} \begin{bmatrix} k_0^2 - \alpha^2 & -\alpha\beta \\ -\alpha\beta & k_0^2 - \beta^2 \end{bmatrix} \begin{bmatrix} A_x(\alpha,\beta) \\ A_y(\alpha,\beta) \end{bmatrix}. \tag{23.182}$$

The inverse transform of $A(\alpha,\beta)$ is

$$\tilde{A}(x,y,z) = \iint A(\alpha,\beta)\bigg|_{z=0} e^{j(\alpha x + \beta y)} e^{\pm j\sqrt{k_0^2 - \alpha^2 - \beta^2}\,z}\, d\alpha d\beta. \tag{23.183}$$

Equation 23.183 can be interpreted as a plane wave expansion. In the above, the upper sign is used for $z<0$ and the lower sign for $z>0$. The relation between $\tilde{A}$ and $\tilde{K}$ in the spatial domain, Equation 23.178, can be converted to the spectral domain and is given by

$$\begin{Bmatrix} A_x(\alpha,\beta) \\ A_y(\alpha,\beta) \end{Bmatrix} = \mu_0 \begin{bmatrix} G(\alpha,\beta) & 0 \\ 0 & G(\alpha,\beta) \end{bmatrix} \begin{bmatrix} K_x(\alpha,\beta) \\ K_y(\alpha,\beta) \end{bmatrix}, \tag{23.184}$$

where

$$G(\alpha, \beta) = \frac{-j}{2\sqrt{k_0^2 - \alpha^2 - \beta^2}}. \tag{23.185}$$

Equation 23.185 is the same as (23.110a) discussed in Section 23.5. We assume that the spatial periodicity of the screen is such that (23.110a) applies.

The FSS screen shown in Figure 23.12 is a periodic structure of rectangular patches, such that the current flowing on the patches can be written as

$$\tilde{K}(x + ma, y + nb) = \tilde{K}(x, y) e^{j(\alpha_m a + \beta_n b)} \tag{23.186}$$

$$\alpha_m = \alpha_0 + \frac{2m\pi}{a} \tag{23.187}$$

$$\beta_n = \beta_0 + \frac{2n\pi}{b}, \tag{23.188}$$

where α_0 and β_0 belong to the incident wave.

Substituting (23.185) in (23.181),

$$\begin{Bmatrix} E_x^s(\alpha_m, \beta_n) \\ E_y^s(\alpha_m, \beta_n) \end{Bmatrix} = \frac{1}{j\omega\mu_0\varepsilon_0} \begin{bmatrix} k_0^2 - \alpha_m^2 & -\alpha_m\beta_n \\ -\alpha_m\beta_n & k_0^2 - \beta_n^2 \end{bmatrix} \mu_0 \begin{bmatrix} G(\alpha_m, \beta_n) & 0 \\ 0 & G(\alpha_m, \beta_n) \end{bmatrix} \begin{Bmatrix} K_x(\alpha_m, \beta_n) \\ K_y(\alpha_m, \beta_n) \end{Bmatrix}. \tag{23.189}$$

Note that the frequency spectrum is discrete (m, n are integers) and we used the Fourier series rather than the Fourier transform because the structure is periodic. The corresponding modes are Floquet modes [32]. See Section 9.7 of Volume 1 for an example of the Floquet modes in one dimension. Thus, we get the relation between the spectrum of the scattered electric field components and the surface currents on the patches. Substituting (23.185) in (23.189) and simplifying we get

$$\begin{Bmatrix} E_x^s(\alpha_m, \beta_n) \\ E_y^s(\alpha_m, \beta_n) \end{Bmatrix} = -\frac{1}{2\omega\varepsilon_0\sqrt{k_0^2 - \alpha_m^2 - \beta_n^2}} \begin{bmatrix} k_0^2 - \alpha_m^2 & -\alpha_m\beta_n \\ -\alpha_m\beta_n & k_0^2 - \beta_n^2 \end{bmatrix} \begin{Bmatrix} K_x(\alpha_m, \beta_n) \\ K_y(\alpha_m, \beta_n) \end{Bmatrix}. \tag{23.190}$$

Since

$$\tilde{E}^s(x, y) = \sum_m \sum_n E^s(\alpha_m, \beta_n)\big|_{z=0} e^{j(\alpha_m x + \beta_n y)}, \tag{23.191}$$

from (23.172) we get

$$-\frac{1}{2\omega\varepsilon_0} \sum_m \sum_n \frac{1}{\sqrt{k_0^2 - \alpha_m^2 - \beta_n^2}} \begin{bmatrix} k_0^2 - \alpha_m^2 & -\alpha_m\beta_n \\ -\alpha_m\beta_n & k_0^2 - \beta_n^2 \end{bmatrix} \begin{Bmatrix} K_x(\alpha_m, \beta_n) \\ K_y(\alpha_m, \beta_n) \end{Bmatrix} e^{j(\alpha_m x + \beta_n y)}$$
$$= -\begin{Bmatrix} \tilde{E}_x^I(x, y) \\ \tilde{E}_y^I(x, y) \end{Bmatrix}. \tag{23.192}$$

Equation 23.192 is the algebraic electric field equation applicable on the metal patches, obtained by using the spectral domain method. Note that (23.192) needs to be enforced on one patch only. It can be solved numerically by using, say, Galerkin method (see Section 15.3.2.3). When K_x and K_y are determined from (23.192), one can obtain $\tilde{E}_x^s$ and $\tilde{E}_y^s$, using (23.190) and (23.191). The singularity due to the expression under the square root in (23.192) can be avoided in the computation by choosing the cell spacings and the wavelength so that the expression does not become zero. This becomes possible since the spectrum is discrete. However, the solution in the spatial domain involves a singularity in the integrand and requires a careful evaluation of this singularity under the integral sign.

The advantage of this method as compared to, say, FEM is that one can choose a small number of basis functions resulting in a small size, fully populated matrix equation. FEM results in a large size, though sparsely populated, matrix equation.

Scott [18] illustrated the use of a combination of *TE* and *TM* waveguide modes as basis functions and developed the numerical computation procedure. He further computed results for three different widths of a rectangular patch FSS for the case of normal incidence of an *x*-polarized plane wave. The results show "that the rectangular plate FSS acts as a simple band-reject filter, the bandwidth of which depends on the width of the plate."

Mittra [31] discussed the application of the SDM for determining the scattering from FSS consisting of circular metal patches.

23.10 Transfer Matrix Method and Layered Media

In Section 7.3.1, we discussed, using ABCD parameters, the computation of the reflection and transmission coefficients for N transmission lines in tandem. In this section we discuss a generalization of such a technique to the propagation of electromagnetic waves in a layered media, called the transfer matrix method (TMM). Let us assume that the layered medium has the z-coordinate as the longitudinal direction. In view of the boundary conditions on the tangential components of the electric and magnetic fields at the interfaces (planes parallel to the x–y plane) of the layers, we assume the following form for the fields:

$$f(x, y, z) = f_1(z)e^{-jk_x x}e^{-jk_y y}. \tag{23.193}$$

Then it is possible to write the relation between the tangential components of the fields in the state variable form [33], where the state variables are the tangential components. For the case of a dielectric layer with the parameters ε_r, μ_r, it is easy to derive the following matrix equation:

$$\frac{d}{dz'}\{\psi\} = [\Omega]\{\psi\}, \tag{23.194}$$

where

$$\{\psi\} = \begin{Bmatrix} E_x(z') \\ E_y(z') \\ H_x(z') \\ H_y(z') \end{Bmatrix} \tag{23.195}$$

$$[\Omega] = \begin{bmatrix} 0 & 0 & \dfrac{k'_x k'_y}{\varepsilon_r} & \mu_r - \dfrac{k'^2_x}{\varepsilon_r} \\ 0 & 0 & \dfrac{k'^2_y}{\varepsilon_r} - \mu_r & -\dfrac{k'_x k'_y}{\varepsilon_r} \\ \dfrac{k'_x k'_y}{\mu_r} & \varepsilon_r - \dfrac{k'^2_x}{\mu_r} & 0 & 0 \\ \dfrac{k'^2_y}{\mu_r} - \varepsilon_r & -\dfrac{k'_x k'_y}{\mu_r} & 0 & 0 \end{bmatrix}$$

(23.196)

$$k'_x = \frac{k_x}{k_0}, \quad k'_y = \frac{k_y}{k_0}, \quad z' = k_0 z. \tag{23.197}$$

In (23.197), the normalized quantities are denoted by a prime.

It can be noted that we used the state variable form (first order derivative of the state variable to the left and no derivative term on the right side) many times before (see Appendix 10B of Volume 1), because in this form, the solution can be written as [33]

$$\Psi(z') = e^{[\Omega]z'}\Psi(0). \tag{23.198}$$

In (23.198), we encounter the evaluation of a function of a matrix. As shown in [33], the solution is given by

$$f([A]) = [W]f([\lambda])[W]^{-1}, \tag{23.199}$$

where
  [W] is the eigenvector matrix
  [λ] is the diagonal matrix of eigenvalues of [A]

From (23.198) and (23.199),

$$\{\Psi(z')\} = [W]e^{[\lambda]z'}[W]^{-1}\{\Psi(0)\}$$

$$\{\Psi(z')\} = [W]e^{[\lambda]z'}\{C\}, \tag{23.200a}$$

where

$$\{C\} = [W]^{-1}\{\Psi(0)\}. \tag{23.200b}$$

For a multiple layer system, we can impose the continuity conditions at the two interfaces of the ith layer and obtain

$$\{C_{i+1}\} = [T_i]\{C_i\}, \tag{23.201a}$$

where

$$[T_i] = [W_{i+1}]^{-1}[W_i]e^{[\lambda_i]k_0 L_i} \tag{23.201b}$$

and L_i is the width of the ith layer.

The $[T_i]$ matrix in (23.201b) is called the transfer matrix for the ith layer, since it transfers the fields from the ith layer to the fields in the $(i + 1)$th layer.

It can be shown that the global relation is given by

$$\{C_T\} = [T]\{C_I\}, \tag{23.202}$$

where

$$[T] = [T_N].[T_{N-1}]...[T_1].[W_1]^{-1}.[W_I]. \tag{23.203}$$

Thus, $[T]$ is the global transfer matrix relating the fields in the incident region (subscript I) to the fields in the transmitted region (subscript T).

TMM, as given in (23.203), can be computationally unstable. Rumpf has shown a method to fix this problem.

The author, Rumpf, R., has shown a method to fix the instability problem in Lecture 4 on Transfer Matrix Method of a downloadable (http://emlab.utep.edu/ee5390cem/Lecture%204%20--%20Transfer%20Matrix%20Method.pdf) e-course with the title *"Computational Electromagnetics"*.

References

1. Balanis, C. A., *Advanced Engineering Electromagnetics*, 2nd edition, Wiley, New York, 2012.
2. Ji, C., Approximate analytical techniques in the study of quadruple-ridged waveguide and its modifications, Doctoral thesis, University of Massachusetts Lowell, Lowell, MA, 2007.
3. Sorrentono, R., *Numerical Methods for Passive Microwave and Millimeter Wave Structures*, in: Sorrentino, R. (Ed.), part 8, IEEE Press, New York, 1989.
4. Wexler, A., Solution of waveguide discontinuities by modal analysis, *IEEE Transactions Microwave Theory and Techniques*, MTT-15(9), 508–517, 1967.
5. Kowalski, G. and Pregla, R., Dispersion characteristics of shielded microstrips with finite thickness, *Arch. Elek. Ubertragung*, 25, 193–196, April 1971.
6. Lee, S. W., Jones, W. R., and Campbell, J. J., Convergence of numerical solution of iris-type discontinuity problems, *IEEE Trans. Microw. Theory Tech.*, MTT-19(6), 528–536, June 1971.
7. Mittra, R. and Itoh, T., in: Mitra, R. (Ed.), *Computer Techniques for Electromagnetics*, Some Numerically Efficient Methods, Pergamon Press, Elmsford, NY, 1973.
8. Bootan Jr., R. C., *Computer Methods for Electromagnetics and Microwaves*, Wiley, New York, 1992.
9. Elliot, R. S., *An Introduction to Guided Waves and Microwave Circuits*, Prentice Hall, Englewood Cliffs, NJ, 1993.
10. Getsinger, W. J., Ridged waveguide field description and application to directional couplers, *IRE Trans. Microw. Theory Tech*, MTT-10, 41–50, January 1962.
11. Helszajn, J., *Ridge Waveguides and Passive Microwave Components*, IEE Electromagnetic Wave Series 49, London, U.K., 2000.

12. Aseltine, J. A., *Transform Method in Linear System Analysis*, McGraw-Hill Book Company, New York, 1958.

13. Bateman, H., *Tables of Integral Transforms*, McGraw-Hill, New York, 1954.

14. Debnath, L. and Bhatia, D., *Integral Transforms and Applications*, 3rd edition, CRC Press, Boca Raton, FL, 2015.

15. Chung, Y.-S., Sircar, T. K., Jung, B. H., and Salazar-Palma, M., An unconditionally stable scheme for a finite-difference time-domain method, *IEEE Trans. Microw. Theory Tech.*, 51(3), 697–704, March 2003.

16. Shvartsburg, A. B., *Impulse Time-domain Electromagnetics of Continuous Media*, Birkhauser, Boston, MA, 1999.

17. Gradshteyn, I. S. and Ryzhik, I. M., *Table of Integrals, Series, and Products*, Academic, New York, 1980.

18. Scott, C., *The Spectral Domain Method in Electromagnetics*, Artech House, Norwood, MA, 1989.

19. Harrington, R. F., *Time-Harmonic Electromagnetic Fields*, IEEE Press, New York, 2001.

20. Namiki, T., A new FDTD algorithm based on alternating-direction implicit method, *IEEE Trans. Microw. Theory Tech.*, 47(10), 2003–2007, October 1999.

21. Fang, Y., Xi, X.-L., Wu, J.-M., Liu, J.-F., and Pu, Y.-R., A J-E collocated WLP-FDTD model of wave propagation in isotropic cold plasma, *IEEE Trans. Microw. Theory Tech.*, 64(7), 1957–1965, July 2016.

22. Zhang, J.-S., Xi, X.-L., Li, Z.-W., Liu, J.-F., and Pu, Y.-R., An unconditionally stable WLP-FDTD model of wave propagation in magnetized plasma, *IEEE Trans. Plasma Sci.*, 44(1), 25–30, January 2016.

23. Lee, J. H. and Kalluri, D. K., Three-dimensional FDTD simulation of EMW transformation in a dynamic inhomogeneous magnetized plasma, *IEEE Trans. Antennas Propagat.*, 47(7), 1146–1151, July 1999.

24. Chen, W.-J., Shao, Y., and Wang, B.-Z., ADE-Laguerre-FDTD method for wave propagation in general dispersive materials, *IEEE Microw. Wireless Compon. Lett.*, 23(5), 228–230, May 2013.

25. Li, Z.-W., Xi, X.-L., Zhang, J.-S., and Liu, J.-F., Unconditionally stable WLP-FDTD method for the modeling of electromagnetic wave propagation in gyrotropic materials, *Opt. Express*, 23(25), 31864, December 2015. DOI: 10.1364/OE.23.031864.

26. Sullivan, D. M., *Electromagnetic Simulation Using the FDTD Method*, IEEE Press, Piscataway, NJ, 2000.

27. Sullivan, D. M., Frequency-dependent FDTD methods using z transforms, *IEEE Trans. Antennas Propagat.*, 40(10), 1223–1230, October 1992.

28. Sullivan, D. M., Z-transform theory and FDTD method, *IEEE Trans. Antennas Propagat.*, 44(1), 28–34, January 1996.

29. Sullivan, D. M., Nonlinear FDTD formulation using z transform, *IEEE Trans. Antennas Propagat.*, 43(3), 676–682, October 1992.

30. Xi, X., Li, Z., and Song, Z., FDTD algorithm for simulation of wave propagation in anisotropic dispersive material based on bilinear transform, *IEEE Trans. Antennas Propagat.*, 63(11), 5134–5138, November 2015.

31. Mittra, R., Spectral-domain analysis of circular patch frequency selective surfaces, *IEEE Trans. Antennas Propagat.*, Ap12, 533–536, May 1984.

32. Ishimaru, A., *Electromagnetic Wave Propagation, Radiation, and Scattering*, Prentice Hall, Englewood Cliffs, NJ, 1991.

33. Derusso, P. M., Roy, R. J., and Close, C. H., *State Variables for Engineers*, Wiley, New York, 1965.

Part VI

Appendices

Appendix 16A: MATLAB® Programs

16A.1 Example of a Main Program and Function Program GLANT to Solve Laplace Equation (see Figure 16.12)

```
function [S,T] = GLANT(Nn,Ne,n1L,n2L,n3L,xn,yn);
% Global Assembly two-dimensional node-based
% triangular elements
for e = 1:Ne;
        n(1,e) = n1L(e);
        n(2,e) = n2L(e);
        n(3,e) = n3L(e);
end
% Initialization
S = zeros(Nn,Nn);
T = zeros(Nn,Nn);
% Loop through all elements
for e = 1: Ne;
        % coordinates of the element nodes
        for i = 1:3;
                x(i) = xn(n(i,e));
                y(i) = yn(n(i,e));
        end
        % compute the element matrix entries
        b(1) = y(2) - y(3);
        b(2) = y(3) - y(1);
        b(3) = y(1) - y(2);
        c(1) = x(3) - x(2);
        c(2) = x(1) - x(3);
        c(3) = x(2) - x(1);
        Area = 0.5 * abs (b(2) * c(3) - b(3) * c(2));
        % Compute the element matrix entries
        for i = 1:3;
                for j = 1:3;
                        Se(i,j) = (0.25/Area) * (b(i) * b(j) + c(i) * c(j));
                        if i == j
                                Te(i,j) =  Area/6;
                        else
                                Te(i,j) =  Area/12;
                        end
% Assemble the Element matrices into Global FEM System
                        S(n(i,e),n(j,e)) = S(n(i,e),n(j,e)) + Se(i,j);
                        T(n(i,e),n(j,e)) = T(n(i,e),n(j,e)) + Te(i,j);
                end
        end
end
```

```
>> Ne = 2;
>> Nn = 4;
>> n1L =[1,2];
>> n2L =[2,3];
>> n3L =[4,4];
>> xn =[0.8,1.4,2.1,1.2];
>> yn =[1.8,1.4,2.1,2.7];
>> [S,T] = GLANT(Nn,Ne,n1L,n2L,n3L,xn,yn);
>> S

S =

  1.2357        -0.7786          0          -0.4571
 -0.7786         1.2500      -0.4571        -0.0143
  0             -0.4571       0.8238        -0.3667
 -0.4571        -0.0143      -0.3667         0.8381

>> T

T =

 0.0583         0.0292          0           0.0292
 0.0292         0.1458       0.0438         0.0729
 0              0.0438       0.0875         0.0438
 0.0292         0.0729       0.0438         0.1458

>> %frn: free node array
>> frn = [1,3];
>> %prn: prescribed node array
>> prn = [2,4];
>> %Vprn: potentials at the prescribed nodes
>> Vprn = [10,-10];
>> Sff = S(frn,frn)

Sff =

1.2357          0
0               0.8238

>> Sfp = S(frn,prn)

Sfp =

-0.7786         -0.4571
-0.4571         -0.3667

>> Vf = -inv(Sff)*Sfp*(Vprn)'

Vf =

2.6012
1.0983

>>
```

16A.2 Example of a Main Program and Function Program PGLANT2 to Solve Poisson's Equation (see Figure 16.14)

```
function [S,T,g] = PGLANT2(Nn,Ne,n1L,n2L,n3L,Rho,Epr,xn,yn);
% Solution of Poisson's Equation
% Laplacian of V = -Rho/(epsilon0*epr)
% Rho is the array of values of volume charge density in each element
% Epr is the array of values of dielectric constant in each element
% Global Assembly two-dimensional node-based
% triangular elements
for e = 1:Ne;
        n(1,e)= n1L(e);
        n(2,e)= n2L(e);
        n(3,e)= n3L(e);
end

% Initialization
S = zeros(Nn,Nn);
T = zeros(Nn,Nn);
g = zeros(Nn,1);
% Loop through all elements
for e = 1: Ne;
        % coordinates of the element nodes
        for i = 1:3;
                x(i)= xn(n(i,e));
                y(i)= yn(n(i,e));
        end
        % compute the element matrix entries
        b(1)= y(2) - y(3);
        b(2)= y(3) - y(1);
        b(3)= y(1) - y(2);
        c(1)= x(3) - x(2);
        c(2)= x(1) - x(3);
        c(3)= x(2) - x(1);
        Area  = 0.5 * abs (b(2)*c(3) - b(3) * c(2));
        % Compute the element matrix entries
        ge = Rho(e)* Area/3;
        for i = 1:3;
                    g(n(i,e))= g(n(i,e)) + ge;
                for j = 1:3;
                    Se(i,j) = (0.25/Area)*(b(i)*b(j) + c(i)
                       *c(j))*Epr(e)*8.854*10^(-12);
                    if i == j
                            Te(i,j)= Area/6;
                    else
                            Te(i,j)= Area/12;
                    end
% Assemble the Element matrices into Global FEM System
                    S(n(i,e),n(j,e)) = S(n(i,e),n(j,e)) + Se(i,j);
                    T(n(i,e),n(j,e)) = T(n(i,e),n(j,e)) + Te(i,j);
                end
        end
end
```

```
>> Nn = 7;
>> Ne = 6;
>> n1L = [1,1,1,1,1,1];
>> n2L = [2,3,4,5,6,7];
>> n3L = [3,4,5,6,7,2];
>> xn = [0.5,0,0,1,1,1,0];
>> yn = [1,1,0,0,1,2,2];
>> Rho = [1,1,1,0,0,0]*10^(-5);
>> Epr = [2.5,2.5,2.5,1,1,1];
>> [S,T,g] = PGLANT2(Nn,Ne,n1L,n2L,n3L,Rho,Epr,xn,yn);
>> frn = [1];
>> Sff = S(frn,frn);
>> gf = g(frn,1);
>> Vf = inv(Sff)*(gf)

Vf =

4.3026e + 004

>>
```

16A.3 Example of a Main Program That Uses Function Program GLANT to Solve the Homogeneous Rectangular Waveguide Problem (see Figure 16.18)

```
>> % Main Program : Rectangular Waveguide
>> Nn = 16;
>> Ne = 18;
>> xn = [0,1/3,2/3,3/3,0,1/3,2/3,3/3,0,0,3/3,3/3,1/3,2/3,1/3,2/3];
>> yn = [0,0,0,0,1/6,1/6,1/6,1/6,2/6,3/6,2/6,3/6,2/6,2/6,3/6,3/6];
>> n1L = [2,2,2,2,3,3,5,5,9,9,7,8,14,11,6,6,13,13];
>> n2L = [5,6,3,7,4,8,6,13,13,15,8,11,11,12,7,14,14,16];
>> n3L = [1,5,7,6,8,7,13,9,15,10,14,14,16,16,14,13,16,15];
>> [S,T] = GLANT(Nn,Ne,n1L,n2L,n3L,xn,yn);
>> ATE = inv(T)* S;
>> [EVTE,kcsqTE] = eig(ATE);
>> kcTE = sqrt(diag(kcsqTE));
>> kcTES = sort(kcTE)

kcTES =

 0.0000
 3.2795
 6.5158
 7.2817
 8.1754
10.3923
12.1165
14.3625
14.9742
17.1434
20.7846
```

```
21.1244
22.4283
23.3670
25.3472
26.6653

>> frn = [6,7,13,14];
>> STM = S(frn,frn);
>> TTM = T(frn,frn);
>> ATM = inv(TTM)* STM;
>> [EVTM,kcsqTM] = eig(ATM);
>> kcTM = sqrt(diag(kcsqTM));
>> kcTMS = sort(kcTM)


kcTMS =

 8.2014
12.1845
17.0024
20.4251


>>
```

16A.4 Example of a Main Program and Function Program GLAN2T Based on Second-Order Triangular Elements to Solve the Homogeneous Waveguide Problem (see Figure 16.23)

```
>> % Example of using Second-order Triangles
>> % Isosceles Triangular Waveguide
>> % 2 elements taken for the purpose of illustration
>> % written by D.K. Kalluri
>> Nn = 9;
>> Ne = 2;
>> n1L = [1 1];
>> n2L = [2 3];
>> n3L = [3 7];
>> n4L = [4 6];
>> n5L = [5 8];
>> n6L = [6 9];
>> x([1 4 6 9]) = [ 0 1 0.5 0];
>> y([1 4 6 9]) = [ 0 0 0.5 1];
>>    [S,T] = GLAN2T(Nn,Ne,n1L,n2L,n3L,n4L,n5L,n6L,x,y);
>> ATE = inv(T)*S;
>> [EVTE,kcsqTE] = eig(ATE);
>> kcTE = sqrt(sort(kcsqTE))

kcTE =

0        0        0        0        0        0        0        0        0
0        0        0        0        0        0        0        0        0
0        0        0        0        0        0        0        0        0
```

```
 0          0          0          0          0          0          0          0          0
 0          0          0          0          0          0          0          0          0
 0          0          0          0          0          0          0          0          0
 0          0          0          0          0          0          0          0          0
 0          0          0          0          0          0          0          0          0
17.7247    14.7641    10.9545    4.6926     6.7702     13.7638    7.7460     0.0000     3.2492

>>
```

```matlab
function [S,T] = GLAN2T(Nn,Ne,n1L,n2L,n3L,n4L,n5L,n6L,xn,yn);
% Global Assembly two-dimensional node-based
% triangular elements of second-order
% written by D. K. Kalluri
for e = 1:Ne;
      n(1,e) = n1L(e);
      n(2,e) = n2L(e);
      n(3,e) = n3L(e);
      n(4,e) = n4L(e);
      n(5,e) = n5L(e);
      n(6,e) = n6L(e);
end
Q1 = (1/6)*[0,0,0,0,0,0;0,8,-8,0,0,0;0,-8,8,0,0,0;
      0,0,0,3,-4,1;0,0,0,-4,8,-4;0,0,0,1,-4,3];
Q2 = (1/6)*[3,0,-4,0,0,1;0,8,0,0,-8,0;-4,0,8,0,0,-4;
      0,0,0,0,0,0;0,-8,0,0,8,0;1,0,-4,0,0,3];
Q3 = (1/6)*[3,-4,0,1,0,0;-4,8,0,-4,0,0;0,0,8,0,-8,0;
      1,-4,0,3,0,0;0,0,-8,0,8,0;0,0,0,0,0,0];
Tek = (1/180)*[6,0,0,-1,-4,-1;0,32,16,0,16,-4;0,16,32,-4,16,0;
      -1,0,-4,6,0,-1;-4,16,16,0,32,0;-1,-4,0,-1,0,6];
% Initialization
S = zeros(Nn,Nn);
T = zeros(Nn,Nn);
% Loop through all elements
for e = 1: Ne;
      % coordinates of the element nodes
      x(1) = xn(n(1,e));
      x(2) = xn(n(4,e));
      x(3) = xn(n(6,e));
      y(1) = yn(n(1,e));
      y(2) = yn(n(4,e));
      y(3) = yn(n(6,e));

% compute the element matrix entries
      b(1) = y(2) - y(3);
      b(2) = y(3) - y(1);
      b(3) = y(1) - y(2);
      c(1) = x(3) - x(2);
      c(2) = x(1) - x(3);
      c(3) = x(2) - x(1);
      Area = 0.5  * abs (b(2) * c(3) - b(3) * c(2));
      COTTH1 = -(0.5/Area)*(b(2) * b(3) + c(2) * c(3));
      COTTH2 = -(0.5/Area)*(b(3)* b(1) + c(3)* c(1));
      COTTH3 = -(0.5/Area)*(b(1)* b(2) + c(1)* c(2));
```

```
        Te = Area*Tek;
        Se = COTTH1 * Q1 + COTTH2 * Q2 + COTTH3 * Q3;
        % Compute the element matrix entries
        for i = 1:6;
                for j = 1:6;
                        % Assemble the Element matrices into Global FEM System
                        S(n(i,e),n(j,e)) = S(n(i,e),n(j,e))+ Se(i,j);
                        T(n(i,e),n(j,e)) = T(n(i,e),n(j,e))+ Te(i,j);
                end
        end
end
```

16A.5 Example of a Main Program and Function Program GLAET (Based on First-Order Vector Shape Functions for Triangles) to Solve the Homogenous Waveguide Problem (see Figure 16.28)

```
>> % Example of first-order edge-based FEM
>> % Two triangular elements for isosceles triangle waveguide problem
>> Nn = 4;
>> Neg = 5;
>> Ne = 2;
>> xn = [0,1,0.5,0];
>> yn = [0,0,0.5,1];
>> n1L = [1 1];
>> n2L = [2 3];
>> n3L = [3 4];
>> n1EL = [1,-3];
>> n2EL = [2,4];
>> n3EL = [3,5];
>>    [E,F] = GLAET(Neg,Nn,Ne,n1L,n2L,n3L,n1EL,n2EL,n3EL,xn,yn);
>> %TM Modes (Htan is the field variable), all edges are free edges
>> ATM = inv(F)* E;
>>    [EVTM,kcsqTM] = eig(ATM);
>> kcTM = sqrt(diag(kcsqTM));
>> kcTMs = sort(kcTM)

kcTMs =

0
0.0000
0 + 0.0000i
6.0000
7.6345
>> %TE Modes, Etan is the field variable, nonconducting edge is 3
>> %Define nce array with nonconducting edge
>> nce = [3];
>> ETE = E(nce,nce);
>> FTE = F(nce,nce);
>> ATE = inv(FTE)* ETE;
>> [EVTE,kcsqTE] = eig(ATE);
```

```
>> kcTE = sqrt(diag(kcsqTE));
>> kcTEs = sort(kcTE)

kcTEs =

3.4641

>>

% GLAET.M
function [E,F] = glaet(Neg, Nn, Ne, n1L, n2L, n3L, n1EL, n2EL, n3EL, xn, yn)
for e = 1:Ne
        n(1,e) = n1L(e);
        n(2,e) = n2L(e);
        n(3,e) = n3L(e);
        ne(1,e) = n1EL(e);
        ne(2,e) = n2EL(e);
        ne(3,e) = n3EL(e);
end

E = zeros(Neg, Neg);
F = zeros(Neg, Neg);

for e = 1:Ne
        for i = 1:3;
                x(i) = xn(n(i,e));
                y(i) = yn(n(i,e));
        end

        b(1) = y(2) - y(3);
        b(2) = y(3) - y(1);
        b(3) = y(1) - y(2);
        c(1) = x(3) - x(2);
        c(2) = x(1) - x(3);
        c(3) = x(2) - x(1);

        Area = 0.5*abs(b(2)*c(3)-b(3)*c(2));
        l(1) = sqrt(b(3)*b(3) + c(3)*c(3));
        l(2) = sqrt(b(1)*b(1) + c(1)*c(1));
        l(3) = sqrt(b(2)*b(2) + c(2)*c(2));

        for i = 1:3
                for j = 1:3
                        ff(i,j) = b(i)*b(j) + c(i)*c(j);
            end
        end

        G(1,1) = 2*(ff(2,2)- ff(1,2)+ ff(1,1));
        G(2,2) = 2*(ff(3,3)- ff(2,3)+ ff(2,2));
        G(3,3) = 2*(ff(1,1) - ff(3,1) + ff(3,3));
        G(2,1) = G(1,2);
        G(1,3) = ff(2,1) - 2*ff(2,3) - ff(1,1) + ff(1,3);
```

```
         G(3,1)  =  G(1,3);
         G(2,3)  =  ff(3,1)  -  ff(3,3)  -  2*ff(2,1)  +  ff(2,3);
         G(3,2)  =  G(2,3);

         for i = 1:3
             for j = 1:3
                     Ee(i,j)  =  (1/Area)*(l(i)*l(j));
                     Fe(i,j)  =  (1/48)*(1/Area)*l(i)*l(j)*G(i,j);
                     if(ne(i,e) < 0)
                     Ee(i,j)  =  -Ee(i,j);
                     Fe(i,j)  =  -Fe(i,j);
                 else;
                 end;

                 if(ne(j,e) < 0);
                         Ee(i,j)  =  -Ee(i,j);
                         Fe(i,j)  =  -Fe(i,j);
                 else;
                 end;

                 ane(i,e)  =  abs(ne(i,e));
                 ane(j,e)  =  abs(ne(j,e));
                 E(ane(i,e),ane(j,e))  =  E(ane(i,e),ane(j,e))  +  Ee(i,j);
                 F(ane(i,e),ane(j,e))  =  F(ane(i,e),ane(j,e))  +  Fe(i,j);
         end
   end
end
```

16A.6 Example of a Main Program Using the Function Program INHWGD to Solve Inhomogeneous Waveguide Problem (see Figures 17.3 and 17.4)

INHWGD is given in Appendix 18C and is based on using first-order node-based scalar shape functions for longitudinal components and first-order edge-based vector shape functions for the transverse components. The triangular elements are used.

```
>> % example of inhomogeneous waveguide
>> Ne = 2;
>> Nn = 4;
>> Neg = 5;
>> xn = [0,1,1,0];
>> yn = [0,0,1,1];
>> n1L = [1,1];
>> n2L = [2,3];
>> n3L = [3,4];
>> mur = [1,1];
>> epr = [4,1];
>> k0 = 5;
>> n1EL = [1,-3];
```

```
>> n2EL = [2,4];
>> n3EL = [3,5];
>> [Att,Btt,Btz,Bzz,C] = INHWGD(epr,mur,k0,Neg,Nn,Ne,n1L,n2L,n3L,n1EL,
   n2EL,n3EL,xn,yn);
>> ncE = 3;
>> nce = 3;
>> C = C(nce,nce);
>>    [EVIN,GAMASQ] = eig(C);
>> GAMA = sqrt(sort(GAMASQ));
>> GAMA
GAMA =

0 + 6.5192i

>>
```

Appendix 16B: Cotangent Formula

16B.1 Cotangent Formula

The elements of the S matrix can be expressed in terms of the coordinates of the vertices of the triangle given by Equation 16.52 and repeated here for convenience:

$$S_{jk}^{(e)} = \frac{1}{4A}\left(b_j b_k + c_j c_k\right).$$

(16.51)

They can also be expressed in terms of the angles of the vertices given by Equation 16.53 and repeated here for convenience:

$$\left[S^{(e)}\right] = \frac{1}{2}\begin{bmatrix} \cot\theta_2 + \cot\theta_3 & -\cot\theta_3 & -\cot\theta_2 \\ -\cot\theta_3 & \cot\theta_1 + \cot\theta_3 & -\cot\theta_1 \\ -\cot\theta_2 & -\cot\theta_1 & \cot\theta_1 + \cot\theta_2 \end{bmatrix}.$$

(16.53)

It is further shown that Equation 16.53 can be written as

$$\left[S^{(e)}\right] = \sum_{i=1}^{3} \cot\theta_i Q_i,$$

(16.107)

where $[Q_i]$ are given by Equations 16.108 through 16.110. In this appendix, we will show that Equation 16.53 is obtained from Equation 16.51 based on the geometry and associated trigonometric identities [1]. Figure 16B.1 shows b's and c's from their definitions in terms of coordinates given by Equations 16.39b, 16.39c, 16.42b, 16.42c, 16.44b, and 16.44c. In the notation of cyclic values, we have

$$b_i = y_{i+1} - y_{i+2}, \quad i = 1,2,3,4,5,$$

(16B.1)

$$c_i = x_{i+2} - x_{i+1},$$

(16B.2)

i being cyclic values of 1, 2, 3, $i = 4$ is same as $i = 1$, and $i = 5$ is same as $i = 2$.

Note the negative signs on some of the b's and c's. For example,

$$c_1 = x_3 - x_2.$$

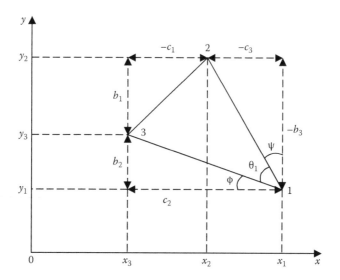

FIGURE 16B.1
Definition of b's and c's in terms of coordinates.

In the diagram, it is marked as $-c_1$ since x_3 is shown to be smaller than x_2. Note also that we maintained the counter-clockwise progression of the nodes 1, 2, 3:

$$\cot\theta_1 = \cot\left(\frac{\pi}{2} - \phi - \psi\right) = \tan(\phi + \psi) = \frac{\tan\phi + \tan\psi}{1 - \tan\phi\tan\psi},$$

$$\therefore \cot\theta_1 = \frac{(b_2/c_2) + (-c_3/-b_3)}{1 - (b_2/c_2)(-c_3/-b_3)} = \frac{(b_2 b_3 + c_2 c_3)b_3 c_2}{(c_2 b_3 - b_2 c_3)b_3 c_2} = \frac{b_2 b_3 + c_2 c_3}{b_2 c_3 - b_3 c_2}.$$

$$(16B.3)$$

Since the denominator of Equation 16B.3 is twice the area of the triangle,

$$\cot\theta_1 = -\frac{b_2 b_3 + c_2 c_3}{2A}.$$

$$(16B.4)$$

Equation 16B.4 can be generalized as

$$b_i b_j + c_i c_j = -2A\cot\theta_k, \quad i \neq j.$$

$$(16B.5)$$

Let us consider next whether any modification is needed if $i = j$. Let $i = j = 1$, we will show that

$$b_1^2 + c_1^2 = 2A\left[\cot\theta_2 + \cot\theta_3\right].$$

Proof

$$\cot\theta_2 = -\frac{b_3 b_1 + c_3 c_1}{2A},$$

$$\cot\theta_3 = -\frac{b_1 b_2 + c_1 c_2}{2A},$$

$$2A\left(\cot\theta_2 + \cot\theta_3\right) = -\left(b_3 b_1 + c_3 c_1 + b_1 b_2 + c_1 c_2\right) = -\left[b_1\left(b_2 + b_3\right) + c_1\left(c_2 + c_3\right)\right]$$
$$= -\left[b_1\left(-b_1\right) + c_1\left(-c_1\right)\right] = b_1^2 + c_1^2.$$

We can generalize the above result as

$$b_i^2 + c_i^2 = 2A\left[\cot\theta_j + \cot\theta_k\right]. \tag{16B.6}$$

From Equations 16.51, 16B.5, and 16B.6, we obtain Equation 16.53.

16B.2 Area of the Triangle

Incidentally, the geometry also shows that the denominator determinant in Equation 16.37 is twice the area. The details are given below. The directed line segment $\overrightarrow{12} = \bar{D}$ is given by

$$\overrightarrow{12} = \bar{D} = \bar{r}_2 - \bar{r}_1 = \hat{x}\left(x_2 - x_1\right) + \hat{y}\left(y_2 - y_1\right) = \hat{x}c_3 - \hat{y}b_3. \tag{16B.7}$$

The directed line segment $\overrightarrow{13} = \bar{E}$:

$$\overrightarrow{13} = \bar{E} = \bar{r}_3 - \bar{r}_1 = \hat{x}\left(x_3 - x_1\right) + \hat{y}\left(y_3 - y_1\right) = -\hat{x}c_2 + \hat{y}b_2.$$

The magnitude of the cross product of $\bar{D}$ and $\bar{E}$ is given by

$$\left|\bar{D} \times \bar{E}\right| = \begin{vmatrix} \hat{x} & \hat{y} & \hat{z} \\ c_3 & -b_3 & 0 \\ -c_2 & b_2 & 0 \end{vmatrix} = \left|\hat{x}\left(0 - 0\right) - \hat{y}\left(0 - 0\right) + \hat{z}\left(c_2 b_2 - c_2 b_3\right)\right| = \left|c_3 b_2 - c_2 b_3\right|.$$

From the definition of the cross product, it is the area of the parallelogram with sides $\bar{D}$ and $\bar{E}$ which is twice the area of the triangle. Therefore, the area A of the triangle is

$$A = \frac{1}{2}\left|c_3 b_2 - c_2 b_3\right|.$$

It is obvious that we should get the same answer for A, if we consider the directed segments $\overrightarrow{23}$ and $\overrightarrow{21}$ as the vectors. Another possibility is to consider $\overrightarrow{31}$ and $\overrightarrow{32}$ as the vectors. Thus, the alternative expressions for the area A are

$$A = \frac{1}{2}\left|c_1 b_3 - c_3 b_1\right| = \frac{1}{2}\left|c_2 b_1 - c_1 b_2\right|.$$

Reference

1. Silvester, P. P. and Ferrari, R. L., *Finite Elements for Electrical Engineers*, 3rd edition, Cambridge University Press, Cambridge, U.K., 1996.

Appendix 16C: Neumann Boundary Conditions: FEM Method

To prove that natural boundary conditions (Neumann type), namely

$$D_{n1} = D_{n2}, \tag{16C.1}$$

$$\varepsilon_1 E_{n1} = \varepsilon_2 E_{n2}, \tag{16C.2}$$

$$\varepsilon_1 \frac{\partial \Phi_1}{\partial n} = \varepsilon_2 \frac{\partial \Phi_2}{\partial n}, \tag{16C.3}$$

or

$$\frac{\partial \Phi}{\partial n} = 0 \tag{16C.4}$$

are enforced in the process of minimizing the functional and need not be explicitly enforced. The proof makes use of Green's theorem of vector calculus. Let us prove the theorems first.

16C.1 First Green's Theorem

Statement:

$$\iiint\limits_{v} f_1 \nabla^2 f_2 \, dv = \oiint\limits_{s} f_1 \frac{\partial f_2}{\partial n} \, ds - \iiint\limits_{v} \left(\bar{\nabla} f_1 \cdot \bar{\nabla} f_2 \right) dv, \tag{16C.5}$$

where f_1 and f_2 are two scalar functions, v is a volume bounded by a closed surface s, and $\hat{n}$ is a unit vector which is the outward normal to the surface. See Figure 16C.1.

Proof:

From vector identity

$$\bar{\nabla} \cdot \left(f \bar{F} \right) = f \bar{\nabla} \cdot \bar{F} + \nabla f \cdot \bar{F}, \tag{16C.6}$$

$$\iiint\limits_{v} \left(f \bar{\nabla} \cdot \bar{F} \right) dv = \iiint\limits_{v} \bar{\nabla} \cdot \left(f \bar{F} \right) dv - \iiint\limits_{v} \bar{\nabla} f \cdot \bar{F} \, dv. \tag{16C.7}$$

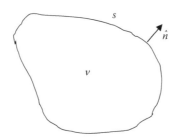

FIGURE 16C.1
Geometry for first Green's theorem.

The first volume integral on the RHS can be converted into a closed surface integral by using the following divergence theorem:

$$\iiint_v \bar{\nabla} \cdot \left(f\bar{F} \right) dv = \oiint_s f\bar{F} \cdot ds = \oiint_s fF_n \, ds. \tag{16C.8}$$

$$\therefore \iiint_v \left(f\bar{\nabla} \cdot \bar{F} \right) dv = \oiint_s fF_n \, ds - \iiint_v \bar{\nabla} f \cdot \bar{F} \, dv. \tag{16C.9}$$

Let

$$f = f_1 \tag{16C.10}$$

and

$$\bar{F} = \bar{\nabla} f_2, \tag{16C.11}$$

where f_1 and f_2 are two scalar functions. Substituting Equations 16C.10 and 16C.11 into Equation 16C.9, we obtain Equation 16C.5.

16C.2 Second Green's Theorem

Statement:

$$\iiint_v \left(f_1 \nabla^2 f_2 - f_2 \nabla^2 f_1 \right) dv = \oiint_s \left(f_1 \frac{\partial f_2}{\partial n} - f_2 \frac{\partial f_1}{\partial n} \right) ds. \tag{16C.12}$$

Proof:
By exchanging f_1 and f_2 in Equation 16C.5, we obtain

$$\iiint_v f_2 \nabla^2 f_1 \, dv = \oiint_s f_2 \frac{\partial f_1}{\partial n} \, ds - \iiint_v \bar{\nabla} f_2 \cdot \bar{\nabla} f_1 \, dv. \tag{16C.13}$$

Subtracting Equation 16C.13 from Equation 16C.5, we obtain Equation 16C.12.

16C.3 Two-Dimensional Versions of Green's Theorems

We can apply two-dimensional versions of the Green's theorems by considering an open surface bounded by a closed curve c as shown in Figure 16C.2. In particular, from Equation 16C.5, we obtain

$$\iint_s f_1 \nabla_t^2 f_2 \, ds = \oint_c f_1 \frac{\partial f_2}{\partial n} \, d\ell - \iint_s \left(\bar{\nabla}_t f_1 \cdot \bar{\nabla}_t f_2 \right) ds, \tag{16C.14}$$

where

$$\bar{\nabla}_t = \hat{x} \frac{\partial}{\partial x} + \hat{y} \frac{\partial}{\partial y}. \tag{16C.15}$$

16C.3.1 Expansion of $W(\Phi + \theta h)$

Let

$$\nabla_t^2 \Phi = 0 \tag{16C.16}$$

and its functional

$$W(\Phi) = I(\Phi) = \frac{1}{2} \iint \varepsilon \left| \bar{\nabla}_t \Phi \right|^2 ds. \tag{16C.17}$$

Let θ be a parameter and h another function with the same Dirichlet boundary condition as Φ. Expanding $W(\Phi + \theta h)$, we obtain:

$$W(\Phi + \theta h) = \frac{1}{2} \iint \left[\varepsilon \bar{\nabla}_t (\Phi + \theta h) \cdot \bar{\nabla}_t (\Phi + \theta h) \right] ds, \tag{16C.18}$$

$$\bar{\nabla}_t (\Phi + \theta h) = \bar{\nabla}_t \Phi + \theta \bar{\nabla}_t h,$$

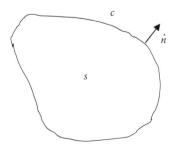

FIGURE 16C.2
Open surface bounded by a closed curve.

$$\bar{\nabla}_t\left(\Phi+\theta h\right)\cdot\bar{\nabla}_t\left(\Phi+\theta h\right)=\left(\bar{\nabla}_t\Phi+\theta\bar{\nabla}_t h\right)\cdot\left(\bar{\nabla}_t\Phi+\theta\bar{\nabla}_t h\right)$$

$$=\bar{\nabla}_t\Phi\cdot\bar{\nabla}_t\Phi+2\theta\left(\bar{\nabla}_t\Phi\cdot\bar{\nabla}_t h\right)+\theta^2\bar{\nabla}_t h\cdot\bar{\nabla}_t h. \qquad (16C.19)$$

By substituting Equation 16C.19 into Equation 16C.18, we obtain

$$W\left(\Phi+\theta h\right)=W\left(\Phi\right)+\theta\iint\varepsilon\left(\bar{\nabla}_t\Phi\cdot\bar{\nabla}_t h\right)ds+\theta^2 W\left(h\right). \qquad (16C.20)$$

From Equation 16C.14 with $f_2=\Phi$ and $f_1=h$,

$$\iint_s\left(\bar{\nabla}_t\Phi\cdot\bar{\nabla}_t h\right)ds=\oint_c h\frac{\partial\Phi}{\partial n}d\ell-\iint_s h\nabla_t^2\Phi\,ds=\oint_c h\frac{\partial\Phi}{\partial n}d\ell. \qquad (16C.21)$$

Since $\nabla_t^2\Phi=0$,

$$\therefore W\left(\Phi+\theta h\right)=W\left(\Phi\right)+\theta\oint_c\varepsilon h\frac{\partial\Phi}{\partial n}d\ell+\theta^2 W\left(h\right). \qquad (16C.22)$$

16C.4 Implication of Minimization of the Functional

$$\frac{W\left(\Phi+\theta h\right)-W\left(\Phi\right)}{\theta h}=\oint_c\varepsilon\frac{\partial\Phi}{\partial n}d\ell+\frac{\theta h}{h^2}W\left(h\right). \qquad (16C.23)$$

In a crude analogy, the LHS may be evaluated in the limit $\theta\to$ and it may be considered as $\partial W/\partial\Phi$, which has to be zero, for the functional to be minimum for the exact choice of Φ. In that analogy, the LHS is zero and the last term on the RHS of Equation 16C.23 is zero. So, we obtain the result that the functional is minimized when

$$\oint_c\varepsilon\frac{\partial\Phi}{\partial n}d\ell=0. \qquad (16C.24)$$

16C.4.1 Error Estimate for an Approximate Choice of Φ

Let Φ_a be an approximate choice for Φ and written in the form

$$\Phi_a=\left(\Phi+\theta h\right). \qquad (16C.25)$$

From Equation 16C.22, we see that the error in the functional is of the order of θ^2. Equation 16C.25 gives θ parameter as the error in estimating Φ and the consequence is an error of the order of θ^2 for the functional W, when Φ_a is used in minimizing the functional. The functional W for the Laplace equation under discussion has the physical significance of electric energy.

An approximate solution for $\Phi = \Phi_a$ with an error of say $\theta = 0.1$ will lead to an error of $\theta^2 = (0.1)^2 = 0.01$ in calculation of electric energy. Although we discussed the solution of the Laplace equation in this appendix, the same kind of argument holds good for the FEM solution of the eigenvalue problem (waveguide problem), discussed in the last section of this appendix.

16C.5 Neumann Boundary Condition

Let us now apply Equation 16C.24 to the elements shown in Figure 16C.3.

The global nodes under consideration are 1, 2, 3, and 4. In doing close contour integration, the terms that remain arise due to integration from $\boxed{2}$ to $\boxed{1}$ in element 1 and $\boxed{1}$ to $\boxed{2}$ in element 2

$$\oint_c \varepsilon h \frac{\partial \Phi}{\partial n} \, d\ell = \int_{\boxed{2}\boxed{1}} \varepsilon_L h \frac{\partial \Phi}{\partial n} \bigg|_L \, d\ell + \int_{\boxed{1}\boxed{2}} \varepsilon_R h \frac{\partial \Phi}{\partial n} \bigg|_R \, d\ell. \qquad (16C.26)$$

The LHS of Equation 16C.26 will be zero only if

$$\varepsilon_L \frac{\partial \Phi}{\partial n} \bigg|_L = \varepsilon_R \frac{\partial \Phi}{\partial n} \bigg|_R, \qquad (16C.27)$$

which is equivalent to

$$D_{n1} = D_{n2} \qquad (16C.28)$$

at the interface $\boxed{1}$ $\boxed{2}$.

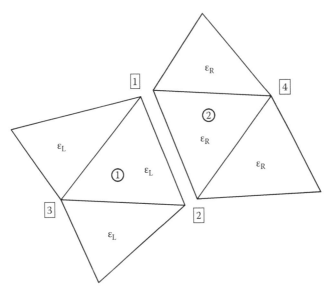

FIGURE 16C.3
Application of Neumann boundary condition to the shown elements.

16C.6 Homogeneous Waveguide Problem

For the homogeneous waveguide problem in Chapter 2, the Helmholtz equation and its functional are given by Equations 2.97 and 2.99 and they are repeated here for convenience:

$$\nabla_t^2 \Phi + k_t^2 \Phi = 0, \tag{16C.29}$$

$$I(\Phi) = \frac{1}{2} \iint_s |\nabla_t \Phi|^2 \, ds - k_t^2 \iint_s \Phi^2 \, ds, \tag{16C.30}$$

where $k_t^2 = \lambda$ is the eigenvalue and k_t is the cutoff wave number. If we multiply Equation 16C.29 by Φ and integrate over the cross section of the waveguide, we obtain

$$\iint_s \Phi \nabla_t^2 \Phi \, ds = -k_t^2 \iint_s \Phi^2 \, ds. \tag{16C.31}$$

Applying Equation 16C.14 with $f_1 = f_2 = \Phi$, the LHS becomes

$$\iint_s \Phi \nabla_t^2 \Phi \, ds = \oint_c \Phi \frac{\partial \Phi}{\partial n} \, d\ell - \iint_s |\bar{\nabla}_t^2 \Phi|^2 \, ds. \tag{16C.32}$$

For a homogeneous waveguide with PEC walls either $\Phi = 0$ (TM modes) or $\partial\Phi/\partial n = 0$ (TE modes), and hence the first term on the RHS is zero. Hence Equation 16C.31 yields

$$-\iint_s |\nabla_t \Phi|^2 \, ds = -k_t^2 \iint_s \Phi^2 \, ds. \tag{16C.33}$$

From Equations 16C.30 and 16C.33, we can conclude that the exact solution will yield a functional $I(\Phi)$ which is not only minimum, but also that minimum is zero. It is tempting to write Equation 16C.23 as

$$\lambda = k_t^2 = \frac{\displaystyle\iint_s |\nabla_t \Phi|^2 \, ds}{\displaystyle\iint_s \Phi^2 \, ds} \tag{16C.34}$$

and interpret an error estimate for an approximate Φ_a as follows: if Φ_a has an error of ε in some sense, the eigenvalue λ will have an error of ε^2 only. This is in agreement with the well-known "stationary" property of the eigenvalues of a matrix. For additional insight, see Section 21.2.

Appendix 16D: Standard Area Integral

16D.1 Standard Area Integral

In Section 16.3.3, Equation 16.78 gives the following standard integral [1]:

$$\iint_{\Delta^{(e)}} \left(\varsigma_1\right)^{\ell} \left(\varsigma_2\right)^{m} \left(\varsigma_3\right)^{n} \, dx \, dy = \frac{\ell! \, m! \, n!}{(\ell + m + n + 2)!} \, 2A^{(e)}. \tag{16.78}$$

We used this integral throughout Chapter 16. In this appendix, we shall prove the result [1]. There are a few preliminary results before we prove the main result.

Show that $dx \, dy = 2A^{(e)} \, d\varsigma_1 \, d\varsigma_2$.

This is essentially a geometrical result obtained from the definition of area coordinates.

Without loss of generality, we can consider one of the sides of the triangle to be along the x-axis as shown in Figure 16D.1.

Let the point P have the coordinates (x, y) or the area coordinates ς_1, ς_2, and ς_3 where

$$\varsigma_i = \frac{h_i}{H_i}. \tag{16D.1}$$

Note that the height of the point P above the line BC is $h_1 = y$:

$$\therefore dy = dh_1. \tag{16D.2}$$

From the triangle PEF, $h_3 = DP$ and

$$PE = dh_3 = dx \sin \theta_2, \tag{16D.3}$$

$$\therefore dx \, dy = \frac{dh_3}{\sin \theta_2} \, dh_1. \tag{16D.4}$$

From Equation 16D.1, $\varsigma_3 = h_3/H_3$ and $\varsigma_1 = h_1/H_1$:

$$\therefore dh_3 = H_3 \, d\varsigma_3, \tag{16D.5}$$

$$dh_1 = H_1 \, d\varsigma_1. \tag{16D.6}$$

From Equations 16D.4 through 16D.6, we obtain

$$dx \, dy = \frac{H_3 d\varsigma_3 H_1 d\varsigma_1}{\sin \theta_2}.$$

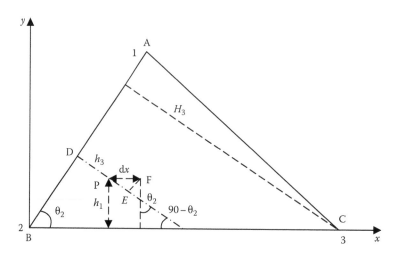

FIGURE 16D.1
Geometry to evaluate $dx\,dy$.

Recognizing $H_3/\sin\theta_2 = BC$ and $(BC)H_1 = 2A^{(e)}$:

$$dx\,dy = 2A^{(e)}\,d\varsigma_1\,d\varsigma_3.$$

Using similar arguments, we can show that $ds = dx\,dy$ can also be written as

$$dx\,dy = 2A^{(e)}\,d\varsigma_2\,d\varsigma_3 = 2A^{(e)}\,d\varsigma_3\,d\varsigma_1.$$

The above result can also be obtained without recourse to geometry. The area element may be written in whatever coordinates, say ς_1 and ς_2, provided the Jacobian of the coordinate transformation is included:

$$d\varsigma_1 d\varsigma_2 = \frac{\partial(\varsigma_1,\varsigma_2)}{\partial(x,y)}dx\,dy, \tag{16D.7}$$

where the Jacobian is the determinant

$$\frac{\partial(\varsigma_1,\varsigma_2)}{\partial(x,y)} = \begin{vmatrix} \dfrac{\partial\varsigma_1}{\partial x} & \dfrac{\partial\varsigma_1}{\partial y} \\[2mm] \dfrac{\partial\varsigma_2}{\partial x} & \dfrac{\partial\varsigma_2}{\partial y} \end{vmatrix}. \tag{16D.8}$$

Since

$$\varsigma_1 = \frac{1}{2A^{(e)}}\left(a_1 + b_1 x + c_1 y\right), \tag{16D.9}$$

$$\varsigma_2 = \frac{1}{2A^{(e)}}\left(a_2 + b_2 x + c_2 y\right), \tag{16D.10}$$

the Jacobian

$$\frac{\partial\left(\varsigma_1,\varsigma_2\right)}{\partial\left(x,y\right)} = \begin{vmatrix} \dfrac{b_1}{2A^{(e)}} & \dfrac{c_1}{2A^{(e)}} \\ \dfrac{b_2}{2A^{(e)}} & \dfrac{c_2}{2A^{(e)}} \end{vmatrix} = \frac{b_1c_2 - b_2c_1}{4\left[A^{(e)}\right]^2} = \frac{2A^{(e)}}{4\left[A^{(e)}\right]^2} = \frac{1}{2A^{(e)}}.$$

From Equation 16D.5, we obtain

$$dx\,dy = 2A^{(e)}\,d\varsigma_1\,d\varsigma_2.$$

16D.2 Limits of Integration

After changing the variables of integration from (x, y) to $(\varsigma_1, \varsigma_2)$, we need to determine next the limits of integration to perform integration over the area of the element $\Delta^{(e)}$. From Figure 16D.2, the area strip shown is parallel to $\varsigma_1 = 0$ and the width of the strip is $d\varsigma_1$. As ς_1 varies from 0 to 1, the strip expands to the triangle.

The limits for ς_2 can now be easily seen as the ς_2 coordinates of the points Q and R. The point Q has $\varsigma_2 = 0$. The value of ς_2 at R can easily be determined from the following relation:

$$\varsigma_1 + \varsigma_2 + \varsigma_3 = 1,$$

and $\varsigma_3 = 0$, at R. Hence,

$$\varsigma_1 + \varsigma_2 = 1 \quad \text{and} \quad \varsigma_2 = 1 - \varsigma_1 \text{ at R.}$$

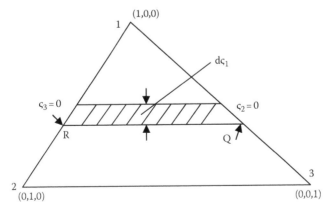

FIGURE 16D.2
Geometry for area integration.

Thus, we have

$$\iint_{\Delta^{(e)}} f(x,y)\,dx\,dy = 2A^{(e)} \int_{\varsigma_1=0}^{1} \int_{\varsigma_2=0}^{1-\varsigma_1} f(\varsigma_1,\varsigma_2)\,d\varsigma_2 d\varsigma_1.$$

Evaluate $\vartheta(m,n) = \displaystyle\int_{0}^{1-\varsigma_1} \varsigma_2^m (1-\varsigma_1-\varsigma_2)^n \, d\varsigma_2$, where m and n are integers.

Let

$$u = (1-\varsigma_1-\varsigma_2)^n, \quad du = n(1-\varsigma_1-\varsigma_2)^{n-1}(-1)\,d\varsigma_2 n,$$

$$dv = \varsigma_2^m d\varsigma_2, \quad v = \frac{1}{m+1}\varsigma_2^{m+1}.$$

Using integration by parts, we obtain

$$\vartheta(m,n) = (1-\varsigma_1-\varsigma_2)\frac{\varsigma_2}{m+1}\bigg|_{\varsigma_2=0}^{\varsigma_2=1=\varsigma_1} - \int \frac{1}{m+1}\varsigma_2^{m+1} n(1-\varsigma_1-\varsigma_2)^{n-1}(-1)\,d\varsigma_2$$

$$= 0 + \int_{0}^{1-\varsigma_1} \frac{n}{m+1}\varsigma_2^{m+1}(1-\varsigma_1-\varsigma_2)^{n-1}\,d\varsigma_2,$$

$$\therefore \vartheta(m,n) = \frac{n}{m+1}\vartheta(m+1,\,n-1).$$

By induction, we can write

$$\vartheta(m,n) = \frac{n(n-1)}{(m+1)(m+2)}\vartheta(m+2,n-2)$$

$$= \frac{n(n-1)\cdots 1}{(m+1)(m+2)\cdots(m+n)}\vartheta(m+n,0) = \frac{m!\,n!}{(m+n)!}\vartheta(m+n,0)$$

$$\vartheta(m+n,0) = \int_{0}^{1-\varsigma_1} \varsigma_2^{m+n}(1-\varsigma_1-\varsigma_2)^0 \, d\varsigma_2 = \frac{1}{m+n+1}\varsigma_2^{m+n+1}\bigg|_{0}^{1-\varsigma_1}$$

Evaluate

$$I(\ell,m,n) = \int_{0}^{1}\int_{0}^{1-\varsigma_1} \varsigma_1^\ell \varsigma_2^m (1-\varsigma_1-\varsigma_2)^n \, d\varsigma_2 d\varsigma_1.$$

Integration by parts gives

$$I(\ell,m,n) = \frac{n}{m+1}\int_{\varsigma_1\varsigma_2=0}^{1}\int^{1-\varsigma_1} \varsigma_1^\ell \varsigma_2^{m+1}(1-\varsigma_1-\varsigma_2)^{n-1}\,d\varsigma_2 d\varsigma_1 = \frac{n}{m+1} I(\ell,m+1,n-1)$$

and by induction

$$I(l,m,n) = \frac{n(n-1)}{(m+1)(m+2)} I(l,m+2,n-2)$$

$$= \frac{n(n-1)\cdots 1}{(m+1)(m+2)\cdots(m+n)} I(l,m+n,0) = \frac{n!m!}{(m+n)!} I(l,m+n,0).$$

Further consideration will show

$$I(\ell,m,n) = \frac{\ell!m!n!}{(\ell+m+n)!} I(0,\ell+m+n,0).$$

$I(0,\ell+m+n,0)$ is easily evaluated since

$$\int_0^1 \int_0^{1-\varsigma_1} (1)\varsigma_2^{l+m+n}(1)d\varsigma_2 d\varsigma_1 = \int_0^1 \frac{1}{l+m+n+1} \varsigma_2^{l+m+n+1}\Big|_0^{1-\varsigma_1} d\varsigma_1$$

$$= \int_0^1 \frac{1}{l+m+n+1}\Big[(1-\varsigma_1)^{l+m+n+1} - 0\Big]d\varsigma_1$$

$$= \frac{-1}{l+m+n+1} \int_0^1 (1-\varsigma_1)^{l+m+n+1} d(1-\varsigma_1)$$

$$= \frac{-1}{(l+m+n+1)} \frac{(1-\varsigma_1)^{l+m+n+2}}{l+m+n+2}\Big|_0^1$$

$$= \frac{1}{(l+m+n+1)(l+m+n+2)}.$$

$$\therefore I(l,m,n) = \frac{l!m!n!}{(l+m+n+2)!}$$

Evaluate $\displaystyle\iint_{\Delta^{(e)}} (\varsigma_1)^{\ell} (\varsigma_2)^m (\varsigma_3)^n dx\, dy.$

$$\iint_{\Delta^{(e)}} (\varsigma_1)^{\ell} (\varsigma_2)^m (\varsigma_3)^n dx\, dy$$

$$= 2A^{(e)} \iint_{\Delta^{(e)}} (\varsigma_1)^{\ell} (\varsigma_2)^m (1-\varsigma_1-\varsigma_2)^n d\varsigma_1 d\varsigma_2$$

$$= 2A^{(e)} I(l,m,n) = \frac{\ell!m!n!}{(\ell+m+n+2)!} 2A^{(e)}.$$

Reference

1. Silvester, P. P. and Ferrari, R. L., *Finite Elements for Electrical Engineers*, 3rd edition, Cambridge University Press, Cambridge, U.K., 1996.

Appendix 16E: Numerical Techniques in the Solution of Field Problems*

Dikshitulu K. Kalluri

16E.1 Introduction

Very few books at the undergraduate level on electric and magnetic fields include a numerical approach to the solution of field problems. While it may not be possible to give a complete account of this aspect of the subject at the undergraduate level, a few lectures devoted toward this end, in the author's opinion, are worthwhile. The general trend to use digital computers for the solution of electrical problems will justify spending a little time on the numerical techniques. Moreover, when boundaries of irregular shape are met with, since exact analytical solutions will not exist, approximate solutions are easily obtained by the numerical techniques.

16E.2 Finite-Difference Equations

As an illustration of this method, Poisson's equation (16E.1) in x- and y-coordinates is chosen:

$$\nabla^2 V = \frac{\partial^2 V}{\partial x^2} + \frac{\partial^2 V}{\partial y^2} = g(x,y). \tag{16E.1}$$

The above partial differential equation can be replaced by a set of algebraic equations of potential by considering a discrete number of points (nodes) in the field region described by the partial differential equation. The resulting set of algebraic equations with the given boundary conditions constitute the starting point of the numerical procedure. The algebraic equations are obtained by the finite-difference method outlined below.

If Equation 16E.1 holds in a region with the rectangle (Figure 16E.1a) as the boundary, on which the potential value is specified as a boundary condition, the discrete points are obtained by superposing a square net on the rectangle. The square net intersection points are the discrete set of points. A typical element of this set is denoted by (j, k). The node (j, k)

---

* Reprinted from the *Bull. Electrical Eng. Educ.*, 28, 1–12, June 1962. Manchester University Press, Manchester. With permission.

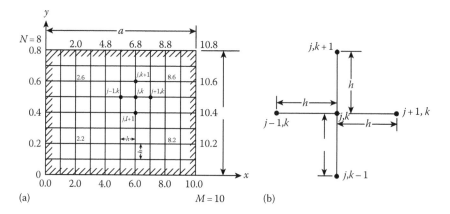

FIGURE 16E.1
(a) A rectangular region with a square net. (b) A typical element of the net.

is surrounded by the four nodes $(j - 1, k)$, $(j + 1, k)$ $(j, k - 1)$, and $(j, k + 1)$. The potential at $(j + 1, k)$ in terms of the potential at (j, k) is obtained by Taylor's series expansion*:

$$V_{j+1,k} = V_{j,k} + h\frac{\partial}{\partial x}V_{j,k} + \frac{h^2}{2!}\frac{\partial^2}{\partial x^2}V_{j,k} + \frac{h^3}{3!}\frac{\partial^3}{\partial x^3}V_{j,k} + \frac{h^4}{4!}\frac{\partial^4}{\partial x^4}V_{j,k}. \tag{16E.2}$$

Similarly, we have

$$V_{j-1,k} = V_{j,k} - h\frac{\partial}{\partial x}V_{j,k} + \frac{h^2}{2!}\frac{\partial^2}{\partial x^2}V_{j,k} - \frac{h^3}{3!}\frac{\partial^3}{\partial x^3}V_{j,k} + \frac{h^4}{4!}\frac{\partial^4}{\partial x^4}V_{j,k}. \tag{16E.3}$$

Adding Equations 16E.2 and 16E.3, we obtain

$$V_{j-1,k} + V_{j+1,k} = 2V_{j,k} + h^2\frac{\partial^2}{\partial x^2}V_{j,k} + \frac{h^4}{12}\frac{\partial^4}{\partial x^4}V_{j,k}.$$

If we allow an error of $(h^4/12)(\partial^4/\partial x^4)V_{j,k}$, called the discretization error, then

$$V_{j-1,k} + V_{j+1,k} - 2V_{j,k} \approx h^2\frac{\partial^2}{\partial x^2}V_{j,k}. \tag{16E.4}$$

Similarly, for the y-coordinate, we have

$$V_{j,k-1} + V_{j,k+1} - 2V_{j,k} \approx h^2\frac{\partial^2}{\partial y^2}V_{j,k}. \tag{16E.5}$$

From Equations 16E.4, 16E.5, and 16E.1, we have

$$V_{j-1,k} + V_{j+1,k} + V_{j,k-1} + V_{j,k+1} - 4V_{j,k} \approx h^2 g(x,y). \tag{16E.6}$$

---

* Higher-order terms than the fourth are neglected.

If, in Equation 16E.1, $g(x, y) = 0$ in the field region, then Poisson's equation reduces to Laplace's equation and Equation 16E.6 reduces to

$$V_{j,k} = \frac{1}{4}\left(V_{j-1,k} + V_{j+1,k} + V_{j,k-1} + V_{j,k+1}\right),$$ (16E.7)

which when stated in words would read that the potential at any point inside the field region is the average of the potentials of the four equidistant surrounding points in the form of a star. For Poisson's equation,

$$V_{j,k} = \frac{1}{4}\left(V_{j-1,k} + V_{j+1,k} + V_{j,k-1} + V_{j,k+1}\right) - \frac{h^2}{4}g(x,y).$$ (16E.8)

The nodes (Figure 16E.1) whose $j = 0$ or M for all k, and whose $k = 0$ or N for all j, are on the boundaries. They will have the potentials, specified by the boundary conditions.

16E.3 Solution by Iteration Versus Matrix Inversion

Thus, the application of finite-difference approximations to the continuous problem results in a set of algebraic simultaneous equations, whose solution can be obtained either by an iterative scheme or by the inversion of the associated matrix. In Figure 16E.1a, the discretization results in 63 simultaneous equations, the special feature of these equations being that each of them contains a few terms only. Reduction of net spacing, in order to minimize the discretization error, results in larger number of simultaneous equations. The order of the matrix to be inverted is the same as the number of simultaneous equations; inversion of such a matrix may involve considerable round-off errors described later.

In the iterative scheme that is a successive approximation method, we start with a set of guessed values for the unknowns. These values are substituted into one side of the equation like Equation 16E.7 and a new value is obtained for $V_{j,k}$. We now move to the next node and the procedure is repeated till we get a complete new set of values at all nodes; thus, we would have completed one iteration. This set of values now forms a starting point for the second iteration. This iterative process is continued till the nth set of values and the $(n + 1)$th set of values are the same, when the algebraic equations are truly satisfied. However, the process is normally terminated when the difference between the n_cth set and the $(n_c + 1)$th set is less than a constant. The constant which is indicative of the error involved in terminating the iterative process may be chosen to suit the accuracy requirements. The process is said to have converged at this stage. A second method of specifying the convergence criterion would be that there should not be any change up to the Ath place in the values of the n_cth set and the $(n_c + 1)$th set.

The difference in the values of the true solution and the terminated set is called the error. This will include the rounding-off one effects while computing. The iteration method is self-correcting; arithmetic errors committed while computing will not affect the final values though they may alter the mode of convergence.

16E.4 Iterative Schemes

Let the iterative scheme described above be carried out in a sequence such that the $(j - 1)$ th row and the $(k - 1)$th column go through the iteration before the jth row and the kth column (in Figure 16E.1a, we proceed from left to right and then from bottom to top). Then while computing the $(n + 1)$th values at (j, k), we would already have the $(n + 1)$th values for $(j - 1, k)$ and $(j, k - 1)$; also $(j + 1, k)$ and $(j, k + 1)$ would have the nth values. However, in the *Richardson method,* one makes use of values as soon as a complete set is available and no sooner. In the *Liebman method,* one makes use of a value as soon as it is available though it may not be part of a complete set. Thus, Equation 16E.8 has to be accordingly modified.
   Richardson method:

$$V_{j,k}^{n+1} = \frac{1}{4}\left(V_{j-1,k}^{n} + V_{j+1,k}^{n} + V_{j,k-1}^{n} + V_{j,k+1}^{n}\right) - \frac{h^2}{4}g(x,y), \tag{16E.8a}$$

where $V_{j-1,k}^{n}$ is the nth value at $j - 1, k$ and so on.
   Liebman method:

$$V_{j,k}^{n+1} = \frac{1}{4}\left(V_{j-1,k}^{n} + V_{j+1,k}^{n} + V_{j,k-1}^{n+1} + V_{j,k+1}^{n}\right) - \frac{h^2}{4}g(x,y). \tag{16E.8b}$$

   Since some of the values used in the Richardson method are old when compared to those used in the Liebman method, it is quite reasonable to expect that the Liebman method would converge faster than the Richardson method for the same error criterion. To further improve the convergence rate, the *extrapolated Liebman method* is used, for which the basic algorithm (equation) used is given in the following equation:

$$V_{j,k}^{n+1} = V_{j,k+\omega}^{n}\left(\frac{1}{4}\left[V_{j-1,k}^{n+1} + V_{j+1,k}^{n} + V_{j,k-1}^{n+1} + V_{j,k+1}^{n} - h^2g(x,y)\right] - V_{j,k}^{n}\right), \tag{16E.8c}$$

where ω is called the relaxation factor.
   To see clearly what this iterative scheme means, one may denote the quantity in square brackets by $V_{j,k}^{(n+1)'}$, the $(n + 1)$th value at node j, k if the Liebman method were used; then Equation 16E.8c may be written as

$$V_{j,k}^{n+1} = V_{j,k}^{n} + \omega\left(V_{j,k}^{(n+1)'} - V_{j,k}^{n}\right)$$

$$= V_{j,k}^{(n+1)'} + (\omega-1)\left(V_{j,k}^{(n+1)'} - V_{j,k}^{n}\right). \tag{16E.9}$$

   The substance of this equation may be stated thus: $V_{j,k}^{n}$ is the value available from the previous iteration and $V_{j,k}^{(n+1)'}$ is the value one would obtain by the Liebman method. $\left(V_{j,k}^{(n+1)'} - V_{j,k}^{n}\right)$ would be the difference between two successive iterations if no relaxation factors were used. By adding a portion of this difference to $V_{j,k}^{(n+1)'}$ to obtain the guessed value for the next iteration, we are trying to eliminate at least one iteration in between by predicting the next value from the previous experience. Thus, it is only logical that by this relaxation process, the convergence rate is improved. The optimum value for ω will give the fastest convergence, while any value of $1 < \omega < 2$ would improve the convergence. It is further known that it is preferable to overestimate rather than underestimate. This is a useful observation, for the reason that the estimation of the optimum relaxation factor, though simple enough in a few cases like a rectangular boundary, is in general very difficult.

16E.5 Convergence Rates of Iterative Schemes [1–5]

Frankel [1] has determined the convergence rates of the above three iterative schemes for Laplace's equation in a rectangular region. The same arguments hold for Poisson's equation also since the algorithms are the same in both except for an additive constant in the latter case. Since we deal with errors rather than potential itself, the additive constant will not matter. Consider a rectangular region having M nodes in a row and N nodes in a column.

Let $V_{j,k}$ be the true solution, $V_{j,k}^n$ be the value of the nth set at j, k, $V_{j,k}^{n_c}$ be the value when the process is terminated after n_c cycles, and $V_{j,k}^0$ be the guessed value.

Also denote $\epsilon_{j,k}^n$ as the error at the nth stage.

For the Richardson method

$$V_{j,k}^{n+1} = \frac{1}{4}\left(V_{j-1,k}^n + V_{j+1,k}^n + V_{j,k-1}^n + V_{j,k+1}^n\right)$$

and

$$\nabla^2 V_{j,k}^n = L V_{j,k}^n = V_{j-1,k}^n + V_{j+1,k}^n + V_{j,k-1}^n + V_{j,k+1}^n - 4V_{j,k}^n \quad \text{for } h = 1.$$

Thus,

$$V_{j,k}^{n+1} = V_{j,k}^n\left(1 + \frac{L}{4}\right),$$

$$\epsilon_{j,k}^{n+1} = V_{j,k}^{n-1} - V_{j,k} = V_{j,k}^n - V_{j,k} + \frac{L}{4}\left(V_{j,k}^n - V_{j,k}\right),$$

for $L V_{j,k} = 0$, $V_{j,k}$ being the true solution.
Thus,

$$\epsilon_{j,k}^{n+1} = \left(1 + \frac{L}{4}\right)^n \epsilon_{j,k} = K \epsilon_{j,k}^n, \quad K = \left(1 + \frac{L}{4}\right).$$

Further,

$$\epsilon_{j,k}^n = K \epsilon_{j,k}^{n-1}, \quad \epsilon_{j,k}^{n-1} = K \epsilon_{j,k}^{n-2}, \quad \therefore \epsilon_{j,k}^n = K^2 \epsilon_{j,k}^{n-2}.$$

In general,

$$\epsilon_{j,k}^n = K^n \epsilon_{j,k}^0. \tag{16E.10}$$

Thus, we have expressed error at the nth stage in terms of the error in guessing. Whatever may be the error in guessing, if $|K|^* < 1$, for sufficiently large $n = n_c$, the error ϵn_c will be smaller than the allowed error. Thus, for convergence to be assured, $K^* < 1$:

$$\epsilon_{j,k}^n = K^{*n0}_{\epsilon_{j,k}}, \quad K^* = \frac{1}{2}\left(\cos\frac{\pi}{M} + \cos\frac{\pi}{N}\right)$$

$$\approx 1 - \frac{1}{4}\left[\left(\frac{\pi}{m}\right)^2 + \left(\frac{\pi}{N}\right)^2\right]. \tag{16E.11}$$

If the error criterion specified is that there should not be any change up to A places of a value and if the guessed value $V_{j,k}^0$ is zero, then n_c the number of iterations required may be calculated as follows:

$$K^{*n_c} = \frac{\epsilon_{j,k}^{n_c}}{0_{\epsilon_{j,k}}} = \frac{.000\ldots0X_PX_QX_R\cdots}{.X_1X_2X_3\ldots X_AX_BX_CX_D\cdots} \quad (A \text{ zeros is in the numerator}).$$

Taking $\log_{10}$ on both sides,

$$n_c\log_{10}K^* \approx (A+1)-(-1), \tag{16E.12}$$

$$n_c \approx \frac{-A}{\log_{10}K^*}.$$

Thus, for the Richardson method,

$$n_c = \frac{A}{-\log_{10}\left[\frac{1}{2}\left(\cos\pi/M + \cos\pi/N\right)\right]} \approx \frac{A}{-\log_{10}\left(1-\frac{1}{4}\left[(\pi/M)^2 + (\pi/N)^2\right]\right)}. \tag{16E.13}$$

For Liebman and extrapolated Liebman method: On the same lines as for the Richardson method, we can write

$$\epsilon_{j,k}^{n+1} = K(\omega)^n_{\epsilon_{j,k}}. \tag{16E.14}$$

---

* K is an operator, but we can speak of the magnitudes of $K_{r,s}$, its eigenvalues. The eigenfunctions of L will be the eigenfunctions of K also. When we express $\epsilon_{j,k}^0$ in terms of such eigenfunctions, for Richardson algorithm, $K_{r,s}$ are shown (Equation 16E.1) to be

$$K_{r,s} = \frac{1}{2}\left(\cos\frac{\pi r}{M} + \cos\frac{\pi r}{N}\right), \quad r = 1,\cdots,M-1, \ s = 1,\cdots,N-1.$$

If $|K_{r,s}| < 1$ we are assured of convergence. The largest value of $|K_{r,s}|$, which however is less than 1, will contribute the most significant and the least convergent term. Thus, the overall convergence rate is more or less determined by K^* the maximum value of $|K_{r,s}|$. For more details, see Reference [1].

However, in this case, K is not simply related to the Laplace operator L. Frankel [1] proved that for $\omega = 1/4$ (Liebman method),

$$K^* = \frac{1}{2}\left(\cos\frac{\pi}{M} + \cos\frac{\pi}{N}\right)^2 \approx \left(1 - \frac{1}{4}\left[\left(\frac{\pi}{M}\right)^2 + \left(\frac{\pi}{N}\right)^2\right]\right)^2 \qquad (16E.15)$$

and thus

$$n_c = \frac{A}{-\log K_{10}^*} = \frac{A}{-2\log_{10}\left(\frac{1}{2}\left[\cos(\pi/M) + \cos(\pi/N)\right]\right)}. \qquad (16E.16)$$

Note that according to Equations 16E.13 and 16E.16, the Liebman method is twice as fast as the Richardson method. It is further proved that the optimum value for ω (relaxation factor in extrapolated Liebman method) is

$$\omega \approx 2 - \sqrt{2}\pi\left(M^{-2} + N^{-2}\right)^{1/2} \qquad (16E.17)$$

and in such a case

$$K^* \approx 1 - \sqrt{2}\pi\left(M^{-2} + N^{-2}\right)^{1/2}, \qquad (16E.18)$$

$$n_c \approx \frac{A}{-\log_{10}\left[1 - \sqrt{2}\pi\left(M^{-2} + N^{-2}\right)^{1/2}\right]}. \qquad (16E.19)$$

On the basis of the above, the convergence rates for the three iterative schemes are calculated for the problem in Figure 16E.1 with $a = 10$, $b = 8$, $h = 1$, and $A = 3$ (convergence up to third place), ignoring the symmetries that exist.

Richardson method: $n_c = 92.4$.

Liebman method: $n_c = 46.2$.

Extrapolated Liebman method: $n_c = 20.2$ *and* $\omega = 1.29$.

16E.6 Solution to the Problem on Poisson's Equation

The problem stated in Section 16E.1 is solved by both Liebman and extrapolated Liebman methods. To start with, the potential is guessed to be zero at all nodes inside the field region. Further $a = 10$, $b = 8$, $h = 1$, and $g(x, y) = -40$ are assumed. Fourfold symmetry that exists is taken into account. Tables 16E.1A and 16E.1B compare the 10th and 20th sets of values obtained by both the methods. Figure 16E.2 shows graphically the mode of convergence to the true value by both the methods for the node (5, 4).

TABLE 16E.1A

10th Sets of Values

$j =$		1	2	3	4	5
$k = 1$	L	36.85	56.40	67.38	73.68	76.88
	EL	43.40	68.69	83.75	92.25	95.73
$k = 2$	L	56.25	89.58	109.3	120.9	126.7
	EL	67.68	111.4	138.5	154.1	160.3
$k = 3$	L	66.50	108.0	133.5	148.6	156.0
	EL	81.46	135.8	170.4	190.4	198.3
$k = 4$	L	70.73	115.7	143.7	160.2	168.1
	EL	86.45	144.7	182.1	203.6	212.0

TABLE 16E.1B

20th Sets of Values

$j =$		1	2	3	4	5
$k = 1$	L	44.83	70.65	85.75	93.90	96.80
	EL	47.48	75.43	91.94	100.8	103.6
$k = 2$	L	70.10	114.3	141.2	155.9	161.2
	EL	74.72	122.5	151.9	167.9	173.1
$k = 3$	L	83.48	138.3	172.5	191.5	198.2
	EL	89.16	148.5	185.8	206.3	212.9
$k = 4$	L	87.95	146.4	183.2	203.7	211.0
	EL	93.77	156.9	196.8	218.8	225.9

L = Liebman method.
EL = Extrapolated Liebman method.

16E.7 Boundary Conditions and Symmetries

The boundary condition which specifies the potential along the boundary is called the Dirichlet condition. On the other hand, if one specifies the normal gradient on the boundary, this boundary condition is known as Neumann condition. If, for example, the normal gradient along the boundary is specified as zero, then the potential of the nodes, adjacent to the boundary and situated at the image points with respect to the boundary, should be the same. Thus, in the computation of the potential at a node on the boundary, the potential of the node adjacent to the boundary and on a line normal to the boundary is doubled. The same situation exists about a symmetry line. The extent

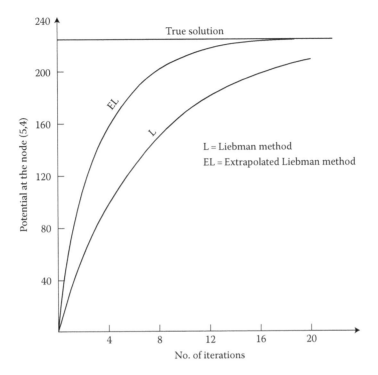

FIGURE 16E.2
Mode of convergence.

of the field is assumed to be only up to the symmetry line and the computation of the potential at the nodes on the symmetry line involves doubling of the potential at the node adjacent to the symmetry line and on a line at right angles (for square and radial nets) to the symmetry line.

16E.8 Grading of the Net and Other Types of Nets

It is shown that decreasing the net spacing (h) decreases the discretization error. However, a more dense network means a greater number of nodes in consequence of which the round-off errors increase in addition to the increased computational time. If more accuracy is desired only in a part of the field region, it is possible to fill up that region with a dense network and the remainder with a coarser one. This is called grading of the network. At the nodes, on the boundary between the two networks, one meets with unequal stars. At such nodes, the potential instead of being the average of the potentials of the surrounding nodes will be the weighted average.

Depending on the shape of the boundary, it may prove more profitable to fill up the field region with triangular, hexagonal, or radial networks. The principle of these methods is the same as that of a square net. To illustrate the algorithms, one meets in such cases, the following example [2] is given.

16E.9 Radial Network: Unequal ARMS [2]

Figure 16E.3 shows an inner circular boundary at 10 V and an outer boundary at 0 V consisting of a part of a circle closed by two straight lines whose vertex is rounded. Figure 16E.4a shows a radial network superposed on the field region.

B is the center of this network. Figure 16E.4b shows a node j, k with the four surrounding nodes. Laplace's equation in polar coordinates holds for the field region:

$$V^2 V = \frac{1}{r}\frac{\partial}{\partial r}\left\{ r \frac{\partial V}{\partial r} \right\} + \frac{1}{r}\frac{\partial^2 V}{\partial \theta^2} = 0 \qquad (16E.20)$$

instead of using Taylor series, a slightly different method is used. The first differential is written as

$$\left\{ \frac{\partial V}{\partial r} \right\}_{j,j+1} \approx \frac{V_{j+1} - V_j}{\Delta r}$$

and similarly the second differential in terms of the first differential.

Thus,

$$\left\{ \frac{\partial V}{\partial r} \right\}_{0-p} = \frac{V_p - V_0}{\Delta r_p}, \quad \left(\frac{\partial V}{\partial r} \right)_{q-0} = \frac{V_0 - V_q}{\Delta r_q},$$

$$r\left\{ \frac{\partial V}{\partial r} \right\}_{0-p} = \left(R_0 + \frac{\Delta r_p}{2} \right)\left(\frac{V_p - V_0}{\Delta r_p} \right), \qquad (16E.21)$$

$$r\left\{ \frac{\partial V}{\partial r} \right\}_{q-0} = \left(R_0 + \frac{\Delta r_q}{2} \right)\left(\frac{V_0 - V_q}{\Delta r_q} \right).$$

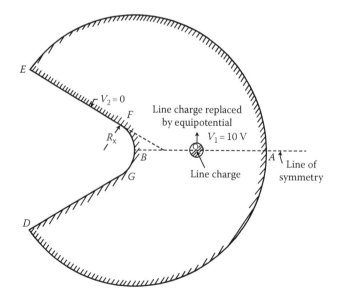

FIGURE 16E.3
A problem on Laplace equation in radial coordinates. (Adapted from Dikshitulu, K., Potential gradients near rounded corners, MS Thesis, University of Wisconsin, Madison, WI, 1959.)

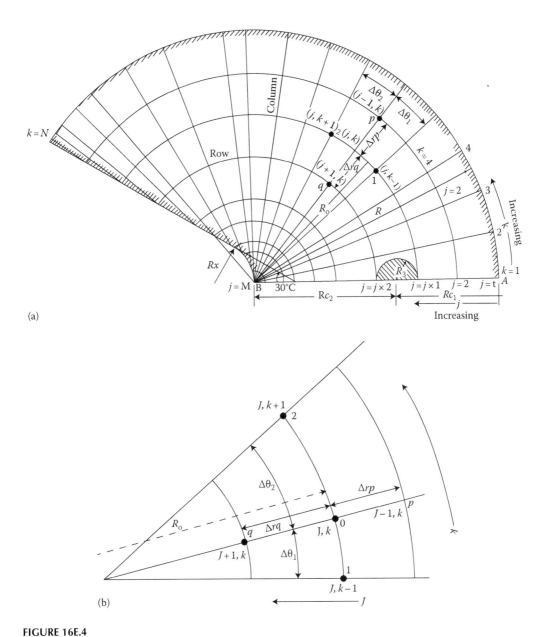

FIGURE 16E.4
(a) Radial network. (b) A typical node of the radial network. (Adapted from Dikshitulu, K., Potential gradients near rounded corners, MS Thesis, University of Wisconsin, Madison, WI, 1959.)

Thus

$$\frac{\partial}{\partial r}\left\{r\frac{\partial V}{\partial r}\right\} = \frac{2}{\Delta r_p + \Delta r_q}\frac{2R_0 + \Delta r_p}{2\Delta r_p}\left(V_p - V_0\right)$$

$$+ \frac{2}{\Delta r_p + \Delta r_q}\frac{2R_0 - \Delta r q0}{2\Delta r_q}\left(V_q - V_0\right).$$

Also

$$\left\{\frac{\partial V}{\partial \theta}\right\}_{0-1} = \frac{V_1 - V_0}{\Delta \theta_1}; \quad \left(\frac{\partial V}{\partial \theta}\right)_{2-0} = \frac{V_0 - V_2}{\Delta \theta_2},$$

$$\left\{\frac{\partial^2 V}{\partial \theta^2}\right\}_0 = \frac{\partial}{\partial \theta}\left(\frac{\partial V}{\partial \theta}\right) = \frac{1}{\Delta \theta_1 / 2 + \Delta \theta_2 / 2}\left[\frac{V_1 - V_0}{\Delta \theta_1} + \frac{V_2 - V_0}{\Delta \theta_2}\right], \tag{16E.22}$$

$$\frac{1}{r}\left\{\frac{\partial^2 V}{\partial \theta^2}\right\}_0 = \frac{2}{R_0\left(\Delta \theta_1 + \Delta \theta_2\right)}\left[\frac{V_1 - V_0}{\Delta \theta_1} + \frac{V_2 - V_0}{\Delta \theta_2}\right].$$

From Equations 16E.20 through 16E.22, for the Liebman algorithm, we can write

$$V_{j,k}^{n+1} = \frac{a_U}{a} V_{j-1,k}^{n+1} + \frac{a_D}{a} V_{j+1,k}^{n} + \frac{a_R}{a} V_{j,k-1}^{n+1} + \frac{a_L}{a} V_{j,k+1}^{n}, \tag{16E.23}$$

where

$$a_U = \frac{2R_0 + \Delta r p}{\Delta r_p\left(\Delta r_p + \Delta r_q\right)}, \quad a_D = \frac{2R_0 - \Delta r_q}{\Delta r_q\left(\Delta r_p - \Delta r_q\right)},$$

$$a_R = \frac{2}{\Delta \theta_1 R_0\left(\Delta \theta_1 + \Delta \theta_2\right)}, \quad a_L = \frac{2}{\Delta \theta_2 R_0\left(\Delta \theta_1 + \Delta \theta_2\right)},$$

and $a = a_U + a_D + a_R + a_L$.

Thus, again the potential at a node is the weighted average of the four surrounding node potentials. Of course, we do not meet with such unequal stars everywhere in the field. But Equation 16E.23 is the most general equation. The results of such computations may be seen in Figure 16E.5. From the node potentials, after interpolating, the equipotentials are obtained. The problem is programmed on IBM 650 and the extrapolated Liebman method is used.

16E.10 Relaxation Methods [3]

The methods described above are well suited for computation on digital computers wherein arithmetical operations are done at great speeds. The mechanical nature of these methods makes it possible to programme them on these machines. Relaxation methods due to Southwell are well suited for hand computation. In these methods, if the algebraic equations are not truly satisfied, the "residues" of the equations are calculated and then these "residues" are "liquidated" (reduced to zero) by giving proper "displacements." However, the operator should be experienced to make a successful use of this method. An excellent account of this method can be found in Reference [3]. The problem on Poisson's equation is solved in Reference [3], with $g(x, y) = -100$, by the relaxation method. One may profitably study this as a complement to the above article.

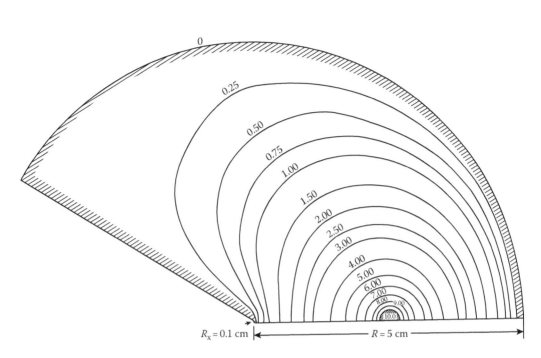

FIGURE 16E.5

Solution to the problem on Laplace equation in radial coordinates. (Adapted from Dikshitulu, K., Potential gradients near rounded corners, MS Thesis, University of Wisconsin, Madison, WI, 1959.)

References

1. Frankel, S. P., Convergence rates of iterative treatments, *Math. Tables Other Aids Comput.*, 4, 65–75, 1950.
2. Dikshitulu, K., Potential gradients near rounded corners, MS thesis, University of Wisconsin, Madison, WI, 1959.
3. Shaw, F. S., *Relaxation Methods*, Dover Publications, Inc., New York, 1953.
4. Crandal, S. H., *Engineering Analysis*, McGraw-Hill Book Company, New York, 1956.
5. Young, D., Iterative methods for solving partial differential equations of elliptic type, *Am. Math. Soc. Trans.*, 76, 92–111, 1954.

Appendix 17A: The Problem of Field Singularities

The problem of field singularities [1,2] can be studied by considering points in the neighborhood of a sharp edge shown in Figure 17A.1.

Very close to the tip of a conducting wedge with an internal angle α, the behavior of the field is the same as for the static field (quasistatic approximation):

$$\bar{E} = -\bar{\nabla}\Phi, \qquad (17A.1)$$

and Φ satisfies the Laplace equation. Thus, the equation for Φ is written as

$$\nabla_t^2 \Phi = \frac{1}{\rho}\frac{\partial}{\partial\rho}\left(\rho\frac{\partial\Phi}{\partial\rho}\right) + \frac{1}{\rho^2}\frac{\partial\Phi^2}{\partial\phi^2} = 0. \qquad (17A.2)$$

The solution can be written as

$$\Phi = A\rho^\nu \sin\nu\left(\phi + \phi_0\right) \qquad (17A.3)$$

for the field region $\alpha < \phi < 2\pi$ and $\rho \sim 0$.

The constants ν and ϕ_0 can be determined from the boundary conditions 1 and 2:

$$\text{BC1:} \quad \phi = \alpha, \quad \Phi = 0. \qquad (17A.4)$$

Thus,

$$\alpha + \phi_0 = 0, \quad \phi_0 = -\alpha. \qquad (17A.5)$$

$$\text{BC2:} \quad \phi = 2\pi, \quad \Phi = 0,$$

$$\sin\nu\left(2\pi - \alpha\right) = 0,$$

$$\nu\left(2\pi - \alpha\right) = \pi,$$

$$\nu = \frac{\pi}{2\pi - \alpha}. \qquad (17A.6)$$

The electric field, from Equation 17A.1, is given by

$$\bar{E} = -A\nu\rho^{\nu-1}\left[\hat{\rho}\sin\nu\left(\phi - \alpha\right) + \hat{\phi}\cos\nu\left(\phi - \alpha\right)\right]. \qquad (17A.7)$$

The corresponding magnetic field can be obtained by substituting Equation 17A.7 into Maxwell's equation:

$$\bar{\nabla} \times \bar{E} = -j\omega\mu\bar{H}, \qquad (17A.8a)$$

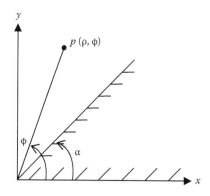

FIGURE 17A.1
A tip of a conducting wedge with an internal angle α.

$$\hat{z}\frac{1}{\rho}\left[\frac{\partial}{\partial\rho}(\rho E_\phi)-\frac{\partial E_\rho}{\partial\phi}\right]=-j\omega\mu\bar{H}, \tag{17A.8b}$$

where

$$E_\rho = -A\nu\rho^{\nu-1}\sin\nu(\phi-\alpha), \tag{17A.9}$$

$$E_\phi = -A\nu\rho^{\nu-1}\cos\nu(\phi-\alpha). \tag{17A.10}$$

$$\frac{\partial}{\partial\rho}(\rho E_\phi)=\frac{\partial}{\partial\rho}\left[-A\nu\rho^\nu\cos\nu(\phi-\alpha)\right]=-A\nu^2\rho^{\nu-1}\cos\nu(\phi-\alpha),$$

$$\frac{\partial E_\rho}{\partial\phi}=-A\nu^2\rho^{\nu-1}\cos\nu(\phi-\alpha), \tag{17A.11a}$$

$$\frac{\partial}{\partial\rho}(\rho E_\phi)-\frac{\partial E_\rho}{\partial\phi}=0$$

as expected.

$$\bar{\nabla}\times\bar{E}=0. \tag{17A.11b}$$

But from the first Maxwell equation, we obtain

$$\bar{\nabla}\times\bar{H}=j\omega\varepsilon\bar{E}. \tag{17A.12}$$

And knowing that $\bar{H}$ has only z-component, we obtain

$$\hat{\rho}\left[\frac{1}{\rho}\frac{\partial H_z}{\partial\phi}\right]=\hat{\rho}\,j\omega\varepsilon E_\rho, \tag{17A.13}$$

$$\hat{\phi}\left[-\frac{\partial H_z}{\partial\rho}\right]=\hat{\phi}\,j\omega\varepsilon E_\phi. \tag{17A.14}$$

From Equation 17A.13, we get

$$H_z = \int j\omega\varepsilon\rho E_\rho \, d\phi$$

$$= j\omega\varepsilon \int -A\nu\rho^{\nu-1} \sin \nu(\phi - \alpha) \, d\phi$$

$$= -j\omega\varepsilon A\nu\rho^\nu [-1]\cos \nu(\phi - \alpha)\left(\frac{1}{\nu}\right). \qquad (17A.15)$$

$$H_z = j\omega\varepsilon A\rho^\nu \cos \nu(\phi - \alpha).$$

From Equation 17A.14, we expect the same answer.

Since $E_z = 0$, but $H_z \neq 0$, we call this a TE case. Let us assume that $\alpha = \pi/2$, that is, we consider the field region outside the right-angled corner. Then from Equations 17A.6, 17A.9, 17A.10, and 17A.15, we obtain

$$\left.\begin{aligned}
\nu &= \frac{\pi}{2\pi - \pi/2} = \frac{\pi}{3\pi/2} = \frac{2}{3}, \\
E_\rho &= -A\nu \frac{1}{\rho^{1/3}} \sin \frac{2}{3}(\phi - \alpha), \\
E_\phi &= -A\nu \frac{1}{\rho^{1/3}} \cos \frac{2}{3}(\phi - \alpha), \quad \frac{\pi}{2} < \phi < 2\pi, \\
H_z &= j\omega\varepsilon A\rho^{2/3} \cos \frac{2}{3}(\phi - \alpha).
\end{aligned}\right\} \qquad (17A.16)$$

Note that the transverse components of $\bar{E}$ become singular, while the tangential component of $\bar{H}$ (H_z) is finite. One can start with magnetostatic potential Φ_m and in the limit of $\rho \to 0$, it satisfies the Laplace equation

$$\nabla_m^2 \Phi = 0. \qquad (17A.17)$$

Proceeding along similar lines as before we can show that (note that BC on $H_{\tan}$ is $\partial H_{\tan}/\partial\phi = 0$:

$$\bar{H} = -A\nu\rho^{\nu-1}\left[\hat{\rho}\cos \nu(\phi - \alpha) - \hat{\phi}\sin \nu(\phi - \alpha)\right],$$

$$E_z = -j\omega\mu A\rho^\nu \sin \nu(\phi - \alpha).$$

For $0 < \alpha < \pi$, $-1/2 < \nu - 1 < 0$ and therefore the transverse components blow up as $\rho \to 0$ since they are proportional $\rho^{\nu-1}$. The smaller the internal angle, the stronger the field singularities.

In conclusion, in the limit $\rho \to 0$, the fields in the vicinity of a right-angled conducting edge (Figure 17A.2) has the following variation:

$$E_\rho \sim \frac{1}{\rho^{1/3}} \sin \frac{2}{3}(\phi - 90°),$$

$$E_\phi \sim \frac{1}{\rho^{1/3}} \cos \frac{2}{3}(\phi - 90°),$$

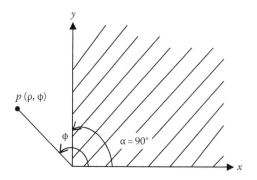

FIGURE 17A.2
Right-angled conducting wedge.

$$H_z \sim \rho^{2/3} \cos \frac{2}{3}(\phi - 90°),$$

$$H_\rho \sim \frac{1}{\rho^{1/3}} \cos \frac{2}{3}(\phi - 90°),$$

$$H_\phi \sim \frac{1}{\rho^{1/3}} \sin \frac{2}{3}(\phi - 90°),$$

$$E_z \sim \rho^{2/3} \sin \frac{2}{3}(\phi - 90°),$$

E_z and H_z are finite at the origin ($\rho \to 0$) but all transverse components E_ρ, H_ρ, E_ϕ, and H_ϕ have singularities at the origin.

Note that we used quasistatic analysis since we were investigating the limit $\rho \to 0$.

References

1. Jin, J., *The Finite Element Method in Electromagnetics*, 2nd edn., Wiley, New York, 2002.
2. Van Bladel, J., *Electromagnetic Fields*, McGraw-Hill, New York, 1964.

Appendix 18A: Input Data

Note: These GID-generated data are different for each problem, for example, the following data are generated from first-order FEM.

18A.1 GID-Generated Input Data

```
ENTITIES
POINT
Num: 1 HigherEntity: 2 conditions: 0 material: 0
LAYER: Layer0
Coord: -10.0 -2.0 0.0
END POINT
POINT
Num: 2 HigherEntity: 2 conditions: 0 material: 0
LAYER: Layer0
Coord: -4.0 -2.0 0.0
END POINT
POINT
Num: 3 HigherEntity: 2 conditions: 0 material: 0
LAYER: Layer0
Coord: -4.0 -5.0 0.0
END POINT
POINT
Num: 4 HigherEntity: 2 conditions: 0 material: 0
LAYER: Layer0
Coord: -10.0 -5.0 0.0
END POINT
STLINE
Num: 1 HigherEntity: 1 conditions: 0 material: 0
LAYER: Layer0
Meshing Info: (num = 19,NOE = 1) Elemtype = 1 IsStructured = 0
  Meshing = Default size = 18
Points: 1 2
END STLINE
STLINE
Num: 2 HigherEntity: 1 conditions: 0 material: 0
LAYER: Layer0
Meshing Info: (num = 17,NOE = 1) Elemtype = 1 IsStructured = 0
  Meshing = Default size = 12
Points: 2 3
END STLINE
STLINE
Num: 3 HigherEntity: 1 conditions: 0 material: 0
LAYER: Layer0
```

```
Meshing Info: (num = 16,NOE = 1) Elemtype = 1 IsStructured = 0
  Meshing = Default size = 18
Points: 3 4
END STLINE
STLINE
Num: 4 HigherEntity: 1 conditions: 0 material: 0
LAYER: Layer0
Meshing Info: (num = 18,NOE = 1) Elemtype = 1 IsStructured = 0
  Meshing = Default size = 12
Points: 4 1
END STLINE
NURBSURFACE
Num: 1 HigherEntity: 0 conditions: 0 material: 0
LAYER: Layer0
Meshing Info: (num = 2,NOE = 1) Elemtype = 2 IsStructured = 0
  Meshing = Default size = 0
NumLines: 4
Line: 1 Orientation: DIFF1ST
Line: 4 Orientation: DIFF1ST
Line: 3 Orientation: DIFF1ST
Line: 2 Orientation: DIFF1ST
Number of Control Points = 2 2 Degree = 1 1
Point 1,1 coords: - 4.0, - 2.0,0.0
Point 1,2 coords: - 10.0, - 2.0,0.0
Point 2,1 coords: - 4.0, - 5.0,0.0
Point 2,2 coords: - 10.0, - 5.0,0.0
Number of knots in U = 4
knot 1 value = 0.0
knot 2 value = 0.0
knot 3 value = 1.0
knot 4 value = 1.0
Number of knots in V = 4
knot 1 value = 0.0
knot 2 value = 0.0
knot 3 value = 1.0
knot 4 value = 1.0
Non rational
IsTrimmed: 0
Center: -7.0 -3.5 0.0
Normal: 0.0 0.0 1.0
END NURBSURFACE
END ENTITIES
MESH dimension 3 ElemType Triangle Nnode 3
Coordinates
```

Global Node #	X_n	Y_n	Z_n
1	−10	−2	0
2	−10	−2.25	0
3	−9.66666667	−2	0
4	−9.66666667	−2.25	0
5	−10	−2.5	0
6	−9.66666667	−2.5	0
7	−9.33333333	−2	0

(Continued)

Global Node #	X_n	Y_n	Z_n
8	-9.33333333	-2.25	0
9	-10	-2.75	0
10	-9.66666667	-2.75	0
11	-9.33333333	-2.5	0
12	-10	-3	0
13	-9	-2	0
14	-9.33333333	-2.75	0
15	-9	-2.25	0
16	-9.66666667	-3	0
17	-9	-2.5	0
18	-9.33333333	-3	0
19	-9	-2.75	0
20	-10	-3.25	0
21	-9.66666667	-3.25	0
22	-8.66666667	-2	0
23	-8.66666667	-2.25	0
24	-9	-3	0
25	-9.33333333	-3.25	0
26	-8.66666667	-2.5	0
27	-10	-3.5	0
28	-8.66666667	-2.75	0
29	-9.66666667	-3.5	0
30	-9	-3.25	0
31	-9.33333333	-3.5	0
32	-8.33333333	-2	0
33	-8.66666667	-3	0
34	-8.33333333	-2.25	0
35	-8.33333333	-2.5	0
36	-10	-3.75	0
37	-9.66666667	-3.75	0
38	-9	-3.5	0
39	-8.33333333	-2.75	0
40	-8.66666667	-3.25	0
41	-9.33333333	-3.75	0
42	-8.33333333	-3	0
43	-8	-2	0
44	-10	-4	0
45	-8.66666667	-3.5	0
46	-9	-3.75	0
47	-8	-2.25	0
48	-9.66666667	-4	0
49	-8	-2.5	0
50	-8.33333333	-3.25	0
51	-9.33333333	-4	0
52	-8	-2.75	0
53	-8.66666667	-3.75	0
54	-8	-3	0
55	-9	-4	0
56	-8.33333333	-3.5	0
57	-10	-4.25	0

(Continued)

Global Node #	X_n	Y_n	Z_n
58	−9.66666667	−4.25	0
59	−9.33333333	−4.25	0
60	−8	−3.25	0
61	−8.66666667	−4	0
62	−8.33333333	−3.75	0
63	−9	−4.25	0
64	−10	−4.5	0
65	−8	−3.5	0
66	−9.66666667	−4.5	0
67	−7.66666667	−3	0
68	−9.33333333	−4.5	0
69	−8.33333333	−4	0
70	−8.66666667	−4.25	0
71	−7.66666667	−3.25	0
72	−8	−3.75	0
73	−9	−4.5	0
74	−10	−4.75	0
75	−9.66666667	−4.75	0
76	−7.66666667	−3.5	0
77	−8.33333333	−4.25	0
78	−8	−4	0
79	−9.33333333	−4.75	0
80	−8.66666667	−4.5	0
81	−7.33333333	−3	0
82	−7.66666667	−3.75	0
83	−9	−4.75	0
84	−7.33333333	−3.25	0
85	−10	−5	0
86	−8.33333333	−4.5	0
87	−8	−4.25	0
88	−9.66666667	−5	0
89	−8.66666667	−4.75	0
90	−7.33333333	−3.5	0
91	−7.66666667	−4	0
92	−9.33333333	−5	0
93	−7	−3	0
94	−9	−5	0
95	−7.33333333	−3.75	0
96	−8	−4.5	0
97	−8.33333333	−4.75	0
98	−7.66666667	−4.25	0
99	−7	−3.25	0
100	−8.66666667	−5	0
101	−7.33333333	−4	0
102	−7	−3.5	0
103	−8	−4.75	0
104	−7.66666667	−4.5	0
105	−8.33333333	−5	0
106	−7	−3.75	0
107	−6.66666667	−3	0

(Continued)

Global Node #	X_n	Y_n	Z_n
108	−7.33333333	−4.25	0
109	−6.66666667	−3.25	0
110	−7	−4	0
111	−8	−5	0
112	−7.66666667	−4.75	0
113	−6.66666667	−3.5	0
114	−7.33333333	−4.5	0
115	−7	−4.25	0
116	−6.66666667	−3.75	0
117	−6.33333333	−3	0
118	−7.66666667	−5	0
119	−7.33333333	−4.75	0
120	−6.33333333	−3.25	0
121	−6.66666667	−4	0
122	−7	−4.5	0
123	−6.33333333	−3.5	0
124	−6	−2	0
125	−6	−2.25	0
126	−7.33333333	−5	0
127	−6.66666667	−4.25	0
128	−6	−2.5	0
129	−6.33333333	−3.75	0
130	−6	−2.75	0
131	−7	−4.75	0
132	−6	−3	0
133	−6.66666667	−4.5	0
134	−6.33333333	−4	0
135	−6	−3.25	0
136	−7	−5	0
137	−6	−3.5	0
138	−6.33333333	−4.25	0
139	−6.66666667	−4.75	0
140	−5.66666667	−2	0
141	−5.66666667	−2.25	0
142	−5.66666667	−2.5	0
143	−6	−3.75	0
144	−5.66666667	−2.75	0
145	−6.33333333	−4.5	0
146	−5.66666667	−3	0
147	−6	−4	0
148	−6.66666667	−5	0
149	−5.66666667	−3.25	0
150	−6.33333333	−4.75	0
151	−5.66666667	−3.5	0
152	−6	−4.25	0
153	−5.33333333	−2	0
154	−5.66666667	−3.75	0
155	−5.33333333	−2.25	0
156	−5.33333333	−2.5	0
157	−6	−4.5	0

(Continued)

Global Node #	X_n	Y_n	Z_n
158	−5.33333333	−2.75	0
159	−6.33333333	−5	0
160	−5.66666667	−4	0
161	−5.33333333	−3	0
162	−5.33333333	−3.25	0
163	−6	−4.75	0
164	−5.66666667	−4.25	0
165	−5.33333333	−3.5	0
166	−5.33333333	−3.75	0
167	−5	−2	0
168	−6	−5	0
169	−5.66666667	−4.5	0
170	−5	−2.25	0
171	−5	−2.5	0
172	−5	−2.75	0
173	−5.33333333	−4	0
174	−5	−3	0
175	−5.66666667	−4.75	0
176	−5	−3.25	0
177	−5.33333333	−4.25	0
178	−5	−3.5	0
179	−5.66666667	−5	0
180	−5.33333333	−4.5	0
181	−5	−3.75	0
182	−4.66666667	−2	0
183	−4.66666667	−2.25	0
184	−4.66666667	−2.5	0
185	−5	−4	0
186	−4.66666667	−2.75	0
187	−5.33333333	−4.75	0
188	−4.66666667	−3	0
189	−4.66666667	−3.25	0
190	−5	−4.25	0
191	−4.66666667	−3.5	0
192	−5.33333333	−5	0
193	−5	−4.5	0
194	−4.66666667	−3.75	0
195	−4.33333333	−2	0
196	−4.33333333	−2.25	0
197	−4.33333333	−2.5	0
198	−4.66666667	−4	0
199	−5	−4.75	0
200	−4.33333333	−2.75	0
201	−4.33333333	−3	0
202	−4.66666667	−4.25	0
203	−4.33333333	−3.25	0
204	−5	−5	0
205	−4.33333333	−3.5	0
206	−4.66666667	−4.5	0
207	−4.33333333	−3.75	0

(Continued)

Global Node #	X_n	Y_n	Z_n
208	-4	-2	0
209	-4.66666667	-4.75	0
210	-4	-2.25	0
211	-4.33333333	-4	0
212	-4	-2.5	0
213	-4	-2.75	0
214	-4	-3	0
215	-4.33333333	-4.25	0
216	-4.66666667	-5	0
217	-4	-3.25	0
218	-4	-3.5	0
219	-4.33333333	-4.5	0
220	-4	-3.75	0
221	-4.33333333	-4.75	0
222	-4	-4	0
223	-4	-4.25	0
224	-4.33333333	-5	0
225	-4	-4.5	0
226	-4	-4.75	0
227	-4	-5	0

end coordinates

Elements

Element#	n1L	n2L	n3L
1	195	196	208
2	196	210	208
3	182	183	195
4	183	196	195
5	167	170	182
6	170	183	182
7	153	155	167
8	155	170	167
9	140	141	153
10	141	155	153
11	124	125	140
12	125	141	140
13	32	34	43
14	34	47	43
15	22	23	32
16	23	34	32
17	13	15	22
18	15	23	22
19	7	8	13
20	8	15	13
21	3	4	7
22	4	8	7
23	1	2	3
24	2	4	3
25	196	197	210
26	197	212	210

(Continued)

Element#	n1L	n2L	n3L
27	183	184	196
28	184	197	196
29	170	171	183
30	171	184	183
31	155	156	170
32	156	171	170
33	141	142	155
34	142	156	155
35	125	128	141
36	128	142	141
37	34	35	47
38	35	49	47
39	23	26	34
40	26	35	34
41	15	17	23
42	17	26	23
43	8	11	15
44	11	17	15
45	4	6	8
46	6	11	8
47	2	5	4
48	5	6	4
49	197	200	212
50	200	213	212
51	184	186	197
52	186	200	197
53	171	172	184
54	172	186	184
55	156	158	171
56	158	172	171
57	142	144	156
58	144	158	156
59	128	130	142
60	130	144	142
61	35	39	49
62	39	52	49
63	26	28	35
64	28	39	35
65	17	19	26
66	19	28	26
67	11	14	17
68	14	19	17
69	6	10	11
70	10	14	11
71	5	9	6
72	9	10	6
73	200	201	213
74	201	214	213
75	186	188	200
76	188	201	200

(*Continued*)

Element#	n1L	n2L	n3L
77	172	174	186
78	174	188	186
79	158	161	172
80	161	174	172
81	144	146	158
82	146	161	158
83	130	132	144
84	132	146	144
85	39	42	52
86	42	54	52
87	28	33	39
88	33	42	39
89	19	24	28
90	24	33	28
91	14	18	19
92	18	24	19
93	10	16	14
94	16	18	14
95	9	12	10
96	12	16	10
97	201	203	214
98	203	217	214
99	188	189	201
100	189	203	201
101	174	176	188
102	176	189	188
103	161	162	174
104	162	176	174
105	146	149	161
106	149	162	161
107	132	135	146
108	135	149	146
109	117	120	132
110	120	135	132
111	107	109	117
112	109	120	117
113	93	99	107
114	99	109	107
115	81	84	93
116	84	99	93
117	67	71	81
118	71	84	81
119	54	60	67
120	60	71	67
121	42	50	54
122	50	60	54
123	33	40	42
124	40	50	42
125	24	30	33
126	30	40	33

(*Continued*)

Element#	n1L	n2L	n3L
127	18	25	24
128	25	30	24
129	16	21	18
130	21	25	18
131	12	20	16
132	20	21	16
133	203	205	217
134	205	218	217
135	189	191	203
136	191	205	203
137	176	178	189
138	178	191	189
139	162	165	176
140	165	178	176
141	149	151	162
142	151	165	162
143	135	137	149
144	137	151	149
145	120	123	135
146	123	137	135
147	109	113	120
148	113	123	120
149	99	102	109
150	102	113	109
151	84	90	99
152	90	102	99
153	71	76	84
154	76	90	84
155	60	65	71
156	65	76	71
157	50	56	60
158	56	65	60
159	40	45	50
160	45	56	50
161	30	38	40
162	38	45	40
163	25	31	30
164	31	38	30
165	21	29	25
166	29	31	25
167	20	27	21
168	27	29	21
169	205	207	218
170	207	220	218
171	191	194	205
172	194	207	205
173	178	181	191
174	181	194	191
175	165	166	178
176	166	181	178

(Continued)

Element#	n1L	n2L	n3L
177	151	154	165
178	154	166	165
179	137	143	151
180	143	154	151
181	123	129	137
182	129	143	137
183	113	116	123
184	116	129	123
185	102	106	113
186	106	116	113
187	90	95	102
188	95	106	102
189	76	82	90
190	82	95	90
191	65	72	76
192	72	82	76
193	56	62	65
194	62	72	65
195	45	53	56
196	53	62	56
197	38	46	45
198	46	53	45
199	31	41	38
200	41	46	38
201	29	37	31
202	37	41	31
203	27	36	29
204	36	37	29
205	207	211	220
206	211	222	220
207	194	198	207
208	198	211	207
209	181	185	194
210	185	198	194
211	166	173	181
212	173	185	181
213	154	160	166
214	160	173	166
215	143	147	154
216	147	160	154
217	129	134	143
218	134	147	143
219	116	121	129
220	121	134	129
221	106	110	116
222	110	121	116
223	95	101	106
224	101	110	106
225	82	91	95
226	91	101	95

(Continued)

Element#	n1L	n2L	n3L
227	72	78	82
228	78	91	82
229	62	69	72
230	69	78	72
231	53	61	62
232	61	69	62
233	46	55	53
234	55	61	53
235	41	51	46
236	51	55	46
237	37	48	41
238	48	51	41
239	36	44	37
240	44	48	37
241	211	215	222
242	215	223	222
243	198	202	211
244	202	215	211
245	185	190	198
246	190	202	198
247	173	177	185
248	177	190	185
249	160	164	173
250	164	177	173
251	147	152	160
252	152	164	160
253	134	138	147
254	138	152	147
255	121	127	134
256	127	138	134
257	110	115	121
258	115	127	121
259	101	108	110
260	108	115	110
261	91	98	101
262	98	108	101
263	78	87	91
264	87	98	91
265	69	77	78
266	77	87	78
267	61	70	69
268	70	77	69
269	55	63	61
270	63	70	61
271	51	59	55
272	59	63	55
273	48	58	51
274	58	59	51
275	44	57	48
276	57	58	48

(Continued)

Element#	n1L	n2L	n3L
277	215	219	223
278	219	225	223
279	202	206	215
280	206	219	215
281	190	193	202
282	193	206	202
283	177	180	190
284	180	193	190
285	164	169	177
286	169	180	177
287	152	157	164
288	157	169	164
289	138	145	152
290	145	157	152
291	127	133	138
292	133	145	138
293	115	122	127
294	122	133	127
295	108	114	115
296	114	122	115
297	98	104	108
298	104	114	108
299	87	96	98
300	96	104	98
301	77	86	87
302	86	96	87
303	70	80	77
304	80	86	77
305	63	73	70
306	73	80	70
307	59	68	63
308	68	73	63
309	58	66	59
310	66	68	59
311	57	64	58
312	64	66	58
313	219	221	225
314	221	226	225
315	206	209	219
316	209	221	219
317	193	199	206
318	199	209	206
319	180	187	193
320	187	199	193
321	169	175	180
322	175	187	180
323	157	163	169
324	163	175	169
325	145	150	157
326	150	163	157

(Continued)

Element#	n1L	n2L	n3L
327	133	139	145
328	139	150	145
329	122	131	133
330	131	139	133
331	114	119	122
332	119	131	122
333	104	112	114
334	112	119	114
335	96	103	104
336	103	112	104
337	86	97	96
338	97	103	96
339	80	89	86
340	89	97	86
341	73	83	80
342	83	89	80
343	68	79	73
344	79	83	73
345	66	75	68
346	75	79	68
347	64	74	66
348	74	75	66
349	221	224	226
350	224	227	226
351	209	216	221
352	216	224	221
353	199	204	209
354	204	216	209
355	187	192	199
356	192	204	199
357	175	179	187
358	179	192	187
359	163	168	175
360	168	179	175
361	150	159	163
362	159	168	163
363	139	148	150
364	148	159	150
365	131	136	139
366	136	148	139
367	119	126	131
368	126	136	131
369	112	118	119
370	118	126	119
371	103	111	112
372	111	118	112
373	97	105	103
374	105	111	103
375	89	100	97
376	100	105	97

(Continued)

Element#	n1L	n2L	n3L
377	83	94	89
378	94	100	89
379	79	92	83
380	92	94	83
381	75	88	79
382	88	92	79
383	74	85	75
384	85	88	75
end elements			

```
PROBLEM DATA
END PROBLEM DATA
INTERVAL NUMBER: 1
INTERVAL DATA
END INTERVAL DATA
```

Appendix 18B: Main Programs

18B.1 First-Order, Node-Based: Main Program

```
Nn = 227; % number of nodes
Ne = 384; % number of elements
inch = 2.54*1e-2./6;
[xn,yn,n1L,n2L,n3L,mt] = meshread('data.txt', Ne, Nn, 0);
[S,T] = GLANT(Nn,Ne,n1L,n2L,n3L,xn*inch,yn*inch);
% TE modes
ATE = inv(T)*S;
[EVTE, kcsqTE] = eig(ATE);
kcTE = sqrt(kcsqTE);
[N,M] = size(kcTE);
for ii = 1 : N
      KCTE(ii) = kcTE(ii,ii);
end
FTE = sort(KCTE*3e8/(2*pi))';
% TM modes
boundary = load('boundary_nodes.txt');% reading the boundary nodes
all_nodes = 1:Nn; % reading all nodes
nce = setxor(boundary,all_nodes); % extracting the free nodes
STM = S(nce,nce);
TTM = T(nce,nce);
ATM = inv(TTM)*STM;
[EVTMET,kcsqTM] = eig(ATM);
kcTM = sqrt(kcsqTM);
[N,M] = size(kcTM);
for ii = 1 : N
      KCTM(ii) = kcTM(ii,ii);
end
disp('Cutoff wavenumbers for the TM modes');
disp(sort(KCTM));
FTM = sort(KCTM*3e8/(2*pi))';
```

18B.2 Edge-Based: Main Program

```
Nn = 227; % number of nodes
Ne = 384; % number of elements
inch = 2.54*1e-2./6;
[xn,yn,n1L,n2L,n3L,mt] = meshread('data.txt', Ne, Nn, 0);
[Neg,n1EL,n2EL,n3EL] = edges(n1L, n2L, n3L);
```

```
[E,F] = GLAET(Neg, Nn, Ne, n1L, n2L, n3L, n1EL, n2EL, n3EL,
    xn*inch,yn*inch);
% TM modes
ATM = inv(F)*E;
[EVTM, kcsqTM] = eig(ATM);
kcTM = sqrt(kcsqTM);
[N,M] = size(kcTM);
for ii = 1 : N
       KCTM(ii) = kcTM(ii,ii);
end
FTM = sort(KCTM*3e8/(2*pi))';

% TE modes
A = intersect(abs(n2EL),abs(n3EL));
B = intersect(abs(n3EL),abs(n1EL));
nce1 = union(A,B);
C = intersect(abs(n1EL),abs(n2EL));
nce = union(nce1,C); % finding the common edges
ETE = E(nce,nce);
FTE = F(nce,nce);
ATE = inv(FTE)*ETE;
[EVTEET,kcsqTE] = eig(ATE);
kcTE = sqrt(kcsqTE);
[N,M] = size(kcTE);
for ii = 1 : N
       KCTE(ii) = kcTE(ii,ii);
end
disp('Cutoff wavenumbers for the TM modes');
disp(sort(KCTE));
FTE = sort(KCTE*3e8/(2*pi))';
```

18B.3 Second-Order, Node-Based: Main Program

```
Nn = 837; % number of nodes
Ne = 384; % number of elements
inch = 2.54*1e-2./6;
[xn,yn,n1L,n2L,n3L,n4L,n5L,n6L,mt] = meshread2T('2ndorder.txt', Ne, Nn, 0);
[S,T] = GLAN2T(Nn,Ne,n1L,n4L,n6L,n2L,n5L,n3L,xn*inch,yn*inch);
% TE modes
ATE = inv(T)*S;
[EVTE, kcsqTE] = eig(ATE);
kcTE = sqrt(kcsqTE);
[N,M] = size(kcTE);
for ii = 1 : N
       KCTE(ii) = kcTE(ii,ii);
end
FTE = sort(KCTE*3e8/(2*pi))';
% TM modes
boundary = load('boundary_nodes.txt'); % reading the boundary nodes
all_nodes = 1:Nn; % reading all nodes
nce = setxor(boundary,all_nodes); % extracting the free nodes
```

```
STM = S(nce,nce);
TTM = T(nce,nce);
ATM = inv(TTM)*STM;
[EVTMET,kcsqTM] = eig(ATM);
kcTM = sqrt(kcsqTM);
[N,M] = size(kcTM);
for ii = 1 : N
      KCTM(ii) = kcTM(ii,ii);
end
disp('Cutoff wavenumbers for the TM modes');
KCTM = sort(KCTM)';
FTM = sort(KCTM*3e8/(2*pi))';
```

18B.4 Inhomogeneous Waveguide: Main Program

```
% Inhomogeneous WG partially filled with dielectric material
clear all
clc
Nn = 227; % number of nodes
Ne = 384; % number of elements
inch = .0256/6;
[xn,yn,n1L,n2L,n3L,mt] = meshread('INHWG_Data.txt', Ne, Nn, 0); % reading
   the data from GID
[Neg,n1EL,n2EL,n3EL] = edges(n1L, n2L, n3L); % generating the Local edges
  for each element
k0 = [303:5:320]; % k0 loop
mur = ones(1,length(n1L));
epsr = ones(1,length(n1L));
e_r = 4; % dielectric constant
diele_elements = (load('dielectric_elements.txt')); % reading dielectric
  filled elements
epsr(diele_elements) = e_r; % set the dielectric elements
for s = 1:length(k0)
       clear C GAMA EVIN GAMASQ diag_GAMA beta Beta
[Att,Btt,Btz,Bzz,C] = INHWGD(epsr,mur,k0(s),Neg,Nn,Ne,...
       n1L,n2L,n3L,n1EL,n2EL,n3EL,xn.*inch,yn.*inch); % calling the INHWG
         function
% free edges or non conductive edges
A = intersect(abs(n2EL),abs(n3EL));
B = intersect(abs(n3EL),abs(n1EL));
nce1 = union(A,B);
CC = intersect(abs(n1EL),abs(n2EL));
nce = union(nce1,CC);
%%%%%%%%%%%%%%%
% free nodes or non conductive nodes
boundary = load('boundary_nodes.txt'); % reading the boundary nodes
all_nodes = 1:Nn; % reading all nodes
ncn = setxor(boundary,all_nodes); % extracting the free nodes
%%%%%%%%%%%%%%%
Attf = Att(nce,nce);
Bttf = Btt(nce,nce);
```

```
Btzf = Btz(nce,ncn);
Bzzf = Bzz(ncn,ncn);
C1f = Bttf-Btzf*inv(Bzzf)*Btzf';
C2f = inv(C1f);
Cf = C2f*Attf;
[EVINf, GAMASQf] = eig(Cf);
GAMA = sqrt((GAMASQf));
diag_GAMA = diag(GAMA);
z = 1;
% finding propagation constant from Gama
for t = 1:length(diag_GAMA)
        if real(diag_GAMA(t)) == 0&&imag(diag_GAMA(t)) > 0
                beta(z) = imag(diag_GAMA(t));
                z = z + 1;
        end
end
Ko = k0(s)
Beta = fliplr(sort(beta))'
Freq = sort(k0(s)*3e8/(2*pi))
end % end of the K0 loop
```

Appendix 18C: Function Programs

18C.1 GLANT MATLAB® Function Program

```
function [S,T] = GLANT(Nn,Ne,n1L,n2L,n3L,xn,yn);
% Global Assembly two-dimensional node-based
% triangular elements
for e = 1:Ne;
        n(1,e) = n1L(e);
        n(2,e) = n2L(e);
        n(3,e) = n3L(e);
end
% Initialization
S = zeros(Nn,Nn);
T = zeros(Nn,Nn);
% Loop through all elements
for e = 1: Ne;
        % coordinates of the element nodes
        for i = 1:3;
                x(i) = xn(n(i,e));
                y(i) = yn(n(i,e));
        end
        % compute the element matrix entries
        b(1) = y(2) - y(3);
        b(2) = y(3) - y(1);
        b(3) = y(1) - y(2);
        c(1) = x(3) - x(2);
        c(2) = x(1) - x(3);
        c(3) = x(2) - x(1);
        Area = 0.5 * abs (b(2)*c(3) - b(3) *c(2));
        % Compute the element matrix entries
        for i = 1:3;
                for j = 1:3;
                        Se(i,j)= (0.25/Area)*(b(i)*b(j) + c(i) * c(j));
                        if i == j
                                Te(i,j)= Area/6;
                        else
                                Te(i,j)= Area/12;
                        end
% Assemble the Element matrices into Global FEM System
                        S(n(i,e),n(j,e)) = S(n(i,e),n(j,e)) + Se(i,j);
                        T(n(i,e),n(j,e)) = T(n(i,e),n(j,e)) + Te(i,j);
                end
        end
end
```

18C.2 Meshread MATLAB® Function

```
function [xn,yn,n1L,n2L,n3L,mt] = meshread(filename, Ne, Nn, mat)
% meshread.m
% This function reads nodal coordinates and connectivity
% matrix from an ASCII mesh file created by the GiD preprocessor
% Inputs: Ne, Nn - numbers of finite elements and nodes
% mat: 1 - read material types, 0 - don't read material types
% Outputs:
% xn, yn - nodal coordinates
% n1L,n2L,n3L - nodes, 3 per element
% mt - material type array [element# element type]
% Andrey Semichaevsky, 08/08/02
M = zeros(Nn,4);
N = zeros(Ne,5);
mt = zeros(Ne,2);
% Open file
[fp, msg] = fopen (filename, 'rt');
if fp == -1
        disp (msg)
        return;
end
% Read data
kk = 0; ii = 1;
while kk == 0
        L = fgetl(fp);
        kk = strncmp(L,'Coordinates',11);
        ii = ii + 1;
end
kk = 0; ii = 1;
while kk == 0
        L = fgetl(fp);
        if ~strcmp(L,'end coordinates')
                MM = str2num(L);
                if ~isempty(MM)
                        M(ii,1:4) = MM(1:4);
                end
        end
        kk = strncmp(L,'end coordinates',15);
        ii = ii + 1;
end
kk = 0; ii = 1;
while kk == 0
        L = fgetl(fp);
        kk = strncmp(L,'Elements',8);
        ii = ii + 1;
end
kk = 0; ii = 1;
while kk == 0
        L = fgetl(fp);
        if ~strcmp(L,'end elements')
                MM = str2num(L);
```

```
                    if ~isempty(MM)
                            if mat == 1
                                    N(ii,1:5) = MM(1:5);
                            else
                                    N(ii,1:4) = MM(1:4);
                            end
                    end
            end
            kk = strncmp(L,'end elements',12);
            ii = ii + 1;
end
% Close file
[msg, errn] = ferror (fp);
if errn ~= 0
        disp ('An error occurred reading from file !')
        disp (msg)
end
fclose (fp);
xn(1:Nn)= M(1:Nn,2);
yn(1:Nn)= M(1:Nn,3);
n1L(1:Ne)= N(1:Ne,2);
n2L(1:Ne)= N(1:Ne,3);
n3L(1:Ne)= N(1:Ne,4);
if mat == 1
mt(1:Ne,1)= N(1:Ne,1);
mt(1:Ne,2)= N(1:Ne,5);
end
```

18C.3 GLAET MATLAB® Function

```
% GLAET.M
function [E,F] = glaet(Neg, Nn, Ne, n1L, n2L, n3L, n1EL, n2EL, n3EL, xn, yn)
for e = 1:Ne
        n(1,e)= n1L(e);
        n(2,e)= n2L(e);
        n(3,e)= n3L(e);
        ne(1,e)= n1EL(e);
        ne(2,e)= n2EL(e);
        ne(3,e)= n3EL(e);
end
E = zeros(Neg, Neg);
F = zeros(Neg, Neg);
for e = 1:Ne
        for i = 1:3;
                x(i)= xn(n(i,e));
                y(i)= yn(n(i,e));
        end
        b(1)= y(2)- y(3);
        b(2)= y(3)- y(1);
        b(3)= y(1)- y(2);
```

```
            c(1) = x(3) - x(2);
            c(2) = x(1) - x(3);
            c(3) = x(2) - x(1);
            Area = 0.5*abs(b(2)*c(3)- b(3)*c(2));
            l(1) = sqrt(b(3)*b(3)+ c(3)*c(3));
            l(2) = sqrt(b(1)*b(1)+ c(1)*c(1));
            l(3) = sqrt(b(2)*b(2)+ c(2)*c(2));
            for i = 1:3
                    for j = 1:3
                            ff(i,j) = b(i)*b(j) + c(i)*c(j);
                    end
            end
            G(1,1) = 2*(ff(2,2) - ff(1,2) + ff(1,1));
            G(2,2) = 2*(ff(3,3) - ff(2,3) + ff(2,2));
            G(3,3) = 2*(ff(1,1) - ff(3,1) + ff(3,3));
            G(2,1) = G(1,2);
            G(1,3) = ff(2,1) - 2*ff(2,3) - ff(1,1) + ff(1,3);
            G(3,1) = G(1,3);
            G(2,3) = ff(3,1) - ff(3,3) - 2*ff(2,1) + ff(2,3);
            G(3,2) = G(2,3);
            for i = 1:3
                    for j = 1:3
                            Ee(i,j) = (1/Area)*(l(i) * l(j));
                            Fe(i,j) = (1/48)*(1/Area) * l(i) * l(j) * G(i,j);
                            if(ne(i,e) < 0)
                            Ee(i,j) = -Ee(i,j);
                            Fe(i,j) = -Fe(i,j);
                    else;
                    end;
                    if(ne(j,e) < 0);
                            Ee(i,j) = -Ee(i,j);
                            Fe(i,j) = -Fe(i,j);
                    else;
                    end;
                    ane(i,e) = abs(ne(i,e));
                    ane(j,e) = abs(ne(j,e));
                    E(ane(i,e),ane(j,e)) = E(ane(i,e),ane(j,e)) + Ee(i,j);
                    F(ane(i,e),ane(j,e)) = F(ane(i,e),ane(j,e)) + Fe(i,j);
                    end
            end
end
```

18C.4 Edges MATLAB® Function

```
function [Neg, n1EL, n2EL, n3EL] = edges(n1L, n2L, n3L)
% edges.m
% This function creates edge elements based on the
% connectivity matrix for a grid of triangular FE.
% Inputs:
```

```
% n1L, n2L, n3L - node # (3 per element)
% Outputs:
% Neg - number of edges,
% n1EL, n2EL, n3EL - edge # (3 per element)
% Andrey Semichaevsky, 07/21/02
N = length(n1L);
n_el = zeros(N,3);
n_el(1:N,1) = linspace(1,N,N)';
n_el(1:N,2) = n1L(1:N)';
n_el(1:N,3) = n2L(1:N)';
n_el(1:N,4) = n3L(1:N)';
ed_array = zeros(N*3,3);
ed_array1 = zeros(N*3,3);
% Build node array
        for i = 1:N
                ed_array((i-1)*3 + 1,2) = n_el(i,2);
                ed_array((i-1)*3 + 1,3) = n_el(i,3);

                ed_array((i-1)*3 + 2,2) = n_el(i,3);
                ed_array((i-1)*3 + 2,3) = n_el(i,4);

                ed_array((i-1)*3 + 3,2) = n_el(i,4);
                ed_array((i-1)*3 + 3,3) = n_el(i,2);
        end
% Check for duplicates
k = 1; cp = 0;
        for i = 1:3*N
                cp = 0;
        for j = 1:3*N

                if ((ed_array (i, 2)) ~= (ed_array (j, 2))) & (((ed_array
                  (i, 3)) ~= (ed_array (j, 3))))
                        % if this condition is true, then copy this edge
                          to the 'refined' table
                        cp = 1;
                end
                if ((ed_array (i, 2)) ~= (ed_array (j, 3))) & (((ed_array
                  (i, 3)) ~= (ed_array (j, 2))))
                        % reverse the order
                        % if this condition is true, then copy this edge
                          to the 'refined' table
                        cp = 1;
                end
        end
        if cp == 1
                ed_array1(k,:) = ed_array(i,:);
                k = k + 1;
                end
        end
clear ed_array;
[N2,M2] = size(ed_array1);
nn = linspace(1,N2,N2);
% Assign numbers to all edges
```

```
ed_array1(1:N2,1) = nn(1:N2)';
ed_array = ed_array1;
[Neg,M] = size(ed_array);
clear ed_array1;
ed_el = zeros(N,4);
ed_el(1:N,1) = n_el(1:N,1);
% Scan through the connectivity matrix and assign edge numbers
% (positive or negative), depending on clockwise/counterclockwise
% element contour
for i = 1:N
        aa(1,1:2) = n_el(i, [2 3]);
        aa(2,1:2) = n_el(i, [3 4]);
        aa(3,1:2) = n_el(i, [4 2]);

        for l = 1:N2
        for k = 1:3
                if aa(k,1:2) == ed_array(l,2:3),
                        ed_el(i,k + 1) = ed_array(l,1);
                end
                if fliplr(aa(k,1:2)) == ed_array(l,2:3),
                        ed_el(i,k + 1) = -ed_array(l,1);
                end
        end
        end
end
ne = zeros(Neg,1);
% Final refinement
for k = 1:Neg
for i = 1:N
        for l = 2:4
                if (ed_el(i,l) == k) | (ed_el(i,l) == -k)
                        ne(k) = 1;
                end
        end
        end
end
tt = zeros(Neg,2);
tt(1:Neg,1) = linspace(1,Neg,Neg)';
% Final numbering for edges
k = 1;
        for i = 1:Neg
                if ne(i) ~= 0
                        tt(i,2) = k;
                        k = k + 1;
                end
        end
% Edge number substitution
        for kk = 1:Neg
        for i = 1:N
                for j = 2:4
                        if ed_el(i,j) == tt(kk,1)
                                ed_el(i,j) = tt(kk,2);
                        end
                        if ed_el(i,j) == (-tt(kk,1))
```

```
                    ed_el(i,j) = -tt(kk,2);
                    end
                    end
            end
end
Neg = k - 1;
n1EL(1:N) = ed_el(1:N,2)';
n2EL(1:N) = ed_el(1:N,3)';
n3EL(1:N) = ed_el(1:N,4)';
```

18C.5 GLAN2T MATLAB® Function

```
function [S,T] = GLAN2T(Nn,Ne,n1L,n2L,n3L,n4L,n5L,n6L,xn,yn);
% Global Assembly two-dimensional node-based
% triangular elements of second-order
% written by D. K. Kalluri
for e = 1:Ne;
        n(1,e) = n1L(e);
        n(2,e) = n2L(e);
        n(3,e) = n3L(e);
        n(4,e) = n4L(e);
        n(5,e) = n5L(e);
        n(6,e) = n6L(e);
end
Q1 = (1/6)*[0,0,0,0,0,0;0,8,-8,0,0,0;0,-8,8,0,0,0;
    0,0,0,3,-4,1;0,0,0,-4,8,-4;0,0,0,1,-4,3];
Q2 = (1/6)*[3,0,-4,0,0,1;0,8,0,0,-8,0;-4,0,8,0,0,-4;
    0,0,0,0,0,0;0,-8,0,0,8,0;1,0,-4,0,0,3];
Q3 = (1/6)*[3,-4,0,1,0,0;-4,8,0,-4,0,0;0,0,8,0,-8,0;
    1,-4,0,3,0,0;0,0,-8,0,8,0;0,0,0,0,0,0];
Tek = (1/180)*[6,0,0,-1,-4,-1;0,32,16,0,16,-4;0,16,32,-4,16,0;
      -1,0,-4,6,0,-1;-4,16,16,0,32,0;-1,-4,0,-1,0,6];
% Initialization
S = zeros(Nn,Nn);
T = zeros(Nn,Nn);
% Loop through all elements
for e = 1: Ne;
        % coordinates of the element nodes
        x(1) = xn(n(1,e));
        x(2) = xn(n(4,e));
        x(3) = xn(n(6,e));
        y(1) = yn(n(1,e));
        y(2) = yn(n(4,e));
        y(3) = yn(n(6,e));
% compute the element matrix entries
        b(1) = y(2) - y(3);
        b(2) = y(3) - y(1);
        b(3) = y(1) - y(2);
        c(1) = x(3) - x(2);
        c(2) = x(1) - x(3);
        c(3) = x(2) - x(1);
```

```
            Area = 0.5 * abs (b(2) * c(3) - b(3) * c(2));
            COTTH1 = -(0.5/Area)*(b(2)* b(3) + c(2)* c(3));
            COTTH2 = -(0.5/Area)*(b(3)* b(1) + c(3)* c(1));
            COTTH3 = -(0.5/Area)*(b(1)* b(2) + c(1)* c(2));
            Te = Area*Tek;
            Se = COTTH1*Q1 + COTTH2*Q2 + COTTH3*Q3;
            % Compute the element matrix entries
            for i = 1:6;
                  for j = 1:6;
                        % Assemble the Element matrices into Global FEM
                          System
                  S(n(i,e),n(j,e)) = S(n(i,e),n(j,e))+ Se(i,j);
                  T(n(i,e),n(j,e)) = T(n(i,e),n(j,e))+ Te(i,j);
                  end
            end
end
```

18C.6 Meshread2T MATLAB® Function

```
function [xn,yn,n1L,n2L,n3L,n4L,n5L,n6L,mt] = meshread2T(filename, Ne,
  Nn, mat)
% meshread.m
% This function reads nodal coordinates and connectivity
% matrix from an ASCII mesh file created by the GiD preprocessor
% Inputs: Ne, Nn - numbers of finite elements and nodes
% mat: 1 - read material types, 0 - don't read material types
% Outputs:
% xn, yn - nodal coordinates
% n1L,n2L,n3L - nodes, 3 per element
% mt - material type array [element# element type]
% Andrey Semichaevsky, 08/08/02

M = zeros(Nn,4);
N = zeros(Ne,8);
mt = zeros(Ne,2);
% Open file
[fp, msg] = fopen (filename, 'rt');
if fp == -1
      disp (msg)
      return;
end
% Read data
kk = 0; ii = 1;
while kk == 0
      L = fgetl(fp);
      kk = strncmp(L,'Coordinates',11);
      ii = ii + 1;
end
```

```
kk = 0; ii = 1;
while kk == 0
      L = fgetl(fp);
      if ~strcmp(L,'end coordinates')
            MM = str2num(L);
            if ~isempty(MM)
                  M(ii,1:4) = MM(1:4);
            end
       end
       kk = strncmp(L,'end coordinates',15);
       ii = ii + 1;
end
kk = 0; ii = 1;
while kk == 0
      L = fgetl(fp);
      kk = strncmp(L,'Elements',8);
      ii = ii + 1;
end
kk = 0; ii = 1;
while kk == 0
      L = fgetl(fp);
      if ~strcmp(L,'end elements')
            MM = str2num(L);
            if ~isempty(MM)
                  if mat == 1
                        N(ii,1:8) = MM(1:8);
                  else
                        N(ii,1:7) = MM(1:7);
                  end
            end
       end
       kk = strncmp(L,'end elements',12);
       ii = ii + 1;
end
% Close file
[msg, errn] = ferror (fp);
if errn ~= 0
       disp ('An error occurred reading from file !')
       disp (msg)
end
fclose (fp);
xn(1:Nn) = M(1:Nn,2);
yn(1:Nn) = M(1:Nn,3);
n1L(1:Ne) = N(1:Ne,2);
n2L(1:Ne) = N(1:Ne,3);
n3L(1:Ne) = N(1:Ne,4);
n4L(1:Ne) = N(1:Ne,5);
n5L(1:Ne) = N(1:Ne,6);
n6L(1:Ne) = N(1:Ne,7);
if mat == 1
mt(1:Ne,1) = N(1:Ne,1);
mt(1:Ne,2) = N(1:Ne,8);
end
```

18C.7 INHWGD MATLAB® Function

```
function [Att,Btt,Btz,Bzz,C] = INHWGD(epr,mur,k0,Neg,Nn,Ne,...
        n1L,n2L,n3L,n1EL,n2EL,n3EL,xn,yn);
% Inhomogeneous waveguide problem, page 73-75 texte
% written by D. K. Kalluri
% Edge elements for transverse fields and node-based
%       elements for longitudinal fields
% Global Assembly of two-dimensional edge-based
% triangular elements
% Neg = number of edge elements
% Nn = number of nodes
% Ne = number of elements
% n1L = array containing global node number
%       of the first local node of e th element
% n2L = array containing global node number
%       of the second local node of e th element
% % n3L = array containing global node number
%       of the third local node of e th element
% n1EL = array containing global edge number
%        of the first local edge of e th element
% n2EL = array containing global edge number
%        of the first local edge of e th element
% n3EL = array containing global edge number
%        of the first local edge of e th element
% a negative value indicates opposite directions
%       of the global edge and the local edge
for e = 1:Ne;
        n(1,e) = n1L(e);
        n(2,e) = n2L(e);
        n(3,e) = n3L(e);
        ne(1,e) = n1EL(e);
        ne(2,e) = n2EL(e);
        ne(3,e) = n3EL(e);
end
% Initialization
Att = zeros(Neg,Neg);
Btt = zeros(Neg,Neg);
Btz = zeros(Neg,Nn);
Bzz = zeros(Nn,Nn);
% Loop through all elements
for e = 1: Ne;
        % coordinates of the element nodes
        for i = 1:3;
                x(i) = xn(n(i,e));
                y(i) = yn(n(i,e));
        end
        % compute the element matrix entries
        b(1) = y(2) - y(3);
        b(2) = y(3) - y(1);
        b(3) = y(1) - y(2);
        c(1) = x(3) - x(2);
```

```
c(2) = x(1) - x(3);
c(3) = x(2) - x(1);
Area = 0.5 * abs (b(2)* c(3) - b(3)* c(2));
l(1) = sqrt(b(3)* b(3) + c(3)* c(3));
l(2) = sqrt(b(1)* b(1) + c(1)* c(1));
l(3) = sqrt(b(2)* b(2) + c(2)* c(2));
K(1) = l(1)/(12*Area);
K(2) = l(2)/(12*Area);
K(3) = l(3)/(12*Area);
% Compute the element matrix entries
for i = 1:3;
        for j = 1:3;
                ff(i,j) = b(i)*b(j) + c(i)*c(j);
                Se(i,j) = (0.25/Area)*(b(i)*b(j) + c(i) *c(j));
                if i == j
                        Te(i,j) =Area/6;
                else
                        Te(i,j) =Area/12;
                end
        end;
end;
G(1,1) = 2*(ff(2,2) - ff(1,2) + ff(1,1));
G(2,2) = 2*(ff(3,3) - ff(2,3) + ff(2,2));
G(3,3) = 2*(ff(1,1) - ff(3,1) + ff(3,3));
G(1,2) = ff(2,3) - ff(2,2) - 2*ff(1,3) + ff(1,2);
G(2,1) = G(1,2);
G(1,3) = ff(2,1) - 2*ff(2,3) - ff(1,1) + ff(1,3);
G(3,1) = G(1,3);
G(2,3) = ff(3,1) - ff(3,3) - 2*ff(2,1)+ ff(2,3);
G(3,2) = G(2,3);
Gtz(1,1) = -K(1)*(b(1)*(b(1)- b(2)) + c(1)*(c(1) - c(2)));
Gtz(2,2) = -K(2)*(b(2)*(b(2)- b(3)) + c(2)*(c(2) - c(3)));
Gtz(3,3) = -K(3)*(b(3)*(b(3)- b(1)) + c(3)*(c(3) - c(1)));
Gtz(1,2) = K(1)*(b(2)*(b(2)- b(1)) + c(2)*(c(2) - c(1)));
Gtz(2,3) = K(2)*(b(3)*(b(3)- b(2)) + c(3)*(c(3) - c(2)));
Gtz(3,1) = K(3)*(b(1)*(b(1)- b(3)) + c(1)*(c(1) - c(3)));
Gtz(1,3) = -K(1)*(-y(1)* b(2)- y(2)*b(1)- y(3)* b(3)...
            + x(1)* c(2) + x(2)* c(1) + x(3)* c(3));
Gtz(2,1) = -K(2)*(-y(2)* b(3)- y(3)*b(2)- y(1)* b(1)...
            + x(2)* c(3) + x(3)* c(2) + x(1)* c(1));
Gtz(3,2) = -K(3)*(-y(3)* b(1)- y(1)*b(3)- y(2)* b(2)...
            + x(3)* c(1) + x(1)* c(3) + x(2)* c(2));
for i = 1:3;
        for j = 1:3;
                Ee(i,j) = (1/Area)*(l(i)*l(j));
                Fe(i,j) = (1/48)*(1/Area)*l(i)*l(j)*G(i,j);
        % Assemble the Element matrices into Global FEM System
        if (ne(i,e) < 0);
                Ee(i,j) = -Ee(i,j);
                Fe(i,j) = -Fe(i,j);
                Gtz(i,j) = -Gtz(i,j);
        else;
        end;
```

```
                          if (ne(j,e) < 0);
                                  Ee(i,j) = -Ee(i,j);
                                  Fe(i,j) = -Fe(i,j);
                          else;
                          end;
                          Atte(i,j) = (1/mur(e))*Ee(i,j) - k0*k0*epr(e)*Fe(i,j);
                          Btte(i,j) = (1/mur(e))*Fe(i,j);
                          Btze(i,j) = (1/mur(e))*Gtz(i,j);
                          Bzze(i,j) = (1/mur(e))*Se(i,j) - k0*k0*epr(e)*Te(i,j);
                          % for storing purposes take absolute value of
                          % ne(i,e) and ne(j,e) which may have negative values
                          ane(i,e) = abs(ne(i,e));
                          ane(j,e) = abs(ne(j,e));
                          Att(ane(i,e),ane(j,e)) = Att(ane(i,e),ane(j,e))+ Atte(i,j);
                          Btt(ane(i,e),ane(j,e)) = Btt(ane(i,e),ane(j,e))+ Btte(i,j);
                          Btz(ane(i,e),n(j,e)) = Btz(ane(i,e),n(j,e))+ Btze(i,j);
                          Bzz(n(i,e),n(j,e)) = Bzz(n(i,e),n(j,e))+ Bzze(i,j);
                          end
                  end
end
C1 = Btt - Btz*inv(Bzz)*Btz';
C2 = inv(C1);
C = C2*Att;
```

18C.8 GLANL MATLAB® Function

```
function [S,T,K,b] = GLANL(Nn,Ne,n1L,n2L,alphae,betae,fe,xn);
% Global Assembly one-dimensional node-based
% first-order elements
% diff. eq: -d/dz(alpha d phi/dz) +beta phi = f(z)
for e = 1:Ne;
        n(1,e) = n1L(e);
        n(2,e) = n2L(e);
        alpha(e) = alphae(e);
        beta(e) = betae(e);
        f(e) = fe(e);
end

% Initialization
S = zeros(Nn,Nn);
T = zeros(Nn,Nn);
K = zeros(Nn,Nn);
b = zeros(Nn,Nn);
% Loop through all elements
for e = 1: Ne;
        % coordinates of the element nodes
        for i = 1:2;
                x(i) = xn(n(i,e));
        end
```

```
% compute the element matrix entries
l(e) = x(2) - x(1);
Se(1,1) = 1/l(e);
Se(1,2) = -1/l(e);
Se(2,1) = Se(1,2);
Se(2,2) = Se(1,1);
Te(1,1) = l(e)/3;
Te(1,2) = l(e)/6;
Te(2,1) = Te(1,2);
Te(2,2) = Te(1,1);
% Compute the Ke and be matrix entries and assemble them
% into Global FEM system
for i = 1:2;
        be(i) = l(e)*f(e)/2;
        b(n(i,e)) = b(n(i,e))+ be(i);
        for j = 1:2;
                Ke(i,j) = alpha(e)*Se(i,j) + beta(e)* Te(i,j);
                S(n(i,e),n(j,e)) = S(n(i,e),n(j,e))+ Se(i,j);
                T(n(i,e),n(j,e)) = T(n(i,e),n(j,e))+ Te(i,j);
                K(n(i,e),n(j,e)) = K(n(i,e),n(j,e))+ Ke(i,j);
        end
end
end
```

Appendix 22A: Complex Poynting Theorem

In a lossless medium,

$$\bar{\nabla} \times \tilde{\bar{H}} = j\omega\varepsilon\tilde{\bar{E}}, \qquad (22A.1)$$

$$\bar{\nabla} \times \tilde{\bar{E}} = -j\omega\mu\tilde{\bar{H}}, \qquad (22A.2)$$

$$\bar{\nabla} \times \tilde{\bar{E}}^* = j\omega\mu\tilde{\bar{H}}^*, \qquad (22A.3)$$

$$\tilde{\bar{H}} \cdot \bar{\nabla} \times \tilde{\bar{E}}^* = j\omega\mu\tilde{\bar{H}} \cdot \tilde{\bar{H}}^*, \qquad (22A.4)$$

$$\tilde{\bar{E}}^* \cdot \bar{\nabla} \times \tilde{\bar{H}} = j\omega\varepsilon\tilde{\bar{E}} \cdot \tilde{\bar{E}}^*. \qquad (22A.5)$$

Subtracting Equation 22A.5 from Equation 22A.4 and using Equation 22A.4, we get

$$\bar{\nabla} \cdot \left(\tilde{\bar{H}} \times \tilde{\bar{E}}^* \right) = j\omega\varepsilon \left| \tilde{\bar{E}} \right|^2 - j\omega\mu \left| \tilde{\bar{H}} \right|^2.$$

Integrating over the volume τ, we obtain

$$\iiint_{\tau} \bar{\nabla} \cdot \left(\tilde{\bar{H}} \times \tilde{\bar{E}}^* \right) d\tau = j\omega \iiint_{\tau} \left[\varepsilon \left| \tilde{\bar{E}} \right|^2 - \mu \left| \tilde{\bar{H}} \right|^2 \right] d\tau. \qquad (22A.6)$$

Using the divergence theorem, we can write

$$\oiint_{s} \left(\tilde{\bar{H}} \times \tilde{\bar{E}}^* \right) \cdot d\bar{s} = j\omega \iiint_{\tau} \left[\varepsilon \left| \tilde{\bar{E}} \right|^2 - \mu \left| \tilde{\bar{H}} \right|^2 \right] d\tau. \qquad (22A.7)$$

Note that Equation 22A.7 shows that in a lossless and sourceless volume bounded by lossless walls, the powerflow is all reactive. If we have lossy medium or we have a source $\tilde{\bar{J}}$ in τ, we add $\tilde{\bar{J}}$ to the RHS of Equation 22A.1 and the complex Poynting theorem becomes

$$\oiint_{s} \left(\tilde{\bar{H}} \times \tilde{\bar{E}}^* \right) \cdot d\bar{s} = j\omega \iiint_{\tau} \left[\varepsilon \left| \tilde{\bar{E}} \right|^2 - \mu \left| \tilde{\bar{H}} \right|^2 \right] d\tau + \iiint_{\tau} \tilde{\bar{E}}^* \cdot \tilde{\bar{J}} \, d\tau. \qquad (22A.8)$$

From Equation 22A.8 follows

$$\mathrm{Re} \oiint_{s} \left(\tilde{\bar{H}} \times \tilde{\bar{E}}^* \right) \cdot d\bar{s} = \iiint_{\tau} \left(\tilde{\bar{E}}^* \cdot \tilde{\bar{J}} \right) d\tau. \qquad (22A.9)$$

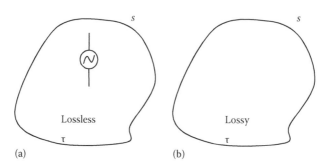

FIGURE 22A.1

Application of complex Poynting theorem to a volume bounded by lossless walls: (a) lossless medium with a source inside and (b) lossy but sourceless medium.

Equation 22A.9 may be interpreted as ohmic loss or/and radiated power. The radiated power from the source in a lossless volume τ coming out of the closed surface over s enclosing the volume τ is given by the RHS of Equation 22A.9. In Figure 22A.1a, the RHS accounts for the active power given by the source and gets out of the volume through the closed surface. In Figure 22A.1b, the RHS accounts for the power loss (ohmic loss) in τ.

*Appendix 23A: Three-Dimensional FDTD Simulation of Electromagnetic Wave Transformation in a Dynamic Inhomogeneous Magnetized Plasma**

Joo Hwa Lee and Dikshitulu K. Kalluri

23A.1 Introduction

Frequency shifting of an electromagnetic wave in a time-varying plasma has been extensively studied [1–7] and some experiments [1–3] were carried out to demonstrate frequency shifting. Analytical studies make various assumptions and use simplified geometries including one-dimensional models, flash ionization, and slow or fast creation of the plasma medium [4–7].

A limited number of theoretical and numerical studies of 3-D models are reported. Buchsbaum et al. [8] examined the perturbation theory for various mode configurations of a cylindrical cavity coaxial with a plasma column and coaxial with the static magnetic field. Gupta used the moment method to study cavities and waveguides containing anisotropic media and compared the results with the perturbation method [9]. A transmission-line-matrix (TLM) method was developed to study the interaction of an electromagnetic wave with a time-invariant and space-invariant magnetized plasma in 3-D space [10]. Mendonça [11] presented a mode-coupling theory in a cavity for a space-varying and slowly created isotropic plasma.

Since the introduction of the FDTD method [12], it has been widely used in solving many electromagnetic problems including those concerned with plasma media [13,14]. For the anisotropic cases [15,16], the equations for the components of the current density vector become coupled and the implementation of the conventional FDTD scheme is difficult. We propose a new FDTD method to overcome this difficulty.

In this paper, we use the FDTD method to analyze the interaction of an electromagnetic wave with a magnetoplasma medium created in a cavity. This paper is organized as follows. The FDTD algorithm is derived first and the implementation of perfect electric conductor (PEC) boundary conditions is investigated. The algorithm is verified by using a mode-coupling theory and a perturbation technique. The application of the algorithm is illustrated by computing the new frequencies and amplitudes of the coupled modes generated due to a switched magnetized time-varying and space-varying plasma in a cavity with an initial TM mode excitation.

---

* C IEEE. Reprinted from *IEEE Trans. Antennas Propag.*, 47 (7), pp. 1146–1151, 1999. With permission.

23A.2 Three-Dimensional FDTD Algorithm

23A.2.1 Maxwell's Equation

Consider a time-varying magnetoplasma medium with collisions. Maxwell's equations and constitutive relation for a cold magnetoplasma are given by

$$\nabla \times \mathbf{E} = -\mu_0 \frac{\partial \mathbf{H}}{\partial t} \tag{23A.1}$$

$$\nabla \times \mathbf{H} = \varepsilon_0 \frac{\partial \mathbf{E}}{\partial t} + \mathbf{J} \tag{23A.2}$$

$$\frac{d\mathbf{J}}{dt} + \nu \mathbf{J} = \varepsilon_0 \omega_p^2 (\mathbf{r},t) \mathbf{E} + \omega_b (\mathbf{r},t) \times \mathbf{J} \tag{23A.3}$$

where ε_0 is the permittivity of free-space, μ_0 the permeability of free-space, ω_p^2 the square of the plasma frequency, $\omega_b = e\mathbf{B}_0/m_e$ the electron gyrofrequency, $\mathbf{B}_0$ the external static magnetic field, and e and m_e are the electric charge and mass of an electron, respectively. The field components in Cartesian coordinate are expressed as

$$\mathbf{E} = E_x \hat{x} + E_y \hat{y} + E_z \hat{z} \tag{23A.4}$$

$$\mathbf{H} = H_x \hat{x} + H_y \hat{y} + H_z \hat{z} \tag{23A.5}$$

$$\mathbf{J} = J_x \hat{x} + J_y \hat{y} + J_z \hat{z} \tag{23A.6}$$

$$\omega_b = \omega_{bx} \hat{x} + \omega_{by} \hat{y} + \omega_{bz} \hat{z}. \tag{23A.7}$$

The substitutions of (23A.4) through (23A.7) in (23A.1) through (23A.3) give the following equations:

$$\frac{\partial H_x}{\partial t} = -\frac{1}{\mu_0} \left(\frac{\partial E_z}{\partial y} - \frac{\partial E_y}{\partial z} \right) \tag{23A.8}$$

$$\frac{\partial E_x}{\partial t} = \frac{1}{\varepsilon_0} \left(\frac{\partial H_z}{\partial y} - \frac{\partial H_y}{\partial z} - J_x \right) \tag{23A.9}$$

$$\begin{bmatrix} \dfrac{dJ_x}{dt} \\[2mm] \dfrac{dJ_y}{dt} \\[2mm] \dfrac{dJ_z}{dt} \end{bmatrix} = \Omega \begin{bmatrix} J_x \\ J_y \\ J_z \end{bmatrix} + \varepsilon_0 \omega_p^2 (r,t) \begin{bmatrix} E_x \\ E_y \\ E_z \end{bmatrix} \tag{23A.10}$$

where

$$\Omega = \begin{bmatrix} -\nu & -\omega_{bz} & \omega_{by} \\ \omega_{bz} & -\nu & -\omega_{bx} \\ -\omega_{by} & \omega_{bx} & -\nu \end{bmatrix}.$$

(23A.11)

The other components of **E** and **H** can be obtained in a similar manner.

23A.2.2 FDTD Equations

Usual grid configuration for **J** is to place J_x, J_y, and J_z [17] at the locations of E_x, E_y, and E_z, respectively. This configuration works fine as long as the equations for the components of **J** are not coupled. When coupled as in (23A.10), J_y and J_z are needed at the position of J_x to update J_x. However, the estimations of J_y and J_z at this position can be very complex; the maintenance of second-order accuracy requires averaging the values at four diagonal positions. Moreover, implementation of the averaging on the boundary is troublesome because some quantities outside the boundary are needed. We overcame these difficulties by placing **J** at the center of the Yee cube as shown in Figure 23A.1. The components of **J** are located at the same space point. It is now possible to solve (23A.10) analytically since all variables are now available at the same space point. Treating **E**, ω_b, ν, and ω_p as constants, each having an average value observed at the center of the time step, the Laplace transform of (23A.10) leads to

$$\mathbf{J}(s) = (s\mathbf{I} - \Omega)^{-1} \mathbf{J}_0 + \varepsilon_0 \omega_p^2 \frac{1}{s} (s\mathbf{I} - \Omega)^{-1} \mathbf{E}$$

(23A.12)

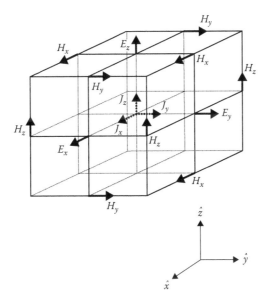

FIGURE 23A.1
Positioning of the electric, magnetic, and current density field vector components about a cubic unit cell.

where $\mathbf{I}$ is the identity matrix. Inverse Laplace transform of (23A.12) leads to the explicit expression for $\mathbf{J}(t)$ as

$$\mathbf{J}(t) = \mathbf{A}(t)\mathbf{J}_0 + \varepsilon_0\omega_p^2\mathbf{K}(t)\mathbf{E} \qquad (23A.13)$$

$$\mathbf{A}(t) = \exp[\Omega t]$$
$$= e^{-vt}\begin{bmatrix} C_1\omega_{bx}\omega_{bx} + \cos\omega_b t & C_1\omega_{bx}\omega_{by} - S_1\omega_{bz} & C_1\omega_{bx}\omega_{bz} + S_1\omega_{by} \\ C_1\omega_{by}\omega_{bx} + S_1\omega_{bz} & C_1\omega_{by}\omega_{by} + \cos\omega_b t & C_1\omega_{by}\omega_{bz} - S_1\omega_{bx} \\ C_1\omega_{bz}\omega_{bx} - S_1\omega_{by} & C_1\omega_{bz}\omega_{by} + S_1\omega_{bx} & C_1\omega_{bz}\omega_{bz} + \cos\omega_b t \end{bmatrix} \qquad (23A.14)$$

$$\mathbf{K}(t) = \Omega^{-1}\left(\exp[\Omega t] - \mathbf{I}\right)$$
$$= \frac{e^{-vt}}{\omega_b^2 + v^2}\begin{bmatrix} C_2\omega_{bx}\omega_{bx} + C_3 & C_2\omega_{bx}\omega_{by} - C_4\omega_{bz} & C_2\omega_{bx}\omega_{bz} + C_4\omega_{by} \\ C_2\omega_{by}\omega_{bx} + C_4\omega_{bz} & C_2\omega_{by}\omega_{by} + C_3 & C_2\omega_{by}\omega_{bz} - C_4\omega_{bx} \\ C_2\omega_{bz}\omega_{bx} - C_4\omega_{by} & C_2\omega_{bz}\omega_{by} + C_4\omega_{bx} & C_2\omega_{bz}\omega_{bz} + C_3 \end{bmatrix} \qquad (23A.15)$$

$$S_1 = \sin\omega_b t/\omega_b \qquad (23A.16)$$

$$C_1 = (1 - \cos\omega_b t)/\omega_b^2 \qquad (23A.17)$$

$$C_2 = (e^{vt} - 1)/v - vC_1 - S_1 \qquad (23A.18)$$

$$C_3 = v(e^{vt} - \cos\omega_b t) + \omega_b\sin\omega_b t \qquad (23A.19)$$

$$C_4 = e^{vt} - \cos\omega_b t - vS_1 \qquad (23A.20)$$

where as shown in (23A.14) through (23A.20) and $\omega_b^2 = \omega_{bx}^2 + \omega_{by}^2 + \omega_{bz}^2$. The FDTD equations for $\mathbf{J}$ are now expressed in terms of their values at a previous time step without having to solve simultaneous equations at each step.

$$\begin{bmatrix} J_x\big|_{i,j,k}^{n+(1/2)} \\ J_y\big|_{i,j,k}^{n+(1/2)} \\ J_z\big|_{i,j,k}^{n+(1/2)} \end{bmatrix} = \mathbf{A}(\Delta t)\begin{bmatrix} J_x\big|_{i,j,k}^{n-(1/2)} \\ J_y\big|_{i,j,k}^{n-(1/2)} \\ J_z\big|_{i,j,k}^{n-(1/2)} \end{bmatrix} + \frac{\varepsilon_0}{2}\omega_p^2\big|_{i,j,k}^n \mathbf{K}(\Delta t)\cdot\begin{bmatrix} E_x\big|_{i+(1/2),j,k}^n + E_x\big|_{i-(1/2),j,k}^n \\ E_y\big|_{i,j+(1/2),k}^n + E_y\big|_{i,j-(1/2),k}^n \\ E_z\big|_{i,j,k+(1/2)}^n + E_z\big|_{i,j,k-(1/2)}^n \end{bmatrix}. \qquad (23A.21)$$

In the above, the time step begins at $n - (1/2)$ and ends at $n + (1/2)$. This formulation is valid for arbitrary values of v and ω_b; the idea is similar to the exponential time stepping [18] for a high conductivity case. Also, if v and ω_b are functions of time and space, (23A.21) can be used by replacing those by $v\big|_{i,j,k}^n$ and $\omega_b\big|_{i,j,k}^n$ in (23A.14) through (23A.20).

Using the grid in Figure 23A.1, the following FDTD equations for the x components of $\mathbf{E}$ and $\mathbf{H}$ can be written as

$$
E_x\big|_{i+(1/2),j,k}^{n+1} = E_x\big|_{i+(1/2),j,k}^{n} + \frac{1}{\varepsilon_0}
$$

$$
\cdot \left[\frac{\Delta t}{\Delta y}\left(H_z\big|_{i+(1/2),j+(1/2),k}^{n+(1/2)} - H_z\big|_{i+(1/2),j-(1/2),k}^{n+(1/2)} \right) - \frac{\Delta t}{\Delta z}\left(H_y\big|_{i+(1/2),j,k+(1/2)}^{n+(1/2)} - H_y\big|_{i+(1/2),j,k-(1/2)}^{n+(1/2)} \right) \right]
$$

$$
- \frac{\Delta t}{2\varepsilon_0}\left(J_x\big|_{i+1,j,k}^{n+(1/2)} + J_x\big|_{i,j,k}^{n+(1/2)} \right) \tag{23A.22}
$$

$$
H_x\big|_{i,j+(1/2),k+(1/2)}^{n+(1/2)} = H_x\big|_{i,j+(1/2),k+(1/2)}^{n-(1/2)} - \frac{1}{\mu_0}
$$

$$
\cdot \left[\frac{\Delta t}{\Delta y}\left(E_z\big|_{i,j+1,k+(1/2)}^{n} - E_z\big|_{i,j,k+(1/2)}^{n} \right) - \frac{\Delta t}{\Delta z}\left(E_y\big|_{i,j+(1/2),k+1}^{n} - E_y\big|_{i,j+(1/2),k}^{n} \right) \right]. \tag{23A.23}
$$

The other components can be written in a similar way.

The stability condition of this method is not easily expressed in a simple form due to the complexity of the algorithm. Nevertheless, it can be said that the standard stability criterion can be applied, that is, the stability is governed by the mode whose phase velocity is fastest in the medium [19,20].

23A.2.3 PEC Boundary Conditions

Figure 23A.2 shows the locations of the fields on $x-y$ plane. The indexes k and $k+(1/2)$ of the planes correspond to those in the FDTD equations. J_x, J_y, J_z, E_x, E_y and H_z are located on k plane, whereas $E_z, H_x,$ and H_y are located on $k+(1/2)$ plane. For a perfect conductor, the tangential components of $\mathbf{E}$ field and the normal component of $\mathbf{H}$ field vanish on the boundaries. Hence, it is natural to choose PEC boundaries to correspond to a k rather than a $k+(1/2)$ plane. For example, the plane $k=1$ may be used for the bottom PEC of a cavity. The boundary conditions are satisfied by assigning zero value to E_x, E_y, J_x, J_y and H_z on $k=1$ plane. Since J_z is not zero on the $k=1$ plane, we need to update its value with time. From (23A.14), the values $E_z\big|_{i,j,(3/2)}^{n}$ and $E_z\big|_{i,j,(1/2)}^{n}$ are needed to obtain the updated value $J_z\big|_{i,j,1}^{n+(1/2)}$. We note that the PEC boundary condition $\partial E_n / \partial n = 0$ will permit us to modify (23A.14) for $k=1$ as given in the following:

$$
\begin{bmatrix} J_x\big|_{i,j,1}^{n+(1/2)} \\[4pt] J_y\big|_{i,j,1}^{n+(1/2)} \\[4pt] J_z\big|_{i,j,1}^{n+(1/2)} \end{bmatrix} = \mathbf{A}\left(\Delta t\right) \begin{bmatrix} J_x\big|_{i,j,1}^{n-(1/2)} \\[4pt] J_y\big|_{i,j,1}^{n-(1/2)} \\[4pt] J_z\big|_{i,j,1}^{n-(1/2)} \end{bmatrix} + \frac{\varepsilon_0}{2}\,\omega_p^2\big|_{i,j,1}^{n}\,\mathbf{K}\left(\Delta t\right) \cdot \begin{bmatrix} E_x\big|_{i+(1/2),j,1}^{n} + E_x\big|_{i-(1/2),j,1}^{n} \\[4pt] E_y\big|_{i,j+(1/2),1}^{n} + E_y\big|_{i,j-(1/2),1}^{n} \\[4pt] 2E_z\big|_{i,j,(3/2)}^{n} \end{bmatrix}. \tag{23A.24}
$$

The algorithm given above leads to explicit computation while maintaining second-order accuracy.

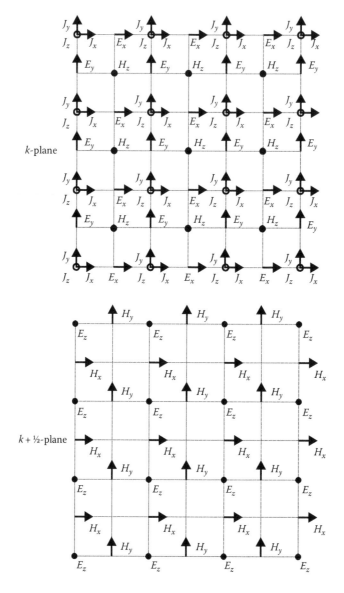

FIGURE 23A.2
Construction of planes for a simple implementation of the PEC boundary conditions.

23A.3 Validation of the Algorithm

The algorithm is validated by considering two test cases, for which results are obtained by other techniques. The first test case involves switching of a plasma medium in a rectangular microwave cavity with dimensions L_x, L_y, and L_z. Several cavity modes are excited due to the interaction of the incident mode and the plasma medium. Hence, the new fields may be written in terms of cavity modes as

$$\mathbf{E}(\mathbf{r},t) = \sum a_l(t)\mathbf{E}_l(\mathbf{r}) \qquad (23A.25)$$

where $\mathbf{E}_l(\mathbf{r})$ is a normalized cavity mode and

$$\int \mathbf{E}_l(\mathbf{r})\mathbf{E}_{l'}^*(\mathbf{r})\mathrm{d}^3r = \delta_{ll'}. \tag{23A.26}$$

In the above, $\mathbf{E}^*$ is the complex conjugate of $\mathbf{E}$. For the space-varying but isotropic plasma creation, the differential equations for a are obtained by the mode-coupling theory as [11]

$$\frac{\partial^2 a_l(t)}{\partial t^2} + \omega_l^2 a_l + \sum C_{ll'}a_{l'} = 0 \tag{23A.27}$$

where

$$\omega_l^2 = c^2 k^2$$

$$= c^2 \left[\left(\frac{m\pi}{L_x} \right)^2 + \left(\frac{p\pi}{L_y} \right)^2 + \left(\frac{q\pi}{L_z} \right)^2 \right] \tag{23A.28}$$

$$C_{ll'} = \int \omega_p^2(\mathbf{r},t)\mathbf{E}_l(\mathbf{r})\mathbf{E}_{l'}^*(\mathbf{r})\mathrm{d}^3r \tag{23A.29}$$

and m, p, q are the mode numbers.

Consider the creation of a time-varying and space-varying plasma medium in the cavity with a plasma frequency profile $\omega_p^2(\mathbf{r},t)$

$$\omega_p^2(\mathbf{r},t) = \omega_{p0}^2\left(1 - e^{-t/T_r}\right)\exp\left[-\left(\alpha_x \frac{x - x_0}{L_x}\right)\right.$$

$$\left. -\left(\alpha_y \frac{y - y_0}{L_y}\right)^2 - \left(\alpha_z \frac{z - z_0}{L_x}\right)^2 \right] \quad t > 0 \tag{23A.30}$$

where ω_{p0}^2 is the saturation plasma frequency, T_r is a rise time, and the spatial variation is Gaussian. If the spatial variation of the plasma is only in x direction, that is, $\alpha_y = \alpha_z = 0$, newly excited modes should have same mode numbers for p and q.

Therefore, we need to consider the changes in mode number m only to describe the fields in the cavity.

Figure 23A.3 shows the comparisons of the results by FDTD with those by the mode-coupling theory for the initial TM_{111} mode excitation. Isotropic and inhomogeneous plasma distribution is considered for comparisons. The following parameters are used: $L_x = 4L_y = 4L_z$, $T_r = 0$, $x_0 = L_x/2$, $\omega_{p0} = \omega_0$, $\nu = 0$, $\alpha_x = 1$, $\alpha_y = \alpha_z = 0$ and ω_0 is the frequency of the initial mode. The results for the mode-coupling theory are obtained by numerically solving the differential equations (23A.27) with the initial conditions obtained from the initial excitation. The results for the FDTD are obtained first by calculating $\mathbf{E}(\mathbf{r},n\Delta t)$ using our algorithm and then by computing $a_l(t)$

$$a_l(t) = a_l(n\Delta t) = \int \mathbf{E}(\mathbf{r},n\Delta t)\mathbf{E}_l^*(\mathbf{r})\mathrm{d}^3r. \tag{23A.31}$$

Figure 23A.3 shows very good agreement.

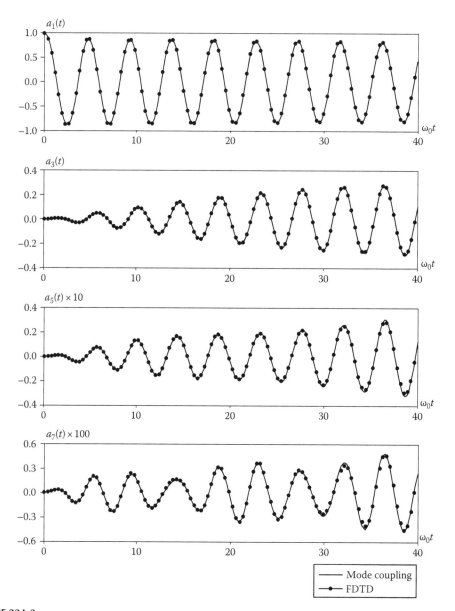

FIGURE 23A.3
Mode amplitudes for the case of a switched isotropic plasma in a cavity. Comparisons of the FDTD and the mode-coupling theory are shown. The coefficients are normalized with respect to $a_1(0)$. $a_m(t)$ indicates $a_{m11}(t)$.

The second validation is for the case of a magnetized plasma. Results based on a perturbation method are available for a magnetized, low-density, plasma-filled cavity in [21]. In the perturbation theory, the frequency shift due to homogeneous magnetized plasma can be obtained as

$$\frac{\Delta\omega}{\omega_0} \approx -\frac{1}{2}\int \varepsilon_0 \mathbf{E}(\mathbf{r})\Delta\varepsilon \mathbf{E}^*(\mathbf{r})\mathrm{d}^3 r \qquad (23\mathrm{A}.32)$$

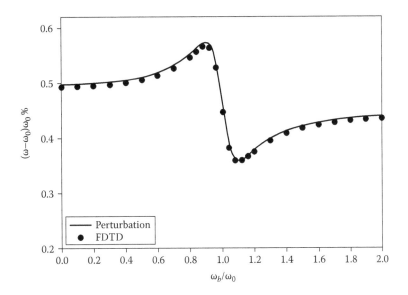

FIGURE 23A.4
Frequency shift due to the creation of a low-density, homogeneous, magnetized plasma in a cavity. Comparisons of the FDTD and the perturbation methods. The ratio of frequency shift (%) versus the cyclotron frequency ω_b/ω_0 is plotted. $\omega_p/\omega_0=0.1$ and $\nu/\omega_0=0.1$.

where $\Delta\varepsilon=\varepsilon-\mathbf{I}$ and ε is the dielectric tensor of the magnetized plasma medium. For the magnetized plasma problem, a differential equation similar to (23A.27) is not available. Nevertheless, the new fields can be approximated by (23A.25) and the empty cavity modes can be used to extract the modes from FDTD simulation for a weak plasma density since the coupling is not strong. During the FDTD computations, $a_l(t)$ is computed using (23A.31) and the frequency change is obtained from the time series of $a_l(t)$ by Prony method [22]. Figure 23A.4 shows the comparison of the perturbation and FDTD methods for an initial TM_{111} mode excitation. A lossy ($\nu/\omega_0=0.1$) homogeneous plasma medium ($\omega_p/\omega_0=0.1$) is used, and the dimensions of the cavity are given by $9L_x=4L_y=3L_z$. The background magnetic field $\omega_b(\mathbf{r})=\omega_b\tilde{z}$ is along z axis. Since we have computed the final frequency shift which is independent of the rise time T_r, we have considered the case of sudden ($T_r=0$) creation. In this figure, the frequency shift ratio is depicted for various background magnetic field intensities.

23A.4 Illustrative Examples of the New FDTD Method for A Dynamic Medium

Figure 23A.5 shows the frequency shifting due to the interaction of the electromagnetic wave with a time-varying and space-varying magnetized plasma. A rectangular cavity with an initial excitation of TM_{111} mode is considered with following parameters: $6L_x=3L_y=2L_z$, $T_r=100/\omega_0$, $x_0=L_x/2$, $y_0=L_y/2$, $z_0=L_z/2$, $\omega_{p0}=\omega_0$, and $\nu=0$. $\alpha_x=\alpha_y=\alpha_z=0$ is used for homogeneous plasma distribution (Figure 23A.5a and b) and $\alpha_x=\alpha_y=\alpha_z=2$ is used for inhomogeneous plasma distribution (Figure 23A.5c and d). The electron cyclotron frequency ω_b is chosen to be zero for Figure 23A.5a and c, and $\omega_b=\omega_0\hat{z}$ for (b) and (d). The results are obtained from the spectrum of the time series E_x of at $x=(1/15)L_x$, $y=(7/15)L_y$,

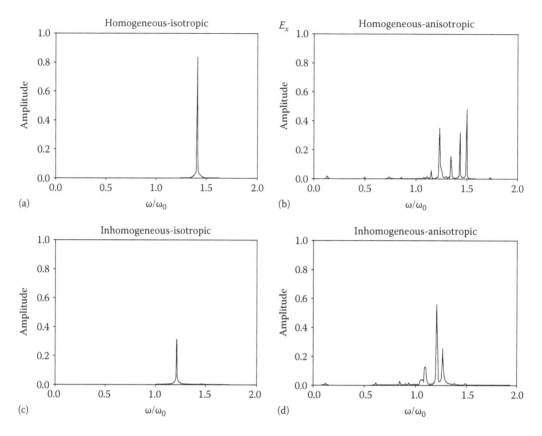

FIGURE 23A.5

Amplitude spectra of the electric field E_x at $x = (1/15)L_x$, $y = (7/15)L_y$, $z = (7/15)L_z$. Parameters are $T_r = 100/\omega_0$, $\omega_{p0} = \omega_0$, $\nu = 0$. (a) Homogeneous-isotropic ($\alpha_x = \alpha_y = \alpha_z = 0, \omega_b = 0$). (b) Homogeneous-anisotropic $\left(\alpha_x = \alpha_y = \alpha_z = 0, \omega_b = \omega_0 \hat{z}\right)$. (c) Inhomogeneous-isotropic ($\alpha_x = \alpha_y = \alpha_z = 2, \omega_b = 0$). (d) Inhomogeneous-anisotropic $\left(\alpha_x = \alpha_y = \alpha_z = 2, \omega_b = \omega_0 \hat{z}\right)$.

$z = (7/15)L_z$ taken after the plasma is almost fully created, that is, after $t = 6.4 T_r$. For a homogeneous and isotropic creation of plasma (Figure 23A.5a), the frequency shift due to the interaction of the electromagnetic wave with the plasma is the same as that of an unbounded plasma. For the cavity with an anisotropic uniform plasma (Figure 23A.5b), several new modes appear. It is noted that some of the modes have higher frequencies than that of the unbounded isotropic plasma. A comparison of Figure 23A.5a with 23A.5c and Figure 23A.5b with 23A.5d shows that nonuniformly filled plasma cavity displays lower values of frequency shifts for the same maximum values of ω_{p0} as predicted in [11].

23A.5 Conclusion

3-D FDTD formulations are constructed for time-varying, inhomogeneous, magneto-plasma medium. An explicit formulation is accomplished by locating **J** at the center of the Yee cube and by using an appropriate time-stepping algorithm.

References

1. Yablonovitch, E., Spectral broadening in the light transmitted through a rapidly growing plasma, *Phys. Rev. Lett.*, 31, 877–879, 1973.
2. Joshi, C. J., Clayton, C. E., Marsh, K., Hopkins, D. B., Sessler, A., and Whittum, D., Demonstration of the frequency upshifting of microwave radiation by rapid plasma creation, *IEEE Trans. Plasma Sci.*, 18, 814–818, 1990.
3. Kuo, S. P. and Faith, J., Interaction of an electromagnetic wave with a rapidly created spatially periodic plasma, *Phys. Rev. E*, 56(2), 1–8, 1997.
4. Kalluri, D. K., Goteti, V. R., and Sessler, A. M., WKB solution for wave propagation in a time-varying magnetoplasma medium: Longitudinal propagation, *IEEE Trans. Plasma Sci.*, 21(1), 70–76, February 1993.
5. Lee, J. H. and Kalluri, D. K., Modification of an electromagnetic wave by a time-varying switched magnetoplasma medium: Transverse propagation, *IEEE Trans. Plasma Sci.*, 26(1), 1–6, February 1998.
6. Kalluri, D. K. and Huang, T. T., Longitudinal propagation in a magnetized time-varying plasma: Development of Green's function, *IEEE Trans. Plasma Sci.*, 26(3), 1022–1030, June 1998.
7. Kalluri, D. K., *Electromagnetics of Complex Media*, CRC, Boca Raton, FL, 1998.
8. Buchsbaum, S. J., Mower, L., and Brown, S. C., Interaction between cold plasmas and guided electromagnetic waves, *Phys. Fluids*, 3(5), 806–819, 1960.
9. R. R. Gupta, A study of cavities and waveguides containing anisotropic media, PhD dissertation, Syracuse University, Syracuse, NY, July 1965.
10. Kashiwa, T., Yoshida, N., and Fukai, I., Transient analysis of a magnetized plasma in three-dimensional space, *IEEE Trans. Antennas Propagat.*, 36, 1096–1105, August 1998.
11. Mendonça, J. T. and Oliveira e Silva, L., Mode coupling theory of flash ionization in a cavity, *IEEE Trans. Plasma Sci.*, 24(1), 147–151, February 1996.
12. Yee, K. S., Numerical solution of initial boundary value problems involving Maxwell's equations in isotropic media, *IEEE Trans. Antennas Propagat.*, AP-14, 302–307, May 1966.
13. Young, J. L., A full finite difference time domain implementation for radio wave propagation in a plasma, *Radio Sci.*, 29, 1513–1522, 1994.
14. Cummer, S. A., An analysis of new and existing FDTD methods for isotropic cold plasma and a method for improving their accuracy, *IEEE Trans. Antennas Propagat.*, 45, 392–400, March 1997.
15. Garcia, S. G., Hung-Bao, T. M., Martin, R. G., and Garcia Olmedo, B., On the application of finite methods in time domain to anisotropic dielectric waveguides, *IEEE Trans. Microwave Theory*, 44, 2195–2205, December 1996.
16. Schneider, J. and Hudson, S., The finite-difference time-domain method applied to anisotropic material, *IEEE Trans. Antennas Propagat.*, 41, 994–999, July 1993.
17. Sano, E. and Shibata, T., Fullwave analysis of picosecond photoconductive switches, *IEEE J. Quantum Electron.*, 26(2), 372–377, February 1990.
18. Taflove, A., *Computational Electrodynamics: The Finite-Difference Time-Domain Method*, Artech House, Boston, MA, 1995.
19. Taflove, A. and Brodwin, M. E., Numerical solution of steady-state electromagnetic scattering problems using the time-dependent Maxwell's equations, *IEEE Trans. Microwave Theory Tech.*, MTT-23(8), 623–630, August 1975.
20. Young, J. L. and Brueckner, F. P., A time domain numerical model of a warm plasma, *Radio Sci.*, 29(2), 451–463, 1994.
21. Harrington, R. F., *Field Computation by Moment Methods*, Krieger, Malabar, FL, 1968.
22. Kannan, N. and Kundu, D., On modified EVLP and ML methods for estimating superimposed exponential signals, *Signal Processing*, 39, 223–233, 1994.

Part VII

Chapter Problems

Problems

Chapter 15

P15.1 Show that

$$\delta = \Delta(1+\Delta)^{-1/2}.$$

P15.2 Show that

$$\delta = 2\sinh\left(\frac{hD}{2}\right),$$

$$\mu = \cosh\left(\frac{hD}{2}\right),$$

$$\mu = \left[1+\frac{\delta^2}{4}\right]^{1/2}.$$

P15.3 Show that

$$h^2 D^2 = \Delta^2 - \Delta^3 + \frac{11}{24}\Delta^4 - \frac{5}{6}\Delta^5 \cdots$$

$$= \delta^2 - \frac{1}{12}\delta^4 + \frac{1}{90}\delta^6 \cdots.$$

Hint: $\sinh^{-1} x = x - x^3/3 + (3/40)x^5 - \cdots$

P15.4 Show that $h^2 D^2 y_n = [1/12][-y_{n+2} + 16y_{n+1} - 30y_n + 16y_{n-1} - y_{n-2} \cdots]$.

P15.5 Given

$$\frac{d^2 V}{dx^2} + V = 2x, \quad 0 < x < 1 \quad \text{and} \quad V(0) = 0, \quad \left[\frac{dV}{dx}\right]_{x=1} = 0.$$

Determine $V(1/2)$ by the method of weighted residuals. Use point matching at $x = 1/2$. Use trial function $V = C_1 \sin Kx$.

P15.6 Given

$$\frac{d^2 V}{dx^2} + V^2 + \frac{4}{9} = 0, \quad 0 < x < 1 \quad \text{and} \quad V(1) = 0, \quad \left[\frac{dV}{dx}\right]_{x=0} = 0.2\, V(0),$$

Determine $V(1/2)$ by the method of weighted residuals. Use point matching at $x = 1/2$.
Hint: try the trial function $V = C_1(1-x)(1-Kx)$ and determine K so that the boundary condition at $x = 0$ is satisfied.

345

P15.7 Solve P15.6 by finite difference method.

P15.8 Given

$$\frac{d^2V}{dx^2} + V = 2x, \quad 0 < x < 1,$$

$$V(0) = 0, \quad \frac{dV}{dx} \text{ at } x = 1 \text{ is zero.}$$

Determine $V(1/2)$ by the method of weighted residuals. Use point matching at $x = 1/2$.

P15.9 Solve the P15.8 by the finite-difference method. Use $h = 1/2$.

P15.10 Solve $-d^2V/dx^2 = 1$, $0 < x < 1$, $V(0) = V(1) = 0$ by using finite-element method. Use three finite elements. Determine $V(1/4)$.

Chapter 16

P16.1 Use finite difference method to compute the lowest cutoff frequency of the TM modes of a rectangular waveguide (Figure 16.1) with $a = 1$ and $b = 1/2$. Use $h = 1/6$. The dimensions are in meters and the medium in the guide is air.

P16.2 Figure P16.2 shows an equilateral triangle. P is the point inside the triangle. The Cartesian coordinates of the vertices and P are marked in the figure. Determine the area coordinates of P and calculate the potential of P approximately. Assume the vertex potentials are: $V_1 = 100$ V, $V_2 = 50$ V, and $V_3 = 25$ V.

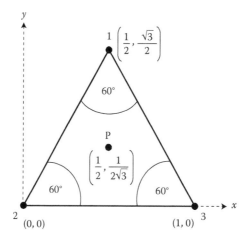

FIGURE P16.2

P16.3 Use first-order node-based FEM to determine the lowest and the next lowest TE mode cut off k_c numbers. Also determine the lowest TM mode cut off k_c. Use reasonable number of elements to illustrate the method rather than obtain accurate answers. You can use the MATLAB® function program GLANT given in Appendix 16A.

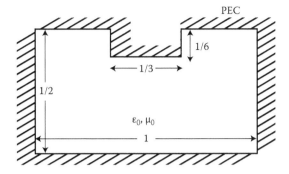

P16.4 Refer to the rectangular waveguide shown in Figure 16.18. Find the lowest cutoff frequency of (a) TM modes and (b) TE modes. Use GLANT to get answers.

P16.5 Refer to Figure 16.9. Calculate the potential at the point $P(1, 1/3)$ given $\varepsilon_{r1} = 1$, $\varepsilon_{r2} = \varepsilon_{r3} = 4$. Use PGLANT2 to obtain answers.

P16.6 Determine the expressions for the following: $S_{11}^{(e)}$, $S_{13}^{(e)}$, $S_{15}^{(e)}$, and $T_{13}^{(e)}$ for the second-order node-based FEM.

P16.7 This problem is of interest in determining the DC resistance of an odd-shaped conductor. Refer to Figure P16.7a. In the solution domain bounded by ABCD, the Laplace equation is satisfied.

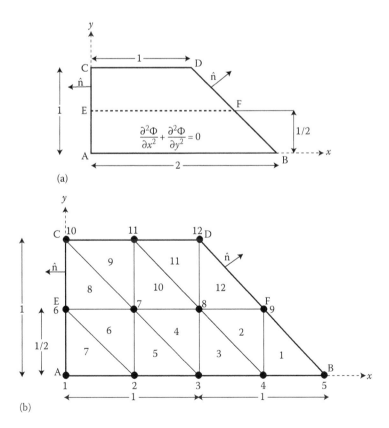

FIGURE P16.7

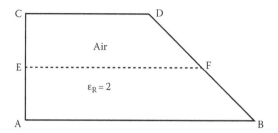

FIGURE P16.8

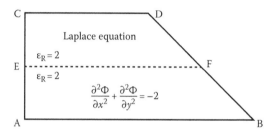

FIGURE P16.9

The boundary conditions are:

On AB: $\phi_0 = 0$;

On CD: $\phi_0 = 1$;

On AC and BD, the homogeneous Neumann BC are satisfied.

Use second-order node-based FEM to determine the potentials at (a) E and (b) F. Use the finite elements shown in Figure P16.7b.

P16.8 Same geometry as in Problem 16.7, but the domain is filled partially with a dielectric as shown in Figure P16.8. Use FEM to determine potentials at (a) E and (b) F.

P16.9 Same geometry as in Problem P16.7, but the Poisson equation given in Figure P16.9 is satisfied on the inside part of the solution domain (Figure P16.9).

Assume homogeneous Dirichlet boundary conditions on AB and CD and homogeneous Neumann boundary conditions on AC and BD. Use FEM to determine potentials at (a) *E* and (b) *F*. Use the elements in figure (b) of Problem P16.7.

P16.10 For vector shape functions α_i, show Properties 2–4 given in Section 16.5.2.

P16.11 For the equilateral triangle shown in Figure P16.2, obtain [E] and [F] matrices.

P16.12 A rectangular waveguide with a PEC baffle is shown in the following figure. Use edge-based FEM to determine the lowest k_c value of (a) TM and (b) TE modes. Use the five elements shown in the following figure. The medium in the guide is air.

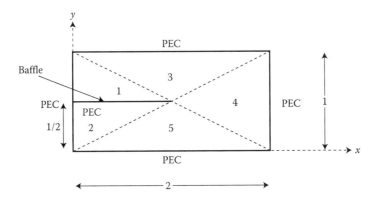

P16.13 Use moment method to find the capacitance with respect to infinity of the metal plate in free space, consisting of two subareas: the left subarea is a square plate of (1 m × 1 m). The right subarea is an equilateral triangle of 1 m side. As discussed in Section 16.6, the shape of the subarea is not terribly important and one can use Equation 16.211 for the computation of l_{nn}. We are limiting to two subareas so that you can do hand computation.

P16.14 a. Use moment method to calculate the capacitance of a conducting plate shown in figure (a) below. Use two subarea approximation shown in the figure.

  b. The shape of the plate is changed to that shown in figure (b). Determine the dimension b so that the capacitance is approximately same as that of the plate in figure (a)

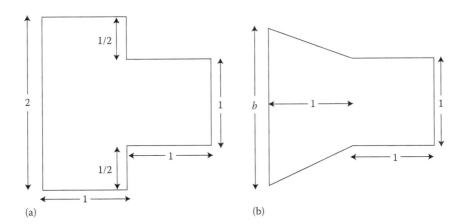

(a)   (b)

P16.15 Refer to the formulation and the solution of the scattering by an infinitely long circular conducting cylinder by moment method in Section 16.7. Plot the magnitude of the total electric field along the x-axis as a function βx in the domain $-20 < \beta x < 60$, for the data $a = 1$ and $\beta = 3.1$, $E_z^i = 1$ using (a) 4 segments and (b) 72 segments. You can use MATLAB for computations.

TABLE P17.1

Node i	1	2	3	4
x_i	0	1	0	0
y_i	0	0	1	0
z_i	0	0	0	1

Chapter 17

P17.1 The purpose of this problem is to illustrate the use of volume coordinates in three-dimensional problems.

A tetrahedral with four nodes is shown in the following figure. The xyz-coordinates of the nodes are given in Table P17.1.

a. Determine the volume of the tetrahedral using volume coordinates.

b. Suppose the volume charge density ρ_v is $= xy \times 10^{-6}$ C/m³. Determine the total charge density inside the tetrahedral.

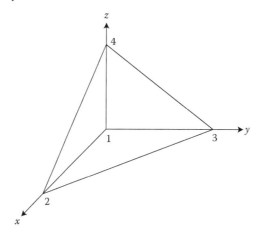

P17.2 Refer to data of the coordinates of the tetrahedra in the figure above.

a. Determine b_i, c_i, d_i, $i = 1, 2, 3, 4$.

b. If the Laplace equation is satisfied in the tetrahedra and the voltages of the nodes 2, 3, and 4 are 10, 20, and 30 V, respectively, determine the voltage of node 1.

Chapter 19

P19.1 The voltage of a nonuniform transmission line satisfies the partial differential equation

$$\frac{\partial^2 V}{\partial x^2} - \frac{1}{c^2}\frac{\partial^2 V}{\partial t^2} - \frac{3}{4x^2}V = 0,$$

in the domain $1 < x < 2$.

The line is short-circuited at both ends Thus, the boundary conditions are: $V(1, t) = 0$ and $V(2, t) = 0$. Let the initial conditions be: $V(x, 0) = (x-1)(2-x)$ and $\frac{\partial V}{\partial t}(x,t)\big|_{t=0} = 0$. Determine $V(1.5, 4\Delta t)$, using FDTD. Choose $\Delta x = 1/4$, $r = c\Delta t / \Delta x = 1$.

P19.2 For a lossy plasma, the frequency domain relative permittivity is complex and is given by

$$\varepsilon_p(\omega) = 1 - \frac{\omega_p^2}{\omega(\omega - jv)}, \tag{P19.1}$$

where v is the collision frequency (rad/s). The above relation is called Drude dispersion relation and in the limit $v \to 0$ reduces to Equation 10.14.

Show that the basic field equations for a cold isotropic lossy plasma are given by Equations 11.4, 11.15 of Volume 1 and

$$\frac{d\mathbf{J}}{dt} + v\mathbf{J} = \varepsilon_0 \omega_p^2 \mathbf{E}. \tag{P19.2}$$

Consider a plasma of plasma frequency $f_p = \omega_p/2\pi = 30$ GHz and a collision frequency $v = 2 \times 10^{10}$ rad/s. Plot the real and imaginary parts of the dielectric constant in the range of 0–90 GHz.

P19.3 Water is an example of materials whose frequency domain relative permittivity exhibits Debye dispersion:

$$\varepsilon_r(\omega) = \varepsilon_\infty + \frac{\varepsilon_s - \varepsilon_\infty}{1 + j\omega t_0}, \tag{P19.3}$$

where ε_∞ is the infinite frequency relative permittivity, ε_s is the static relative permittivity at zero frequency, and t_0 is the relaxation time.

Show that the basic field equations in this medium may be described by

$$\nabla \times \mathbf{E} = -\mu_0 \frac{\partial \mathbf{H}}{\partial t}, \tag{P19.4}$$

$$\nabla \times \mathbf{H} = \frac{\partial \mathbf{D}}{\partial t}, \tag{P19.5}$$

$$t_0 \frac{d\mathbf{D}}{dt} + \mathbf{D} = \varepsilon_s \varepsilon_0 \mathbf{E} + t_0 \varepsilon_\infty \varepsilon_0 \frac{d\mathbf{E}}{dt}, \tag{P19.6}$$

and the one-dimensional field equations are

$$\frac{\partial E}{\partial z} = -\mu_0 \frac{\partial H}{\partial t}, \tag{P19.7}$$

$$-\frac{\partial H}{\partial z} = \frac{\partial D}{\partial t}, \tag{P19.8}$$

$$t_0 \frac{dD}{dt} + D = \varepsilon_s \varepsilon_0 E + t_0 \varepsilon_\infty \varepsilon_0 \frac{dE}{dt}. \tag{P19.9}$$

Plot the real and imaginary parts of complex relative permittivity of water in the frequency domain 0–80 GHz. Assume $\varepsilon_s = 81$, $\varepsilon_\infty = 1.8$, and $t_0 = 9.4 \times 10^{-12}$ s.

P19.4 A second-order Lorentz dispersive material has a relative permittivity

$$\varepsilon_r(\omega) = \varepsilon_\infty + \frac{(\varepsilon_s - \varepsilon_\infty)\omega_R^2}{\omega_R^2 + 2j\omega\delta - \omega^2} \tag{P19.10}$$

where ω_R is the resonant frequency and δ is the damping constant.

a. Show that the basic field equations in this medium are given by

$$\nabla \times \mathbf{E} = -\mu_0 \frac{\partial \mathbf{H}}{\partial t}, \tag{P19.11}$$

$$\nabla \times \mathbf{H} = \frac{\partial \mathbf{D}}{\partial t}, \tag{P19.12}$$

$$\omega_R^2 \mathbf{D} + 2\delta \frac{d\mathbf{D}}{dt} + \frac{d^2\mathbf{D}}{dt^2} = \omega_R^2 \varepsilon_s \varepsilon_0 \mathbf{E} + 2\delta\varepsilon_\infty\varepsilon_0 \frac{d\mathbf{E}}{dt} + \varepsilon_\infty\varepsilon_0 \frac{d^2\mathbf{E}}{dt^2}, \tag{P19.13}$$

and the one-dimensional field equations are

$$\frac{\partial E}{\partial z} = -\mu_0 \frac{\partial D}{\partial t}, \tag{P19.14}$$

$$-\frac{\partial H}{\partial z} = \frac{\partial D}{\partial t}, \tag{P19.15}$$

$$\omega_R^2 D + 2\delta \frac{dD}{dt} + \mu_0 \frac{d^2D}{dt^2} = \omega_R^2 \varepsilon_s \varepsilon_0 E + 2\delta\varepsilon_\infty\varepsilon_0 \frac{dE}{dt} + \varepsilon_\infty\varepsilon_0 \frac{d^2E}{dt^2}. \tag{P19.16}$$

Plot the real and imaginary parts of $\varepsilon_r(\omega)$ using the parameters $\varepsilon_s = 2.25$, $\varepsilon_\infty = 1$, $\omega_R = 4 \times 10^{16}$ rad/s, and $\delta = 0.28 \times 10^{16}$/s. This Lorentz medium has a resonance in the optical range.

Chapter 21

P21.1 Obtain (21.17).
P21.2 Obtain (21.18).
P21.3 Show that Equations 21.35 and 21.36 can be obtained from (21.37) and (21.38) provided the stretch factor s_i is a function of the i coordinate only, where i stands for x or y or z.
P21.4 Obtain (21.39) and (21.40).
P21.5 Derive (21.53).

Chapter 23

P23.1 Derive (23.31) and show that it is the same as (3.45) if $k_c = \pi/a$.

P23.2 (a) Show that $f_2(y)$ has only even harmonics as given in (23.41b). (b) Fill in the steps leading to (23.44) and (23.47).

P23.3 Obtain (23.110) from (23.109) using (23.111).

P23.4 Derive (23.117) starting from (23.116).

P23.5 Show that for the trapezoidal approximation of the convolutional integral, the rule for replacing $j\omega$ is given by (23.155).

P23.6 Determine f_1 and f_2 in (23.170).

P23.7 Starting from (23.171), obtain the relation between E_z at the current time step in terms of D_z at the current and the previous time steps and E_z at the previous time steps by using bilateral z-transform.

P23.8 Starting from Maxwell's equations, obtain (23.194).

P23.9 Derive the transfer matrix for the ith layer given by (23.200b).

P23.10 Refer to Figure 2C.20 in Volume 1, Appendix 2C of this book. Determine the reflection coefficient using the TMM method. You may use the MATLAB® to get the eigenvalues and eigenvectors and other math calculations.

P23.11 Reference 37 gives a technique of fixing the inherently unstable TMM method of calculation by rearranging the columns [W] and [l] to group together the forward-propagating modes in the first two columns and the backward-propagating modes in the last two columns, etc. Perform such a fix for the data of problem P23.10.

P23.12 Use mode matching technique to obtain the resonant frequencies of a cigar-shaped cavity and compare your answers with those given on page 475 of the reference, Bladel, V., *Electromagnetic Fields*, McGraw-Hill, New York, 1964.

Index

Milton Keynes UK
Ingram Content Group UK Ltd.
UKHW052021071024
449327UK00027B/2364

9 780367 873868